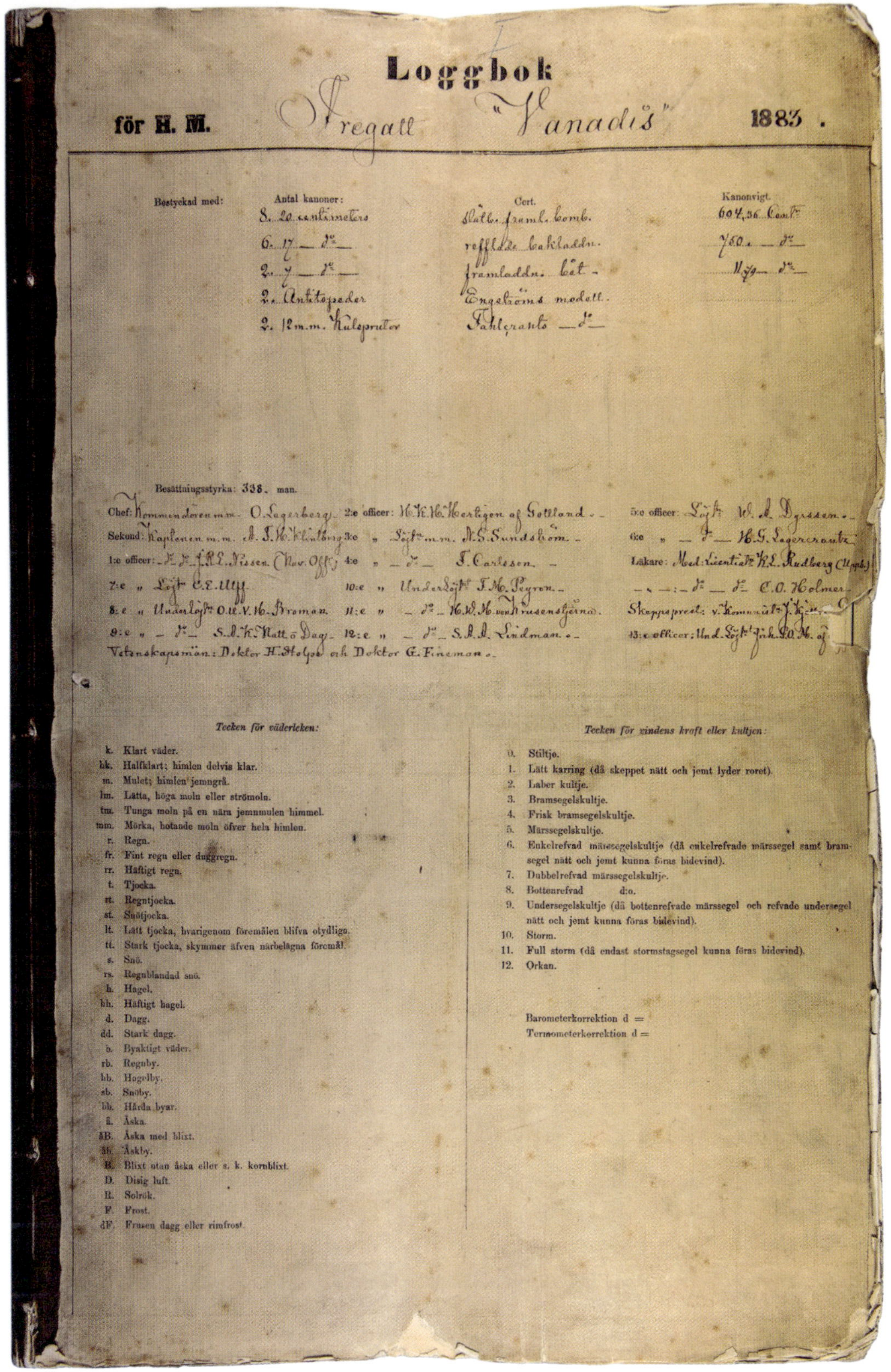

Loggbok

för H. M. Fregatt "Vanadis" 1883.

Bestyckad med:	Antal kanoner:	Cert.	Kanonvigt.
	8. 20 centimeters	slätb. framl. comb.	604,36 Cent:r
	6. 17 — d:o —	refflade bakladdn.	750. — d:o
	2. 7 — d:o —	framladdn. bet.	11,79 — d:o
	2. Antitopeder	Engströms modell.	
	2. 12 m.m. Kulsprutor	Fahlcrantz — d:o —	

Besättningsstyrka: 338. man.

Chef: Kommendören m.m. O. Lagerberg. — 2:e officer: H. K. H. Hertigen af Gottland — 5:e officer: Löjtn. W. A. Dyrssen —
Sekond: Kaptenen m.m. A. F. K. Klintberg 3:e „ Löjtn. m.m. N. S. Sundström — 6:e „ — d:o — H. G. Lagercrantz
1:e officer: — d:o — J. R. L. Nissen (Nav. Off.) 4:e „ — d:o — F. Carlsson — Läkare: Med: Licentiat K. L. Rudberg (Uppb.)
7:e „ Löjtn. C. E. Uggla 10:e „ Underlöjtn. T. M. Peyron — „ — : — d:o — d:o — C. O. Holmer
8:e „ Underlöjtn. O. U. V. H. Broman 11:e „ — d:o — H. W. H. von Krusenstjerna. Skeppspredikant: v. Kommunist J. K[illegible]
9:e „ — d:o — S. A. K. Natt o Dag — 12:e „ — d:o — S. A. A. Lindman. — 13:e officer: Und. Löjtn. Frih. C. O. H. af [illegible]
Vetenskapsmän: Doktor H. Stolpe och Doktor G. Fineman.

Tecken för väderleken:

- k. Klart väder.
- hk. Halfklart; himlen delvis klar.
- m. Mulet; himlen jemngrå.
- lm. Lätta, höga moln eller strömoln.
- tm. Tunga moln på en nära jemnmulen himmel.
- mm. Mörka, hotande moln öfver hela himlen.
- r. Regn.
- fr. Fint regn eller duggregn.
- rr. Häftigt regn.
- t. Tjocka.
- rt. Regntjocka.
- st. Snötjocka.
- lt. Lätt tjocka, hvarigenom föremålen blifva otydliga.
- tt. Stark tjocka, skymmer äfven närbelägna föremål.
- s. Snö.
- rs. Regnblandad snö.
- h. Hagel.
- hh. Häftigt hagel.
- d. Dagg.
- dd. Stark dagg.
- b. Byaktigt väder.
- rb. Regnby.
- hb. Hagelby.
- sb. Snöby.
- bb. Hårda byar.
- å. Åska.
- åB. Åska med blixt.
- åb. Åskby.
- B. Blixt utan åska eller s. k. kornblixt.
- D. Disig luft.
- R. Solrök.
- F. Frost.
- dF. Frusen dagg eller rimfrost.

Tecken för vindens kraft eller kultjen:

0. Stiltje.
1. Lätt karring (då skeppet nätt och jemt lyder roret).
2. Laber kultje.
3. Bramsegelskultje.
4. Frisk bramsegelskultje.
5. Märssegelskultje.
6. Enkelrefvad märssegelskultje (då enkelrefvade märssegel samt bramsegel nätt och jemt kunna föras bidevind).
7. Dubbelrefvad märssegelskultje.
8. Bottenrefvad d:o.
9. Undersegelskultje (då bottenrefvade märssegel och refvade undersegel nätt och jemt kunna föras bidevind).
10. Storm.
11. Full storm (då endast stormstagsegel kunna föras bidevind).
12. Orkan.

Barometerkorrektion d =
Termometerkorrektion d =

THE VANADIS EXPEDITION

An Ethnographic Voyage Around the World 1883–1885

Bo G Erikson Oscar Ekholm Bengt Wanselius

BOKFÖRLAGET STOLPE

AXEL AND MARGARET AX:SON JOHNSON FOUNDATION FOR PUBLIC BENEFIT

CONTENTS

PREFACE

On 5 December 1883, the steam frigate Vanadis left the naval port in Karlskrona to sail round the world. On board was the 22-year-old photographer Oscar Ekholm. Eighteen months later he stepped ashore in Stockholm with 700 unique photographs in the form of glass negatives, mainly featuring ethnographic motifs. Ekholm (1861–90) had thus become a pioneering documentary photographer in the emerging field of Swedish ethnography – only Louis Palander could be deemed to have vied for first place in this branch of knowledge, with his images from Spitsbergen and the Vega's voyage along the Siberian coast in the 1870s, both expeditions having been led by Adolf Erik Nordenskiöld.

Palander went down in the history of photography. Ekholm disappeared into oblivion. When a selection of Ekholm's photos were published for the centenary of the Vanadis's voyage, his first name was not even known, and he was referred to as 'O.B'. – now interpreted as Oscar Birger. In *Den svenska fotografins historia* (The history of Swedish photography), published in 1983, he is not mentioned.

This silence is strange. Oscar Ekholm was not only, after all, one of our first documentary photographers, with a global range of motifs, but he was also one of the first global travellers and circumnavigators to have practised the young art of photography. The person who facilitated his journey was Hjalmar Stolpe, the renowned field archaeologist who excavated our first town, Birka, and now an ethnographer. After an unrelenting struggle he had managed to secure a berth on board the Vanadis, and was officially dispatched and commissioned by the government with support from both the Royal Swedish Academy of Letters and the Royal Swedish Academy of Sciences.

The ethnographic team of Stolpe and Ekholm returned with a unique collection of photographs and nearly 10,000 ethnographic objects (7,500 museum items) – the result of almost manic collection work. In addition to Ekholm's photographs, the pictorial material in the present book comprises drawings by Stolpe and photographs purchased during the Vanadis voyage. He was a keen collector and amateur photographer whose archive of images came to comprise a thousand items.

Stolpe's agreement with Oscar Ekholm prescribed that five paper copies be made of each glass negative during the course of the journey, thus there was a small photographic laboratory on board the Vanadis when they left Sweden. This gave Stolpe an overview of the stock of pictures. It in time amounted to thousands of paper copies and a number of photo albums, which explains why to this day the images are to be found in the possession of so many institutions and individuals. The original plates, however, are kept in the Museum of Ethnography's archive in Stockholm. They were originally Hjalmar Stolpe's private property, as they were paid for with the 10,000 kronor donated to Stolpe in person by the patron Pontus Fürstenberg.

Of that plethora of paper copies, Stolpe's family kept around 400 examples, which have lain packed away in Copenhagen for over a century. Hjalmar Stolpe's grandson, the late Björn Fabricius, was their main keeper, and he eventually decided to place the collection at the disposal of the undersigned, together with the entire Stolpe family archive, comprising thousands of letters and other documents.

This material forms the basis of my book *Kungen av Birka – Hjalmar Stolpe, arkeolog och etnograf* (The King of Birka – Hjalmar Stolpe, archaeologist and ethnographer), published in 2015. For reasons of space, however, justice could not be done to Oscar Ekholm's images, nor to the information on the individual motifs. It transpired that Stolpe himself had sometimes written brief explanatory texts on the backs of the pictures, and had left descriptions in the various lists of images and commented on them in letters sent home to his wife. Thanks to this, it has been possible to furnish many of the glass negatives with fresh information, which I have also transferred to the Museum of Ethnography's image archive.

This new knowledge about the photographs combined with their first-class quality convinced me that Oscar Ekholm now deserves attention in the form of a volume that does full justice to his pictures, in terms of both format and print quality. It shows what a pioneering and skilled documentary photographer was able to achieve in the 1880s during a voyage around the world, despite cumbersome and unruly technology. But without Hjalmar Stolpe there would not have been a voyage in the first place. And without the mild-mannered and discerning archivist, Björn Fabricius, there would not have been a book about the Vanadis voyage.

Stockholm, August 2020

Bo G Erikson

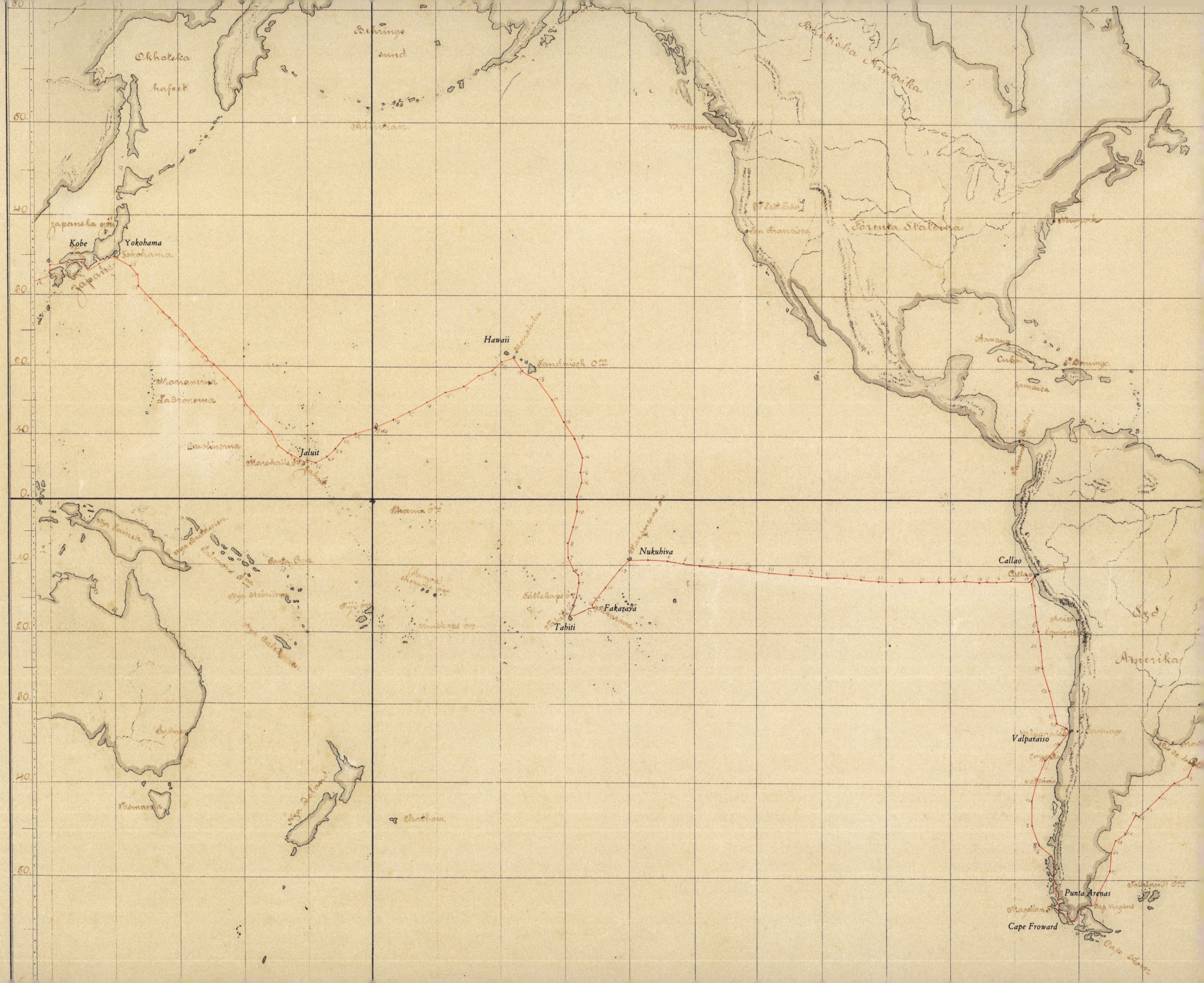
Kobe
Yokohama
Hawaii
Jaluit
Nukuhiva
Callao
Fakarava
Tahiti
Valparaiso
Punta Arenas
Cape Froward

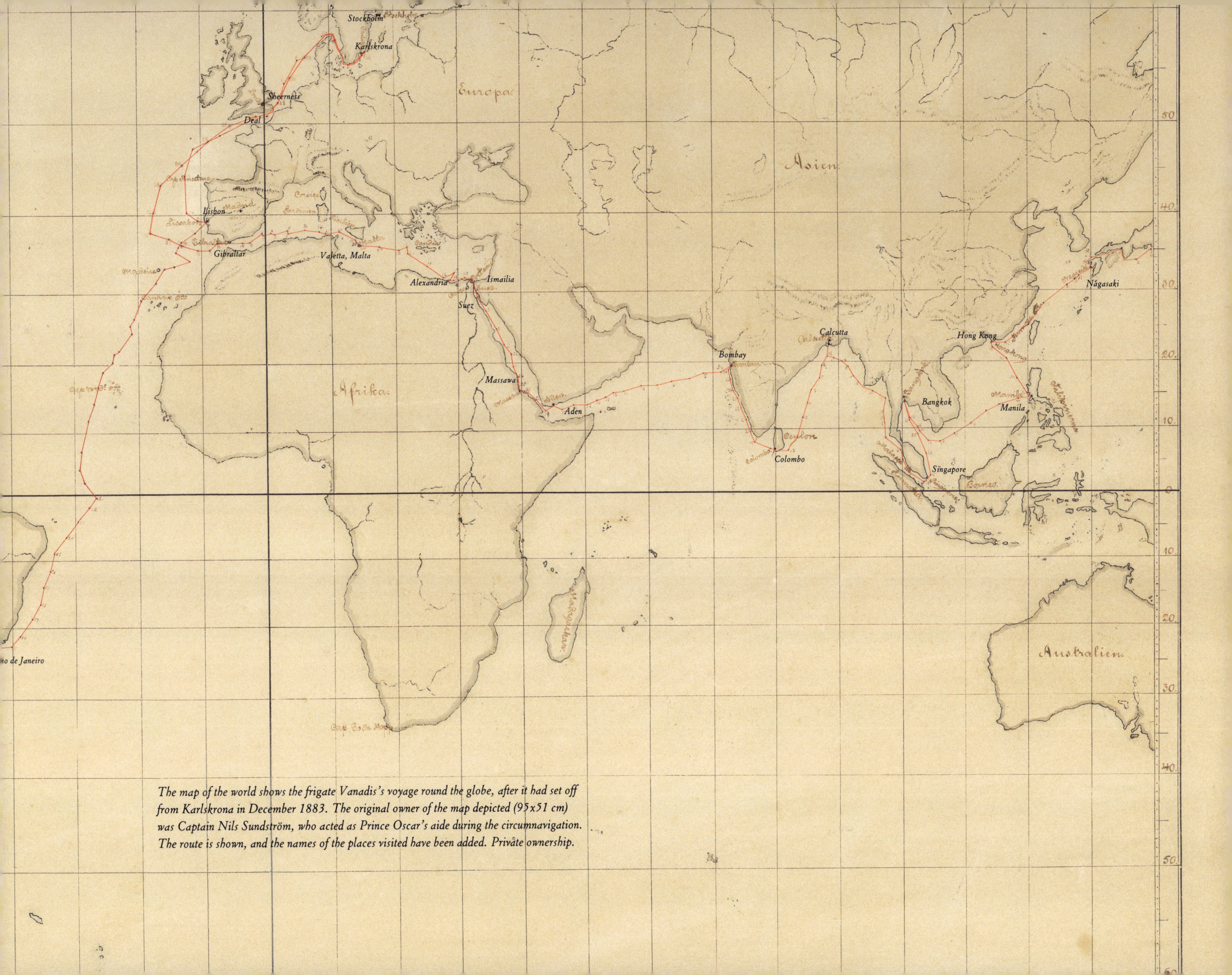

The map of the world shows the frigate Vanadis's voyage round the globe, after it had set off from Karlskrona in December 1883. The original owner of the map depicted (95x51 cm) was Captain Nils Sundström, who acted as Prince Oscar's aide during the circumnavigation. The route is shown, and the names of the places visited have been added. Private ownership.

THE VANADIS EXPEDITION TIMETABLE 1883–1885

PORTS AND ANCHORAGES

1883	
5/12	Karlskrona, departure
5/12	Helsingborg
10–13/12	Sheerness, England
13–15/12	Deal
22–29/12	Lisbon, Portugal
1884	
1–7/2	Rio de Janeiro, Brazil
29/2–1/3	Punta Arenas, Chile
2/3	passes Cape Froward
4–6/3	Columbino Cove
6–7/3	Gray Harbour
14–25/3	Valparaiso
3–14/4	Callao, Peru
8–12/5	Nukuhiva, Marquesas Islands
15–17/5	Fakarava, Tuamoto Islands
19/5–2/6	Papeete, Tahiti, Society Islands
20/6–10/7	Honolulu, Hawaiian Islands
26/7–2/8	Jaluit, Marshall Islands
26/8–1/10	Yokohama, Japan
3/10–5/10	Kobe
8–11/10	Nagasaki
18–28/10	Hong Kong
2–6/11	Manila, Philippines
17–23/11	Bangkok
28/11–3/12	Singapore
18/12–4/1	Calcutta, India
1885	
12–18/1	Colombo, Sri Lanka
26/1–4/2	Bombay
18/2	Aden
21–25/2	Massawa (Mitsiwa), Eritrea
4–6/3	Suez, Egypt
6–7/3	Ismailia
7–9/3	Port Said
11–15/3	Alexandria
24–28/3	Valetta, Malta
11–15/4	Gibraltar
28/4	Deal
2/5	Malmö
2–4/5	Skanör
5–7/5	Karlskrona
8/5	Vigsö, Stockholm archipelago
9/5	Stockholm

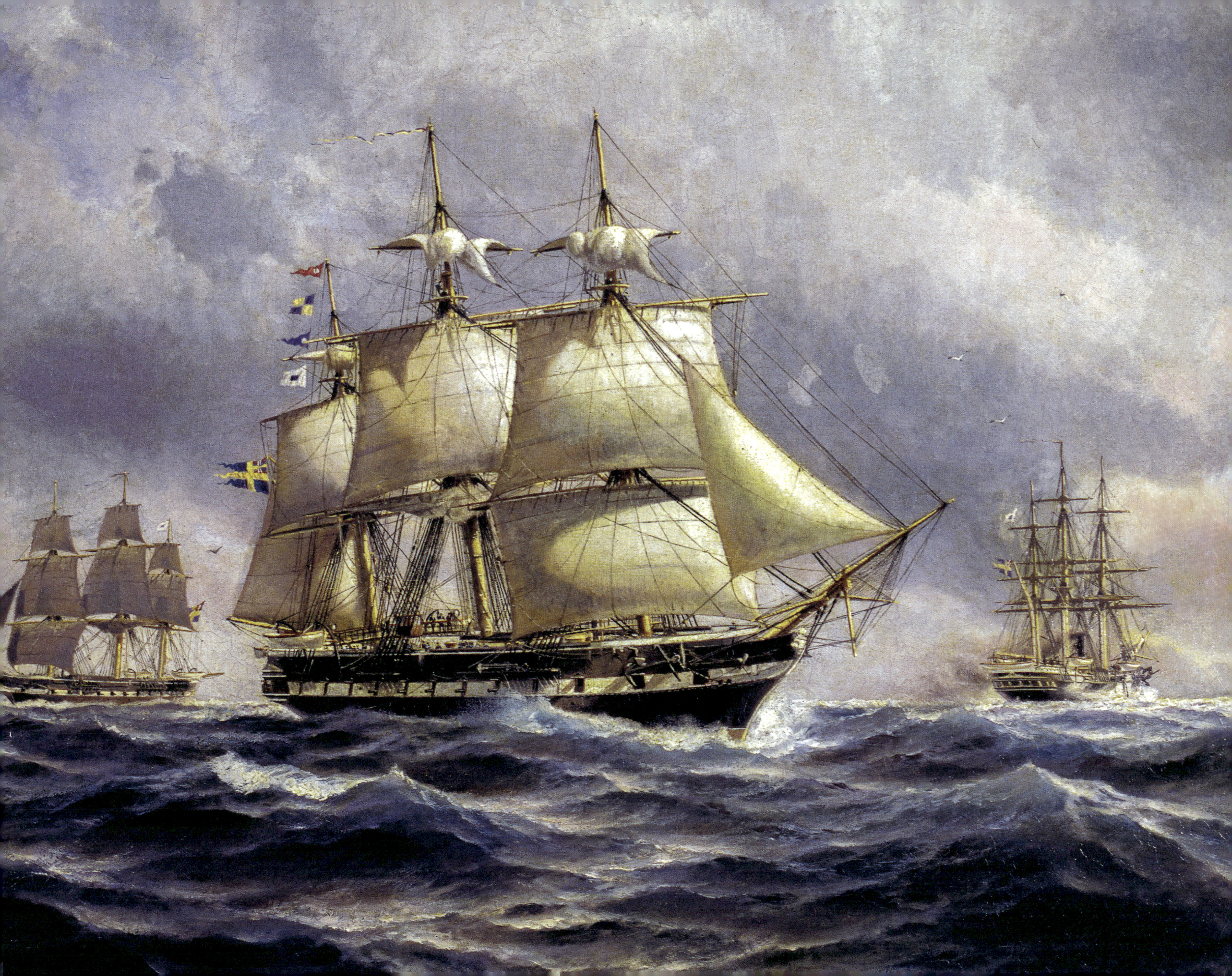

THE VANADIS EXPEDITION

The voyage around the world of the steam frigate Vanadis was followed with great interest both at home and abroad. The beautiful ship and its global circumnavigation aroused strong feelings of national pride, demonstrating that we Nordic peoples were also present on the world's oceans, where Europe's major powers closely monitored their colonial interests and where global trade exploded thanks to ever improving vessels of ever increasing size. In the mid-1880s the world's merchant fleets comprised nearly 44,000 sailing vessels and over 8,000 steamers.

The little double monarchy of Sweden/Norway was certainly not a colonial power. Some years before – in 1878 – the Swedish colony Saint Barthélemy in the West Indies had been sold back to France for 300,000 kronor. But there was great interest in burgeoning world trade, and it was precisely this interest that was the main reason for the Vanadis expedition. The nation's sparse, dormant networks of ambassadors and consulates across the globe needed bolstering, inspecting and supplementing. The government deemed this approach to be the prerequisite for smooth, successful foreign trade and tradepolicy agreements at state level. The instructions established 'protection and promotion of trade' as the main purpose of the voyage. Swedish/ Norwegian exports chiefly comprised wooden products and iron, the quality of Swedish iron being deemed unparalleled.

But this journey of inspection was also to act as a promotional tour. The decision was thus made to despatch the Swedish navy's most beautiful vessel, His Majesty's warship Vanadis, a slender frigate equipped with both sails and a steam engine. In order to lend the expedition additional prestige His Majesty Oscar II placed one of his sons – Prince Oscar – on board. The prince's participation was to make the expedition into a major diplomatic success and open doors to the world's crowned and uncrowned heads. In order to further justify the expensive expedition, which cost 35 million Swedish kronor in today's money, a huge amount at the time, the diplomatic assignment was combined with a military one. For 18 months, 300 men were to be drilled in martial activities and marine seamanship.

There was also a third element to the mission – a scientific one. During the voyage across the oceans and to the Pacific islands, systematic meteorological and magnetic observations were to be made, so as to support future trade trips. There were scientific instruments on board, and specialists had received training at Uppsala University.

But the specialists who were to make the Vanadis expedition memorable and important to this day were two differing pioneers: the archaeologist and ethnographer Hjalmar Stolpe, and the photographer Oscar Ekholm. Their remit was to collect ethnographic objects from the threatened cultures of indigenous peoples and document them with images. Their participation had been questioned by the navy; the state's financial support was meagre, and the assignment was so underfunded that Ekholm had to travel as a paying passenger. But the results of the voyage amazed everybody – nearly 10,000 objects and over 700 photographic plates were brought back to Stockholm.

OSCAR EKHOLM

Hjalmar Stolpe had quickly realised the significance of the nascent technology of photography in scientific contexts, and worked hard to get a photographer on board. Six months before the departure he had found Oscar Ekholm, and wrote as follows to his father Carl Johan Stolpe, the juridical burgomaster in Norrköping:

> I'd like to propose a very skilled one, 21 years old, who has sailed for 16 months as a deckhand on merchant vessels. He is willing to agree with whatever the conditions may be, and Captain Nissen has hit upon the idea of his enrolling under the heading of valet for Nissen. The young chap is delighted. He will be invaluable to me. He has done a lot of photography for my ethnographic album, and is excellent at his job. Is furthermore a decent sort and looks presentable.

Oscar Ekholm was then a totally unknown photographer, and to this day knowledge about him is scant. He was born in Södermalm in Stockholm on 14 June, 1861. His parents were the stonemason Per Ekholm and Hedvig, née Lindberg. They owned a property in Södermalm that provided them with rental income. They had five boys, the youngest of whom died before the age of one. A few years later Per died of tuberculosis at the age of 46. Hedvig remarried, at 41, but her new husband, just 24, suddenly disappeared to America. Hedvig's life was beset with a chain of tragedies. She had to bury one son after another, one of the causes of death being tuberculosis. She alone survived.

Hedvig did, however, get to experience her son Fredrik (born 1857) train as an architect and his brother Oscar, who was four years younger, go on to establish himself

The attractive frigate Vanadis was the fleet's pride and joy. From her launch in 1862 and for the next 30 years she represented the Swedish/Norwegian double monarchy on the world's oceans on 16 long voyages. She was also the favourite ship of the brilliant maritime painter Jacob Hägg. This painting of 1891 with surrounding sister ships is slightly cropped.

as a photographer. The latter went to sea at an early age, and then learned the craft of photography at Robert and Ernst Roesler's studio in Stockholm. It was on Roesler's recommendation that Stolpe hired Oscar as a photographer on the Vanadis, after many people had contacted Stolpe and applied for the job.

But Oscar Ekholm also had a sound photographic lineage. His maternal uncle was Carl Fredrik Lindberg (1830–1893), the Royal Swedish Academy of Letters' first professional photographer, who was also an excellent draughtsman and artist. He was hired in 1867, and over the years he criss-crossed Sweden in order to photograph and draw the country's monuments – archaeological, sacred and profane. He often travelled together with the national antiquarian Hans Hildebrand, who in the Academy's monthly newsletter talked about the important assignment he had given Lindberg: 'to photographically depict the more prominent ancient monuments. Both on his own initiative and at the behest of others Mr Lindberg has produced major series of photographs, on the basis of which one can with good reason give a complimentary testimonial'.

The business-minded Lindberg sold his photographs directly from the Royal Academy of Letters, a 'whole plate' costing 1 krona and 25 öre (1.25 kronor), and a half plate 1 krona. Lindberg was the man behind the often reproduced picture of a profusely bearded national antiquarian sitting by the famous Rök runestone in the county of Östergötland, which dates from the ninth century and features the world's longest runic inscription. With a matchless feeling for surprising compositions, he photographed the ruins of Borgholm Castle dimly visible behind two old women and a lighthouse. Carl Fredrik Lindberg was one of the great pioneers of photography, now unfortunately as forgotten as his nephew Oscar.

Before the Vanadis voyage, Ekholm had thus already done some photography for Hjalmar Stolpe for an ethnographic album – which in time would be rewarded with a prize. Stolpe appreciated the young Oscar's efforts, and throughout the arduous circumnavigation he praised his photographer: 'In the long run he's perfect company – quiet, decent, diligent and interested in his assignment. I'm most satisfied with him'. In a travel letter to the daily paper *Stockholms Dagblad*, he gave Ekholm public praise: 'On a voyage such as this, the good fortune of having the assistance of a skilled photographer cannot sufficiently be extolled'. After returning home, Stolpe observed that Ekholm's pictures represented 'by no means the least significant scientific benefit of the voyage'.

There is no further information on Ekholm in Stolpe's notebooks or correspondence. The special accounts book for the photographer only lists expenditure on hotel rooms, transportation of heavy equipment, the hire of a horse for riding, purchase of glass plates and various chemicals such as gold chloride, bromine chloride and iron sulphate. In his notebook from India, Stolpe records the purchase of '44 dozen half plates at 5.50 kronor. 26 dozen in visiting-card format at 3.00 kronor. Sought in vain to sell the impractical plates in Calcutta'. The most-used format during the voyage was whole plates measuring 18x24 cm. But sometimes the stock did not suffice, as, for example, in Benares in India: 'Unfortunately I had none of my large plates left, so I was unable to take any full-length photographs'.

But Oscar Ekholm was not hired just as a photographer. He was also engaged to 'assist in the ethnographic and anthropological investigations I've been commissioned with carrying out during the expedition', Stolpe wrote to the Swedish Ministry for Naval Affairs immediately prior to departure. The Ministry reserved for Ekholm a 'berth and ship portion' – which Stolpe would pay for – plus daily remuneration. The additional role of valet for the navigation officer Rudolf Nissen was frozen. To mark the photographer's special position on board, the Ministry decided he should be given the rank of non-commissioned officer (Stolpe held the rank of officer), which meant Oscar's hammock was suspended amongst those of the non-commissioned officers. Meals were taken in the so-called Gunroom – a significant advantage compared with the conditions for the other ordinary crew.

In accordance with Stolpe's agreement with Ekholm, during the voyage the latter was to make five paper copies from each glass plate at no extra charge. This did not always happen, however, and after they returned their relationship deteriorated. Ekholm attempted to cash in on his participation in the world voyage, opening his own studio in Stockholm, though without success.

With bankruptcy threatening, Ekholm moved to Sundsvall in 1886 and the next year to Härnösand, where he set up a new studio. In 1888 he married Anna Vogel. Their happy family life was cut tragically short, as only two years later (on 5 January 1890) the 28-year-old Ekholm died of tuberculosis. A month later his nearly one-year-old son expired from the same disease. His widow – who survived him for over four decades – filed for bankruptcy, and Hjalmar Stolpe was incorrectly stated as being in debt to the estate. In a letter to the town bailiff in Härnösand, Stolpe pointed out that it was, on the contrary, Ekholm who was indebted to him:

> From several plates I have never received a copy! There are around 700 plates, and according to the register just 655 copies were provided. I have not had the opportunity to check whether this figure is correct. As Ekholm did not meet his copying obligation during the course of the expedition, it was of course incumbent on him to do so afterwards. His imprudence in immediately setting up his own studio without any assets landed him in a precarious situation, which presumably hindered him. When I ultimately learnt (in 1886, I think) that he was on the verge of going bankrupt and leaving town, I enjoined him to hand over the plates, the apparatus and the relevant utensils.

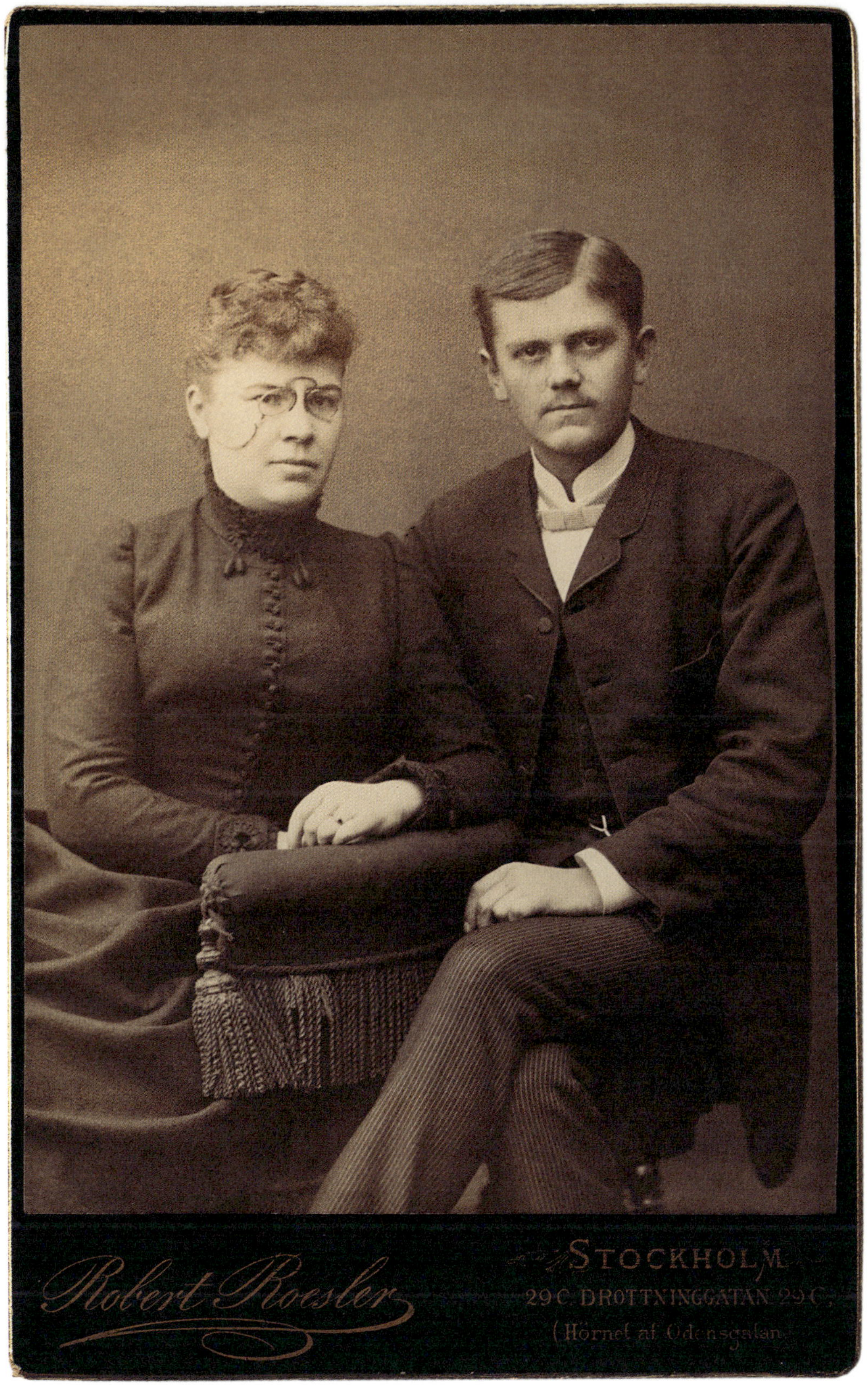

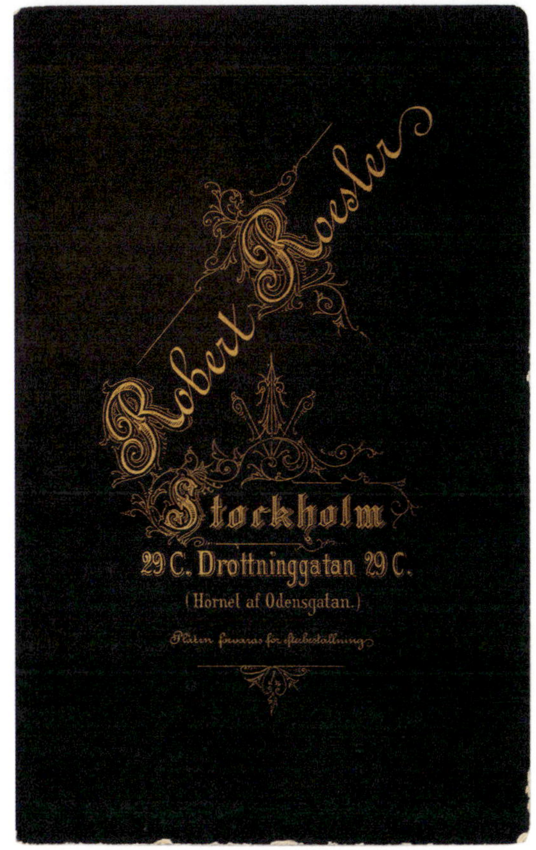

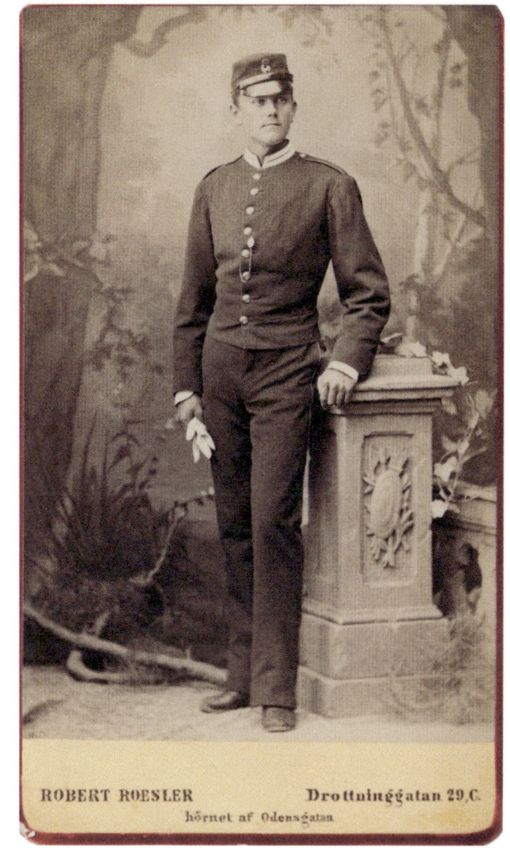

After the Vanadis voyage and the return to Stockholm, Oscar Ekholm (1861–90) tried to establish himself as a photographer with his own studio in the capital. This venture failed, and he moved to Sundsvall, and subsequently Härnösand. In 1888 he married Anna Vogel. Both Oscar and his newborn son died of tuberculosis in 1890; Anna survived them for over four decades. Oscar did his photographic training at Roesler's photographic studio on Drottninggatan, where a portrait photo of him as a young recruit was also taken.

Stolpe thus took charge of the camera equipment when Ekholm left Stockholm. And nearly 300 of the glass-plate negatives ended up in Roesler's photographic studio, where they were numbered and provided with brief descriptions. Over the years they were copied and delivered to friends and backers, and to the Vanadis crew members. Where and how the original plates were subsequently stored before ending up in today's Museum of Ethnography is unclear, though it is evident that they did at times suffer from their century-long storage in sheafs of newspaper and other bundles.

HJALMAR STOLPE

Hjalmar Stolpe (1841–1905) was twice the age of his young photographer, and already a respected scientist. He was someone who was assertive, as often happy as he was angry, uncompromising, a perfectionist in his research and a good-for-nothing in terms of his private finances. At times a hypochondriac, he was also a draughtsman and a prominent singer, one of his appearances being as Papageno in Mozart's *Magic Flute*. During the 1870s he excavated Sweden's oldest town, Birka, on the island of Björkö in Lake Mälaren, and gained a doctorate in the subject. The investigations lasted ten years, and in the course of time the self-taught excavator became our first major field archaeologist, whose excavation and documentation techniques were historic and who is to this day admired for his precision and elegance.

During the arduous Birka years, however, Stolpe became increasingly fascinated by the young science of ethnography, which was tentatively developing in association with central European museums – primarily those in Berlin, Vienna, Paris and London. For more than a year he travelled around the continent, studying pretty much all the ethnographic collections – a good 80 of them – and their multitudes of objects from the cultures of so-called primitive peoples. He often found the collections in disarray and the objects incorrectly labelled – rather like in those old cabinets of curiosities that well-to-do collectors set up in their chambers, where stuffed lizards were displayed alongside stone axes and coconuts.

Back from his European trip, Stolpe was intent on showing how the objects should be organised and displayed. During the period 1878–79 he thus organised the country's first major ethnographic exhibition in the Palace of the Hereditary Prince in Stockholm, based around a selection from the 10,000 ethnographic objects available in Sweden. He carried out this huge project on his own, and in parallel with the Björkö excavations – a well-nigh implausible amount of work. Having trained himself as an archaeologist, he was now a successful self-taught ethnographer – Sweden's first.

His special interest was the ornamentation embellishing objects from the Pacific islands. At the museums, thousands of copies of the ornaments had been made, and using them he created a new method of revealing the ornaments' meaning and the objects' origin, both temporally and spatially. This was Stolpe's so-called theory of ornamentation – an ethnographic variant of Darwin's theory of evolution, he emphasised. Everything on earth was in a state of development, including ornamentation:

> It is a curious spectacle that the laws of ornamentation ... apply right across the globe to all peoples, be they prehistoric or modern – to primitive peoples with the lowest standing just as to the most highly developed cultural civilisations. Everywhere the same course of development.

But he made no secret of the fact that he borrowed the idea from his archaeological colleagues at the Museum of National Antiquities in Stockholm, Hans Hildebrand and Oscar Montelius, and their groundbreaking typological and comparative method, which brought order to the former archaeological chaos. Internationally, Hjalmar Stolpe was considered a pioneering ethnographer, and in due course he was offered a job at the British Museum in London, no less. At home in Sweden, Viktor Rydberg described him as 'one of our most meticulous experts on remote civilisations'.

During the Vanadis voyage Stolpe had many opportunities to study the ornamentation of 'primitive' peoples and expand his archive of copies.

After his archaeological career, Hjalmar Stolpe's goal – his big aim in life – was to create an ethnographic museum based on scientific principles, with him in charge. And he saw the Vanadis voyage as a big step towards this goal. But despite his well-documented proficiency he had to fight hard to attain it. His ardent determination is evident in the way he campaigned for a berth on the ship, even at one point writing: 'The proposed ethnographer Dr H Stolpe is willing to undertake not to seek any convenience at all on board; he is willing to live and spend time in the coal room, in other words subject himself to any conditions at all, as long as he can be involved'. Upon closer consideration, however, he deleted this passage.

In the end it was Oscar II who personally sanctioned his involvement, following a successful visit to Stockholm Palace. Stolpe maintained a special relationship with the royal family throughout his career – slightly surprisingly, considering that he was neither noble nor rich, but a subordinate amanuensis. As early as in the Björkö years the king seems to have taken to the handsome young archaeologist, with his fine singing voice and graced with pleasant social skills. He sent his princes to the Birka excavations with important guests. Stolpe was also invited to stay the night and eat at the royal table at Drottningholm Palace, in the park of which the monarch could be seen walking with his arm around Stolpe's shoulders in conversation with him. The king was also to intervene when the Vanadis voyage's ethnographic remit was threatened by financial disaster, stumping up 6,000 kronor of private money (equivalent to 380,000 kronor today).

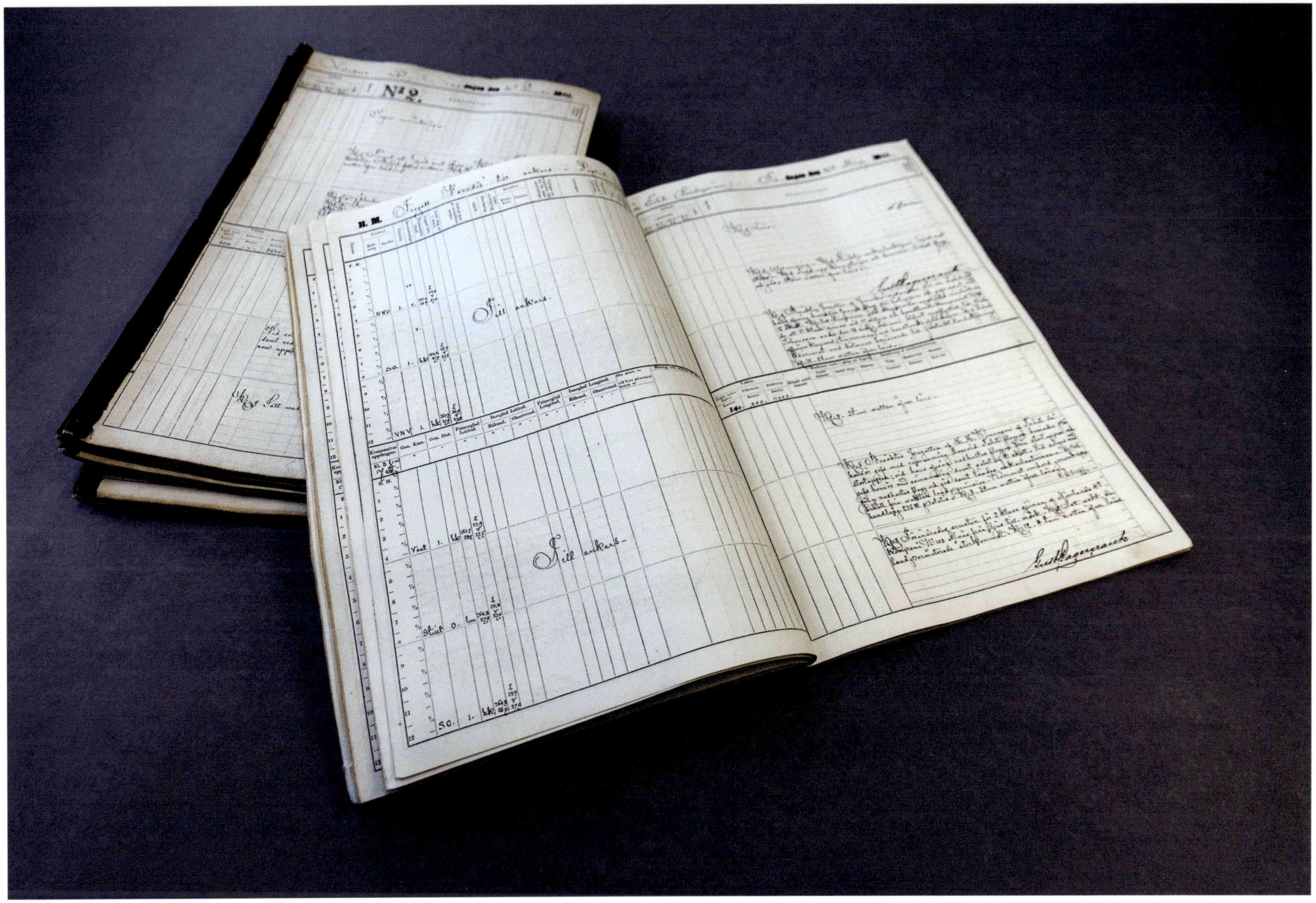

On 5 December 1883 the Vanadis started her voyage round the world, and for 18 months a detailed daily log was kept of the journey and life on board. The top of the first page of the logbook lists the armaments, i.e. the number of cannons and their weight. The captain of the nearly 340-strong crew was the commander-in-chief Otto Lagerberg, and the officers included the Duke of Gotland, Prince Oscar. One of the scientists was Doctor Hjalmar Stolpe. The logbook was an important legal document that provided hour-by-hour details of what happened during the journey.

The practical conditions for participation in this voyage around the world were not exactly lavish. Hjalmar's cabin offered an area of four square metres, and was located furthest abaft, where, as he acerbically stated, the heaving of the sea was at its greatest. Space was to be made here for several metres of reference literature from the Royal Library (including on James Cook's famous Pacific voyages a hundred years prior), clothes, a rifle, a revolver, knives for ethnographic bartering with indigenous peoples, writing utensils, glass plates and cigars. Hjalmar accepted the small space – the officers didn't have it any better – but not the rocking. He had previously only embarked on calm outings on Lake Mälaren to Birka, on the island of Björkö. It now transpired that he was simply unable to cope with the swelling of the sea and the movement of the waves. He couldn't work in the cabin, and this was why the planned book on the Vanadis voyage never got written.

The financial conditions meant that Stolpe was to travel at the state's expense in his own cabin, his living costs covered at the level of 'officers with the rank of captain and their equals'. This did not mean he received a captain's salary – something many people have believed. The modest monthly salary for his amanuensis job at the Museum of National Antiquities continued to apply – though as the captain's equal, he did receive nine kronor per day by way of 'subsistence money and sea wage'.

Stolpe found these conditions to be reasonable and acceptable. Unfortunately the same could not be said about the budget for the ethnographiccollection remit. Stolpe's foremost supporters had always been the country's two most prominent scientific institutions: the Royal Swedish Academy of Sciences and the Royal Swedish Academy of Letters. They had refrained from requesting state support for the ethnographic purchases and the photographer Ekholm. A committee had instead been formed, to raise 18,000 kronor of private funding (approximately 1 million kronor in today's money).

But the money-raising went dismally, and immediately before the start of the expedition a mere thousand kronor had been collected. Stolpe was furious. He realised it was an impending fiasco, but was unable to drop out such a short time before the departure date. The circumnavigation was being mentioned in every newspaper, and he was treated as a celebrity when he and his wife Emmy appeared at the theatre. He was courted by publishing houses and newspapers; international contacts had been established and equipment had been bought. He would have lost face, and the scandal would have affected him alone, as the leading figure in the venture – and not the committee members, headed up by the national antiquarian Hans Hildebrand, who had failed to raise the money. A guardian angel then appeared:

> Dear Doctor Hjalmar Stolpe,
> I have had the honour of receiving the Doctor's esteemed communication, and it was with great pleasure that I subsequently received further confirmation that my gift was so welcome and so abundantly promises to be of great benefit to the scientific part of the expedition. Warm thanks for your both friendly and flattering words. I have this day made the amount (10,000) available to you through the bank Skandinaviska Kredit Aktiebolaget.
> Yours faithfully, Pontus Fürstenberg

It was Stolpe himself who in desperation had sought the support of the Gothenburg art collector and patron Fürstenberg, in order to hire Oscar Ekholm as a photographer. At the same time a bank loan for the ethnographic purchases was also arranged. Without this money, which materialised just prior to departure, Stolpe and Ekholm would have had to stay at home and the ethnographic expedition would have been called off at the eleventh hour, and that would have been a huge fiasco.

THE NEW PHOTOGRAPHIC TECHNOLOGY

Despite Fürstenberg and the bank loan, the ethnographic expedition was still underfunded, and was beset by financial problems that Hjalmar Stolpe needed to resolve. But on one matter he was adamant – there had to be a photographer. Stolpe was extremely interested in the new photographic technology, and liked taking photographs himself, as was evidenced by a number of family photos from Björkö and the summer paradise retreat of Kolmården. And according to the card index he left, his home library included the handbook *Kompendium i praktisk fotografi* (Compendium of practical photography) and *Priskurant å Fotografiska Artiklar* (Price list for Photographic Articles).

Ahead of the Vanadis voyage, Stolpe had doubtless been inspired by the photo collection Louis Palander brought back with him from the coast of Siberia, where during the winter of 1878–79 Adolf Erik Nordenskiöld's ship Vega had been frozen in, before managing to continue its historic voyage through the Northeast Passage. The extreme cold – sometimes as low as -40°C – affected exposures and damaged the pictures of the local population. But despite these difficulties, photographic technology demonstrated unique possibilities as an ethnographic documentation method. And, thought Stolpe, if Palander's camera could work in deep-frozen Siberia, it would of course be even more viable in the tropical South Pacific. Oscar Ekholm would become a new Palander.

And despite the enormous difference in temperature, the similarity between their photographs of indigenous populations is striking. The subjects of the portraits stand in front of a basic backdrop of canvas or cloth. Palander's Arctic photographs of fur-clad people have been called character portraits. Ekholm's South

Pacific photographs of scantily clad or unclad individuals have been adjudged differently.

At the same time as Nordenskiöld was on his expedition, Stolpe was cultivating his interest in photography at the 1878–79 ethnographic exhibition in Stockholm – his first major ethnographic project. The exhibition was temporary, and Hjalmar thus wanted to document it for posterity, with the help of Ekholm's uncle, the Royal Swedish Academy of Letters' photographer Carl Fredrik Lindberg. On the latter's initiative and at his expense, a thousand objects were photographed, with Stolpe as the director. He noted: 'I did all the work, involving the groupings (but if they are ugly and asymmetrical, they are the work of Lindberg himself), the organisation and the cataloguing'. The result was a collection of 278 photographs with texts by the director, divided into five albums. They were awarded a prize at a conference in Venice, and received glowing reviews in international journals, including the esteemed English journal *Nature*: 'This photographic album must be regarded as the best existing ethnological atlas'.

As evidenced by the quoted lines to Hans Hildebrand, Stolpe was very active in terms of documenting the exhibition. He not only selected the ethnographic objects but also arranged them the way he wanted them in front of Lindberg's camera. Hjalmar was thus already extremely interested in photography, and was also very keen to influence the content and design of the images. This would become increasingly clear during the voyage around the world with Oscar Ekholm.

The success of the ethnographic exhibition and the subsequent photographic version of this exhibition meant a lot to Hjalmar Stolpe, showing, as it did, that photographic technology could be used as a method of documentation in major contexts. And Stolpe of course realised that the Vanadis voyage would allow him to repeat the exercise, but on a global level – a world exhibition involving objects from exotic locations combined with photographs from these very sites, ideally with himself in the picture as a globally active ethnographer. The pictorial material could be used directly as part of an exhibition – and this happened – and also in an accompanying book about the Vanadis voyage that would hopefully be worthwhile. Nordenskiöld's Vega book with Palander's illustrations had, after all, been a commercial success.

This is the background to and explanation for Stolpe's uphill struggle to include a photographer on board the Vanadis, and it enables an understanding of the staggering cost he was prepared to take on. He actually invested a total of 10,000 kronor (640,000 in today's money) in the photographer, constituting about half of the budget for the voyage. The camera equipment purchased from the company Johannes Schmidt alone cost just over 80,000 kronor in today's money.

In the decade before the Vanadis voyage photography underwent an important change in technology. Up to that time cameras had been loaded with so-called wet glass plates – a cumbersome method requiring the plate to be prepared and made light-sensitive immediately before exposure. Development had to be equally fast, and performed in total darkness. During the years around 1880, wet plates were being ousted by the new factory-made dry plates, which were ready for instant use, were more light-sensitive and did not require immediate development. The new technology had of course been noted by Stolpe, who invested in a 'photographic apparatus using dry plates'. The format was chiefly 18x24 cm, but half-plates and so-called visitingcard format were also used.

As already mentioned, Stolpe was highly satisfied with Ekholm's work during the circumnavigation, but it was doubtless Stolpe who selected the motifs. He also firmly intervened as regards the composition of the pictures, e.g. the way the portrait subject was to pose. 'People happily defer to my wishes,' he noted in India. When photographing Indian women, however, this caused problems. 'You're not allowed to touch them in order to create the pose you want – you should see me standing there posing, to show how I want them to look,' Hjalmar told his wife. He thus made demands of his portrait subjects, but he also paid them, as can be seen from the cash book for Ekholm. For 'Photography of Negritos and Tagalas', a set of images from the Philippines, 40 kronor was paid. And 'Photography of live naked individuals' on Nuku Hiva in the Pacific Ocean cost 30 kronor. This readiness to pay is evidence of Stolpe's well-known respect for 'primitive peoples'.

Stolpe ensured that he was visible in the pictures. He appears in a good number of photographs in this book, and they are by no means normal portraits. If you look closely at the big overview pictures of holy sites and monumental buildings, a small, well-nourished figure in a bowler hat crops up surprisingly often. His image is that of the gentleman research traveller typical of the period, always impeccably dressed, though clothed in appropriate field attire when visiting grave holes.

This was of course a conscious marketing on the part of the global ethnographer Hjalmar Stolpe, as he intended to appear at the forthcoming Vanadis exhibition in Stockholm and in the planned Vanadis book – frequently mentioned in the correspondence and eagerly awaited by the publisher Karl Kruhs at the bookshop Fritzes Kongl. Hofbokhandel in Stockholm. These were the professional and commercial reasons for the many pictures of Stolpe, but there was also a private one. His beloved wife Emmy constantly wanted to see pictures of her husband during their long separation, thus he often sent her paper copies of his 'selfies'. On the back he wrote 'Sp', meaning 'Sparrow's collection', which has been preserved to this day, Sparrow being his somewhat unromantic term of endearment for his wife.

THE ETHNOGRAPHIC REMIT

As regards the main remit for the ethnographic voyage – the collection of objects and the accompanying photography – Stolpe and Ekholm were of course totally dependent on where the commander Otto Lagerberg chose to anchor his ship. The stocks of coal and provisions needed to be constantly replenished, but he also had to follow the government's instructions and specified schedule for mail handling.

The Vanadis voyage was thus a cross between a military exercise and a commercial public relations trip for the union of Sweden/Norway, during which the expedition was to 'fly the flag' in important cities, inspect consulates and pay respects to those in positions of power. The main role in this diplomatic remit was held by Prince Oscar, who concluded his marine training on board the Vanadis, gaining the rank of captain. He bestowed a royal lustre on the expedition, thereby increasing awareness of it. And the 24-year-old Oscar also became a close friend of Hjalmar Stolpe, whose cabin was situated just below the prince's.

Stolpe's instructions from the Royal Swedish Academy of Sciences were at odds with those issued by the government. The ship's commander Lagerberg needed to spend long periods at sea, with only brief sojourns on land, whilst Stolpe's assignment required the opposite. But it was not just the instructions that were incompatible – the two formidable gentlemen also collided, the aversion being mutual. The Vanadis typically stopped for only a couple of days or weeks at the same port, with the exception of Japan and India, where Stolpe was able to spend months collecting thousands of ethnographic objects. Another source of irritation was the commander's scant interest in scientific matters in general, and Stolpe's ethnographic remit in particular.

The worst thing, however, was the ban imposed on keeping the collected objects on board the Vanadis for any prolonged period of time. There was room for them, but during the course of the voyage Stolpe had to regularly send his collections home with the help of consulates and local businessmen. Via various merchant vessels the wooden packing cases and their unique ethnographic contents ended up alternately in London, Hamburg and Copenhagen, or sometimes went straight to Stockholm. And some of them never arrived. Stolpe was unprepared for these transportation costs, which were a strain on the already limited budget. It is hard to understand the commander's motives in this regard, especially when comparisons are made with an earlier Swedish expedition. On a similar voyage 30 years previously, the frigate Eugenie had returned with some 10,000 'specimens' from the mineral, plant and animal kingdoms. She was a mere 46 metres long, while the length of the Vanadis was 64 metres.

The instructions of the Royal Swedish Academy of Sciences stated that priority was to be given to the collection of ethnographic items, and Hjalmar Stolpe devoted nearly all his time to this task. But these instructions also prescribed the collection of 'casts, crania, skeletons, entire bodies etc.', as well as measurements of the bodies and skulls of the 'primitive' peoples ('the races'), all in accordance with the requirements of physical anthropology. The Academy furthermore emphasised that 'the art of photography, now widespread throughout the world' should be used in these ethnographic and anthropological studies. The 'photograms of persons' that were for sale were of 'little value in scientific terms', and were 'largely ethnographically/anthropologically unreliable', wrote the Academy, which in addition to personal portraits identified 'buildings and various other major ethnographic items' as being suitable photographic motifs.

Ekholm's personal-portrait photographs were thus taken over a hundred years ago on behalf of the Royal Swedish Academy of Sciences. Some of them are now described as being racist, though there is no consensus as to which of them are, the degree of nakedness appearing to be the crucial factor. Is it right or wrong to find them racist? Take a look and make your own judgement.

The instructions of the Royal Swedish Academy of Sciences had been clarified by the professors Gustaf von Düben and Gustaf Retzius, the country's leading physical anthropologists. Their race research and that of others are often seen as a breeding ground for racism. So, was Hjalmar Stolpe a racist? No, he wasn't. He was of course a child of his time, but in the thousands of letters he wrote there is nothing to indicate racist tendencies – something to be deemed quite unusual at a time when the world of European research was seething with racists. It is rather the opposite of racism that emerges – a burning and inquisitive interest in indigenous peoples based on respect, as is clear from his own observation: 'In each small trodden path one reads a small piece of the history of a people who are engaged in an arduous struggle with nature – a people whose standing is not in the slightest lower than once that of our forefathers'.

These were Stolpe's basic values, both during the Vanadis voyage and throughout his ethnographic career. He had no time for the fashionable 'skull-measuring' sciences. And it was thus logical that, after just a few weeks in the Pacific Ocean, he informed the Royal Swedish Academy of Sciences that he no longer wished to carry out their instructions – a courageous act of disobedience against the mighty Royal Academy's instructions by a simple amanuensis (as he called himself).

THE STEAM FRIGATE

In December 1883, the Vanadis awaited Oscar Ekholm and Hjalmar Stolpe in a cold and windy Karlskrona. The ship was now more than 20 years old; it was partially made of oak and was 64 metres long, and the three masts were capable of bearing nearly 1,700 square metres of sail. The engineering company Motala Mekaniska Verkstad had delivered four new boilers that could produce 1,400 horsepower, generating a maximum speed of about 12 knots. The two-bladed propeller was a clever design, and could be raised during sailing – at the same time as the tall chimney was lowered into the bowels of the ship.

The coal stocks sufficed for a week, while the crew had provisions for three months and water for 30 days. The armaments comprised 20 or so cannons, machine guns, plenty of bullets, gunpowder and fireworks. Running along the sides of the vessel was a white-painted belt, broken up by the 22 black-painted gunports – creating the so-called 'Nelson chequer'. Furthest fore was one of the powder kegs, the project room, water cisterns, six water closets, sick bays, the pharmacy and bathrooms. And furthest abaft were the officers' cabins and mess, plus the bread store. Stolpe also brought on board a novel item in the shape of a piano, which he had been commissioned with acquiring. It was clearly to be understood that a great singer had embarked, equipped with the requisite sheet music

For communication with the outside world there was a so-called signal book, containing hundreds of short ciphers for transmission by telegraph in the ports they called at. The number *129* meant 'Prince Oscar is well'. The cipher for Stolpe was *FP*, whilst HK stood for 'Man-eater'. Two huge electric searchlights – 'dynamo-electric light apparatuses' – made in Berlin by Siemens & Halske inspired awe on festive occasions. They had been placed up on the bridge, and were run by two separate generators.

The Vanadis was the first and last sailing Swedish steam frigate. Her name was drawn from Old Norse mythology, and was an alternative designation for the goddess Freya, the goddess of fertility, who presided over life and death, love and combat, fertility and black magic. She was deemed to be the fleet's most handsome vessel, and had a proud history. After her launch in Karlskrona in April 1862, the Vanadis was used by the government for a number of diplomatic and commercial assignments. During the years 1863–93 she was sent on 16 long voyages: the majority to the Mediterranean, one circumnavigation and the remainder to the West Indies (often Saint Barthélemy), with detours to Brazil and New York. At the historic opening of the Suez Canal in 1869 she had the honour of representing Sweden and saluting the principal guests of honour – the Empress of France, the Emperor of Austria, the Viceroy of Egypt and the Crown Prince of Prussia. Another prominent assignment was when, in September 1872, she took the ailing King Charles XV from Kiel to Malmö, where he passed away that very month. When, on 16 March 1878, Sweden returned the colony of Saint Barthélemy to France, the frigate was present and saluted the French naval vessels with its cannons.

THE CREW

When the Vanadis commenced the voyage, the crew comprised 340 men. The majority were listed as seamen (125) and boatswains (146), and there were also a dozen ship's boys. A number of professional groups were represented: carpenters, sailmakers, gunsmiths, divers, medical orderlies and even a few stewards. The Vanadis band, comprising five horn players and a drummer, was responsible for matters musical, and often played when the ship put into port and when distinguished guests were on board.

Some of the crew became successful entrepreneurs after their voyage around the world. The stoker Johan Victor Svenson became managing director of his own 'paraffin-stove factory', AB Primus, and later very successfully manufactured the paraffin-powered 'Avance' ignition-bulb engine. The seaman K J Flink came to be involved in the manufacture and set-up of electronic bell-ringing apparatuses in churches, and later with the manufacture of the 'Sparglöd' sawdust stove. And the ship's boy Ölander became the director of a cinema company. Boatswain No. 100 from the Tjust Company, Johan Oskar Humbla, wrote memos during the journey that in an edited state were in due course printed as a book.

There were 20 officers, who together with their 'equals' (including Hjalmar Stolpe and the ship's doctor) had their cabins in a so-called gunroom, and in the battery one level up. At the top amongst the cannons, the commander Otto Lagerberg (commodore) – a real seadog – had his quarters. His closest neighbour was Prince Oscar (lieutenant) and the navigation officer Rudolf Nissen (captain). Through a skylight one could look down on the officers' mess, which was surrounded by eight cabins. They accommodated Stolpe, the ship's doctor Karl Rudberg, the second-in-command Hjalmar af Klintberg (captain and deputy commanding officer), the consulate inspector Harald Ehrenborg, and the lieutenants Wilhelm Dyrssen, Gustaf Lagercrantz and Henning von Krusenstierna, as well as Nils Sundström, the religious prince's highly religious adjutant, whom Stolpe viewed with a critical eye.

Further down in the bowels of the frigate, a number of second lieutenants were to be found, including Arvid Lindman and Svante Natt och Dag. One could also find the assistant doctor Carl Otto Holmer here, Gottfrid Fineman, a meteorologist and physicist from Uppsala who had recently gained his Ph.Lic, and the person Hjalmar Stolpe found the most difficult of all the crew, the ship's priest Immanuel Kiellman-Göranson, a 'gasbag who wants to stick his nose into all things, be witty and understand everything, and is excessively inquisitive, in other words insufferable in my opinion'.

The officer group boasted an abundance of talent, which would lead to distinguished careers, both military and political. No fewer than seven of them were promoted to the rank of admiral, and three of them ended up as a secretaries of state and managers at the Swedish Ministry for Naval Affairs. The one who attained by far the greatest renown was Arvid Lindman (1862–1936), who embarked as an exuberant 21-year-old second lieutenant. He concluded his military career as a rear admiral, subsequently became a successful captain of industry and then had a brilliant political career as minister of naval defence, minister of foreign affairs and finally prime minister. But his marine background had left its mark on him, and he remained an *Admiral* throughout his life, both inwardly and outwardly.

With several of the officers and their equals, Stolpe was to have differences of opinion, both major and minor. They also proved to be amateur collectors of ethnographic objects, making them into competitors during the buying forays. He often complained of their driving up prices and buying objects right under his nose. Some of the gentlemen were also to become rivals in terms of courting the favour of publishers back home in Stockholm, and would compete as correspondents for a number of newspapers. It was not only Hjalmar Stolpe who was invited to write up his travels for books and the press. Within the four walls of the little gunroom, priceless intrigues flourished, and secret letters were written that only saw the light of day when the Vanadis had returned to Sweden.

The only person who was informed of the social complications of the long voyage was Emmy, Stolpe's wife, who was exhorted not to pass on what she knew. Those on board the Vanadis soon realised that information sent home was reported back remarkably quickly, and became known on board, especially if it concerned something that could be interpreted as being critical of the expedition. The ship's priest fared particularly badly when the weekly newspaper *Karlskrona Weckoblad* accused him of having gossiped about an accident involving a lost lifeboat. This was viewed most unfavourably at the highest levels in Stockholm. In a letter to the Vanadis's commander the minister of naval defence urged the tattling priest 'not to write scandalous articles'. But the latter proved to be innocent, it being a non-commissioned officer who had aroused the wrath of the touchy authorities.

As a result of this, as well as the reporting of a near-accident at the port of Sheerness near London, Hjalmar came to the conclusion that he should be sparing with details about the conditions on board the ship. And this is why his articles in *Stockholms Dagblad* reveal so little about life on board. But he was open in the letters he sent home, both about life at sea and concerning his travel companions, though it was only at sea that despondency and discontent developed. During the periods ashore he was transformed into an enthusiastically energetic ethnographer.

PRINCE OSCAR

The most exciting person on board was of course Prince Oscar, about whom Emmy and her friends wanted to know more. The 24-year-old prince was quiet and religious, read the Bible and the lyric poet Snoilsky and had a set dining place at the officers' table not far from Hjalmar. His cabin and bed were right above Hjalmar's, and now and then he looked in on him for a chat:

> It would seem I have to put up a picture of the King too, as the prince appreciates such things. He's abandoned titles with me, so he now addresses me using the informal 'du', and I say Prince; I thus get out of having to resort to Your Royal Highness, other than on festive occasions. He's pleasant and unassuming. Very friendly, and people don't differentiate between him and others. And he does his dog watch (12 midnight to 4 am) just like everyone else.

Despite his youth, Prince Oscar was an experienced seaman. At the tender age of 13 he was on board the steamer corvette Balder, and later went on long voyages to the Mediterranean and the US. During the Vanadis voyage he assiduously kept a diary – a total of '464 closely written pages', according to one of his biographers. Unfortunately the diary disappeared without trace from the Bernadotte Archive at Stockholm Palace a decade ago. The extant fragments, however, show that his mother, Queen Sophia, made sure the commander 'arranged a solitary, quiet cabin, where the Prince would be able to carry out his worship', and that he promised to read the Bible every day. He suffered from conjunctivitis, and the ship's doctor at times forbade him to carry out duties. The prescribed treatment was that he 'should be burnt with bluestone (vitriol of copper)' in order to get better. For some of the journey the prince had with him his own valet, Svante Segerman, though the latter was found to be 'consumptive', had to leave the vessel in Honolulu and later died in Chicago.

The prince was the second-eldest son of Oscar II, who showed great interest in the Vanadis voyage. In the salons, however, it was whispered that this interest was not solely based on his concern for his son's marine education and the kingdom's trading. The fact was that the prince had fallen in love with an insufficiently blue-blooded woman, Ebba Munck af Fulkila, the daughter of a postal inspector/colonel. As this misalliance might cause constitutional problems for the kingdom, the government and the royal family had hoped – according to malicious reports – that his feelings would cool during the circumnavigation. But this hope proved in vain. Ebba and Oscar were to marry a few years later, and the prince thus had to renounce his title of Royal Highness.

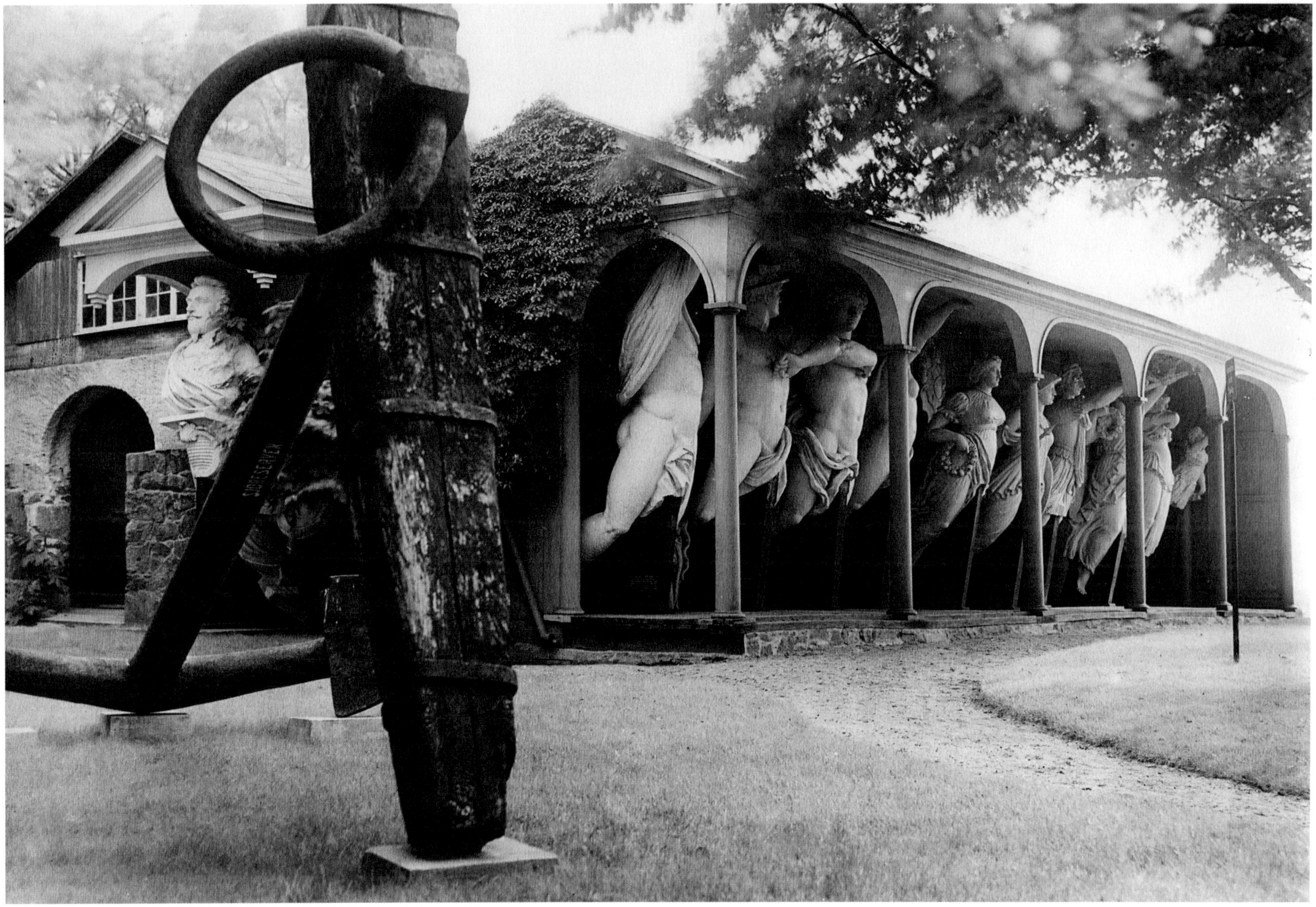

The picture from the so-called Galjonslunden ('figurehead grove') at Karlskrona's naval shipyard dates from the late 19th century, and shows a collection of well-turned figureheads. A sculpted figurehead also adorned the Vanadis's prow. It represented the Norse goddess Freya, who was a member of the Vanir and was thus called Vanadís. The sculptor was Henrik Nerpin from the Handicraft School in Stockholm. Before the figurehead was sculpted in wood, a maquette in the form of a wax model was made. Both the original and the model are to be found at the Maritime Museum in Stockholm.

Stolpe found in Prince Oscar an ally in his struggle against the commander's unwillingness to stop along the South American coast. If he refuses, Hjalmar wrote to his wife,

> I will submit a written protest. I've not been sent on this trip just to be at sea! The Prince for his part reflected likewise today. Then I'll go home, he said; I immediately requested that I might join him. Yes, dear Sparrow – being landed with such a bigoted idiot as our captain was most unfortunate! Stupid, conceited and arrogant about having gained this command – these are the three qualities the rest of us discuss and comment on. He's also a social fool who just wants to go to ports where he can mingle with high society and meet the ladies ... At 4 o'clock I'm to have dinner with the Captain, so it'll be a tedious day that will rob me of even more time! These dinners are insufferable. Poor wines, suffocating heat, dull conversation. The only saving grace is that the Captain usually has apples. We have to go dressed in white today, but alas I don't have a clean jacket.

LIFE ON BOARD

When the Vanadis called at Pacific ports the local bigwigs plus ladies were invited to parties on board in order to promote trade relations. The Vanadis band struck up for dancing, and the ship's priest noted contentedly that 'Swedish naval officers should without a doubt be deemed some of the finest dancers in the world'.

But this was the sunny side of life on board. It was not the life of the Vanadis's 300-strong crew. During the months at sea there was grumbling about the gloominess and total lack of any opportunity for relaxation, as was noted by Prince Oscar in his diary: 'In the evening I obtained permission from the second-in-command to have music played on the forecastle for the people. But this was the first time they'd heard this kind of thing for six months, so they never brought themselves to do any dancing. You could both see and hear that they were dissatisfied, and that they saw life on board as drudgery'.

In Rio de Janeiro the officers were granted permission to go ashore but not the rest of the crew, despite the health situation in the city having been reported as being good.

Commodore Lagerberg called for strict discipline and issued harsh punishments – so harsh that one crew member attempted suicide. Prince Oscar was most concerned: 'Seaman Nelson was reported for disobedience. Hearing with the captain, after which he was sentenced to 8 days' close arrest. He was both angry and distressed, and at 12:30pm attempted to jump into the sea from a porthole, but was caught by his hair and hung below a cannon'.

The Vanadis's logbook – in which records were entered every hour, 24 hours a day – sometimes gives the impression that there was a queue of crewmembers waiting to be arrested. The culprit then had to sit for six to eight days in conditions of 'close', 'dark', or 'light' arrest. Arrests are frequently mentioned, for the offences of drunkenness, 'assault on a senior', improper conduct, noise, embezzlement and desertion. In Lisbon three boatswains had fled the Vanadis. But this did not cause any great fuss, as desertions were a common feature of life at sea at this time – which makes what occurred in Valparaíso in Chile all the more unusual. No fewer than eight boatswains and one seaman deserted en masse, indicating great dissatisfaction with life on board. Ultimately the mass defection was a setback for the commander and his handling of the crew. A total of 21 men were to desert the Vanadis during the circumnavigation.

The daily routine for the crew was regulated down to the last detail. It was a matter of keeping 300 men busy and in reasonably good spirits for 18 months on a shared deck area of 700 square metres.

The crew was divided into two watches, starboard and port, who worked alternately on deck. Wakeup call 6.30am, meal around 12, coffee break in the afternoon when 'positions for combat' was sounded, which meant an inspection to check everyone was present. At the same time the lashing of the cannons was to be checked – very important for the crew's safety and the balance of the vessel. At about 6pm, evening meal, and two hours later, prayers – God and the King were important. Monday and Friday cannon drill, Tuesday and Thursday sail drill, Wednesday 'domestic-economy day' with washing of linen and patching and mending of clothes. And on Saturdays, big cleaning. In addition there was training for boarding, call-out and fireextinguishing, and sometimes at night without warning a 'ready for battle' call, when the crew practised for combat. But the armaments were fairly outdated, and the cannons were unable to shoot either ahead or astern. The boatswain Humbla commented as follows on the training:

> In the morning we had weapons inspection, then training in preparation: Ship ready for battle, which is the worst drill. In the afternoon we practised fire-extinguishing and pot inspection, plus carpet and tarpaulin washing for the watch. Saturday the 21st was cleaning with sand and stones ... In the morning there was instruction for both watches, sail drill on the mizzen-top and cannon drill with the breech-loading cannons. In the afternoon I was 2nd corporal on the top of the prow, where we did a drill involving taking down and putting up topgallant masts and spars.

The sometimes tropical heat naturally affected the training, as Prince Oscar wrote out in the Pacific:

> We have had beautiful weather the whole time, and a constantly good speed. The heat has risen above 28 degrees in the shade. Drills have been strenuous and very tiring at these temperatures, but the people have been granted some amelioration in recent days. Bathing after drills and sun tents on the forecastle [prow]. Thanks to these arrangements the mood amongst them has also become happier. The drill yesterday went excellently. The foretopsail was hoisted, and it started to smoke from the block.

The diet was probably the same as during other Vanadis voyages: herring pudding, crispbread in the form of 'ship's bread' so hard that it had to be soaked in the mealtime drink, peas (spiced with ginger), salt pork, pancakes, potatoes, fruit soup with sago and prunes called 'manoeuvre soup', rice pudding, cod pudding, and porridge. Once or twice putrid pork was served, which meant the crew got angry – 24 barrels of pork had to be thrown overboard. On festive occasions beer, wine, and generous drams of rum were handed out.

Throughout the circumnavigation the ship's company were free from epidemics and febrile illnesses. The vessel's doctor Karl Rudberg's explanation for this was a combination of good hygiene and the fact that their stocks of fresh water were often supplemented with distilled water produced by using the boilers. When Rudberg in due course completed his doctorate, it involved a dissertation on the state of health on board the fleet's vessels during the period 1865–85.

When the Vanadis was in the middle of the Atlantic Hjalmar drew up a summary of his everyday routines for his wife:

> 8am breakfast, 12 noon luncheon, 5 o'clock dinner and 8pm tea. 10 o'clock bedtime, but this does not prevent one's being up on deck for ages. Last night the purser and I walked until half past midnight in the moonlight. It was beautiful beyond description ...
>
> Do you know what is the most pleasant thing of the whole day? It's my morning shower. At 7 o'clock I walk clothed only in my long women's coat up in the battery, where a canvas curtain is suspended behind which you can get a shower of water at a temperature of nearly 30 degrees, pumped straight from the sea. And yet it is cooling for a while.

There was periodically a small collection of pets on board, writes the flagship officer C G Mårtensson:

> We have a monkey, three cats and a few score parrots, who greatly amuse us. Markus and Lukas [two goats] followed the crew up and down the steps, and they also got permission to go ashore, just like the crew, but made no attempt to escape, and the geese also followed their keeper, Trogen. The most difficult time for the black pigs that came on board in Honolulu and on the small islands was every morning when they were showered and washed. Every Saturday, when all the cages had to be cleaned, they wandered around freely, each and every one of them expressing their joy at existence in their own language.

FROM KARLSKRONA TO LISBON

On Sunday 2 December 1883, the Vanadis was inspected in Karlskrona by King Oscar and Crown Prince Gustaf, who were greeted with a 21-gun salute and manning of the yards, i.e. the crew had climbed up into the rigging. The future prime minister Arvid Lindman noted: 'Every man on the yards fired a blank. The whole time the royal personages were on board, the electric lights were on'.

Hjalmar Stolpe wrote to his wife:

> My cabin is now ready. All that remains is for me to put up a few bookshelves, as the bunk is still full of books. I've been out this morning, and have finalised the other arrangements. I now just have to get hold of a curtain to hang in front of my clothes, sou'wester and various other items of sailing gear. The weather is vile. It's snowing incessantly, but it turns into water.
>
> It's now starting to dawn on me that I'm actually going; up until now I've always forgotten it, or to be more precise have not had the time to think about it. And there's another reason why I'm starting to get my head round it, namely the dreadful storm raging right now. It's no fun, Sparrow!
>
> It's an awful storm – I don't really think I've heard anything like it. My only hope is that after this it will be tolerably decent again, but the sea is rough, and we will be rolling like anything. The Vanadis is said to roll valiantly, according to Pantzerhjelm. And my cabin is the worst. Today I happened on a Vanadis painting in a bookshop window, and bought it on your behalf. The round porthole furthest abaft below the white strip is my cabin window.

On the day the king inspected the Vanadis in Karlskrona, Oscar Ekholm wrote to his brother Fredrik:

> I'm feeling alright, and am very satisfied. When I came on board I was welcomed by the commodores Lagerberg, Klintberg and Nissen, and the officers. But any unpacking and laboratory work are out of the question in the current commotion – they will have to wait until we're at sea or in Lisbon.
>
> You would not believe the commanding screeching that goes on from dawn to dusk,

> but I'm living life to the full, go walking in the battery (below deck, where the cannons are) and smoke nearly all day. Our Pastor Göranson has no doubt caught a cold, as he is very ill at the moment.

On 5 December 1883, Arvid Lindman wrote in his diary: 'To the sound of folk songs we slowly left the inner roadstead watched by a huge crowd, who bade us their last farewell from Kungsbron'. An excited Hjalmar Stolpe ran into his cabin and scribbled down his final lines to his wife, which the pilot was to take ashore with him: 'Be strong!!! God protect you and the child! Courage! The circumnavigation is now under way. Whomsoever God wishes to show great favour is being sent by God into the wide world! Now, at 2.55, Raise Anchor is being called out! The royal anthem 'From the depths of Swedish hearts' is being played. Cannons are thundering. Emmy! Farewell!'

In the North Sea the Vanadis encountered angry waters, and Stolpe wrote:

> Strong northerly storm. Towards evening I became dizzy, and when I finally went below and lied down things got serious in my rocking cabin, so that my valet, captain of the after-guard Olsson, had to help me get undressed. As soon as I had got into my bunk everything passed, and I slept well, apart from the fact that I sometimes woke up from being on the verge of falling out of my bunk ... On the dinner table the sugar bowls fell over, and the soup flooded over the tablecloth. The meteorologist Dr Fineman went as white as a sheet and disappeared, the priest looked unhappy, and I doubtless did too.

On the same day that the soup was flowing in torrents, the Vanadis experienced a serious near-accident. When the manned starboard lifeboat was in the process of being launched it fell down the side of the ship and was crushed. Miraculously the crew escaped serious injury. And the drama was to continue. At the first port of call, Sheerness, at the mouth of the Thames, the voyage around the world nearly came to an abrupt end. The Vanadis had anchored in the roadstead and was taking in coal from two barges alongside when a violent storm blew up, and the frigate together with the coal barges started to drift towards a stone pier. Before they could reach it, however, the whole flotilla became lodged against the English monitor Hydra (John Ericsson's invention) and managed to get their anchors to gain a hold. The Vanadis then finally managed to free itself from the monitor, and the perils of the pier, under its own power.

In the English Channel the Vanadis band played marches and operetta excerpts, despite the seas, and Stolpe noted the compass courses and wind strengths: 'It got worse and worse. When I went down I found that two boxes of photographic plates had come adrift in the gunroom. Called the photographer down and chastised him for not having secured them. We may have sailed a few times with the lee gunwale under water, but this is just for a moment, and the boat immediately lifts and pitches over to the other side'. In a letter to Emmy, Hjalmar continued:

> Either go and walk on deck or sit and hold tight to a pillar in the battery – or lie in my cabin ... The days are monotonous, the nights long; hard to sleep with the constant creaking and squeaking of every joint – I think it's worst in my cabin and by my headrest in particular. The ship's movements stretch all the planks in the fittings, and it is especially the dividing walls between the cabins that create these sounds. I hope it will be better when all the joints have opened up and the boards have dried out so we can look in on each other. Don't think it's the actual ship that's making the noise – just the lighter joinery work in the fittings.
>
> And so the 18th arrived. The wind had switched to a westerly, and was getting fresher and fresher, and finally a storm broke out. We were a little northwest of Cape Finisterre. The sea was crashing into the battery. I sat in my usual place on a small rack for conical cannonballs, holding onto a pillar. Water was rushing to and fro, and it was too slippery to walk on a surface that was one moment sloping in one direction and the next moment in the opposite direction.
>
> One man was given the job of mopping the gunroom all night, to stop the water damaging our things. He had a bucket in which to squeeze out the mop, and when the bucket was full of dirty water it flew into the second-in-command's cabin and released its contents all over his face and soaked the bunk.
>
> I had another worry. The valet had neglected to empty [drawing of chamber pot], and I lay thinking about all the splashes that would trickle out onto my new carpet, which I had paid a lot for in London. But it turned out worse than that. The oaf in charge of the lamps (the Prince's impudent servant) had left a pitcher of lamp oil standing open in the galley, and the pitcher was of course smashed to bits and the oil got into my cabin, soaking the carpet, which I'd planned to use as my desk carpet (at home). I fear it's ruined.

This stormy episode when everything went awry and ran amok was dubbed the 'horrendous night'.

On Saturday 22 December, the Vanadis anchored in the Lisbon roadstead, and Hjalmar noted in his diary an odd expression of military life at sea – attentive shots from salute cannons. When the Swedish/Norwegian minister in Lisbon, Count Stenbock, and the legation secretary Baron Kantzow left the ship they were given a salute of 11 shots, and the Portuguese crown prince one of 21 shots.

Hjalmar put in at the Hotel Braganza. From his window he looked down onto the Vanadis, and for a week he would now work day and night appeasing his bad conscience regarding the overdue report about his last great archaeological excavation, which should have been submitted before his departure. The previous year saw the discovery of boat graves in Vendel in Uppland – referred to by Stolpe as the archaeological sensation of the century. Starting in the sixth century AD and for the following 500 years, a family of chieftains had been buried there in their gig-shaped boats –14 generations in succession in 14 graves containing swords, helmets and sacrificed horses unlike anything seen before.

His boss, the national antiquarian Hans Hildebrand, was angry with him. When Stolpe complained about his meagre travel funds and tried to get Hildebrand to contact Alfred Nobel in Paris, his boss replied that he did not intend to do anything at all until he had got the Vendel essay and the inventory of finds. So Hjalmar had with him his grave drawings – the best ever seen by the world of archaeology up to that time – on board the sometimes flooded Vanadis. Hjalmar had taken a big risk with the irreplaceable material. He wrote a few apologetic lines to Hildebrand: 'I, a poor sinful person etc. The papyrus is, however, on its way. What I have I've sent with Captain Appelgren'. It was only after again crossing the Atlantic and the North Sea that the Vendel chieftains were thus returned to Sweden.

Stolpe could now devote himself to ethnography with an eased conscience, and he quickly solicited a collection from the Norwegian sea captain J Amundsen on the bark Borghild from Frederikshald. The collection also included a geological curiosity – volcanic ash from the huge eruption earlier that year at Krakatoa in the sound between Sumatra and Java. The eruption had created 40-metre waves that had killed 36,000 people and caused atmospheric disturbances all over the world. Amundsen had taken on board samples of the volcanic ash, which Stolpe now sent home to Adolf Erik Nordenskiöld.

Hjalmar otherwise ran from shop to shop in Lisbon in search of finds, with moderate success. The city's colonial museum was full of rubbish, and Lisbon was a dismal city, he wrote to Emmy. Its population was found to be of a dubious quality from an aesthetic/anthropological standpoint: 'The people are ugly. I've barely seen a single attractive face. I never fail to check all the photographers' displays, where one has the best chance of seeing a number of facial types in one place, and the ugliness is striking. Even the children are ugly. There's something ape-like about the nation'.

Neither did the Portuguese royal family meet with approval. But it was now Prince Oscar who thought that 'the entire royal family apart from Don Fernando [the king's father] seemed to me to be real *mosucks*. The crown prince is a dreadful snob. – They had a disgusting custom whereby anyone greeting a member of the royal family went down on one knee and kissed the royal hand'.

Christmas in Lisbon started with the boatswains Flur, När and Hassel fleeing the vessel. On Christmas Eve the captain issued an invitation for dinner. The Vanadis had been decorated with transparencies – illuminated transparent paintings – and down in the depths of the machine room between the boilers stood a Christmas tree lit up with candles. A double quartet sang Hosanna, and the crew performed the play *Jernbäraren* (The Iron Carrier) by August Blanche. Between Christmas and New Year, Oscar Ekholm wrote to his brother Fredrik:

> My hammock is in the non-commissioned officers' mess, where we occupy ourselves with [the card games] Priffe, Knekt, Hund etc., as well as stories and anecdotes, because I must say they are nice, reasonable, smart chaps. I eat in the gunroom, as you will see from my letter to mama. How the photography is going I cannot yet tell you, as I've not yet been given a laboratory, but that will happen in the trade wind, when we will have plenty of time.
>
> How our journey has gone you will find, as I've seen, in the newspapers from Stockholm. If things continue this way I'll be on an excursion around the world.

A few days before the New Year of 1884, the frigate left Lisbon and commenced its journey across the Atlantic, towards South America.

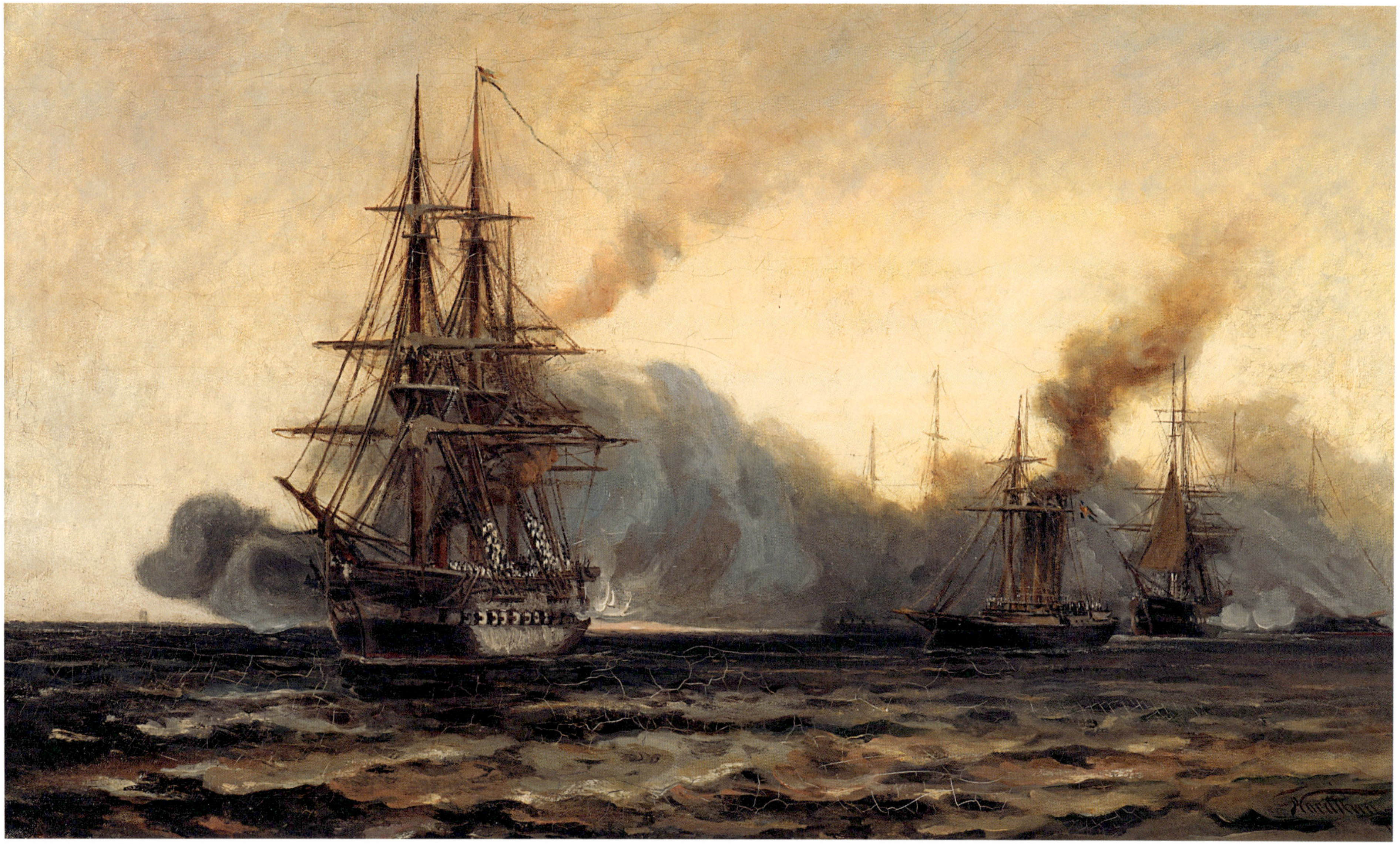

A dramatic painting showing the frigate Vanadis dating from the 1880s. The frigate (on the left) bears the royal command flag on the mainmast (naval flag and pennant). The salute from the vessels at the back is thus a tribute to royalty on board the Vanadis. The location is unknown, and the artist is Adolf Nordling.

Opposite page: The non-commissioned officers, gathered on the deck of the Vanadis.

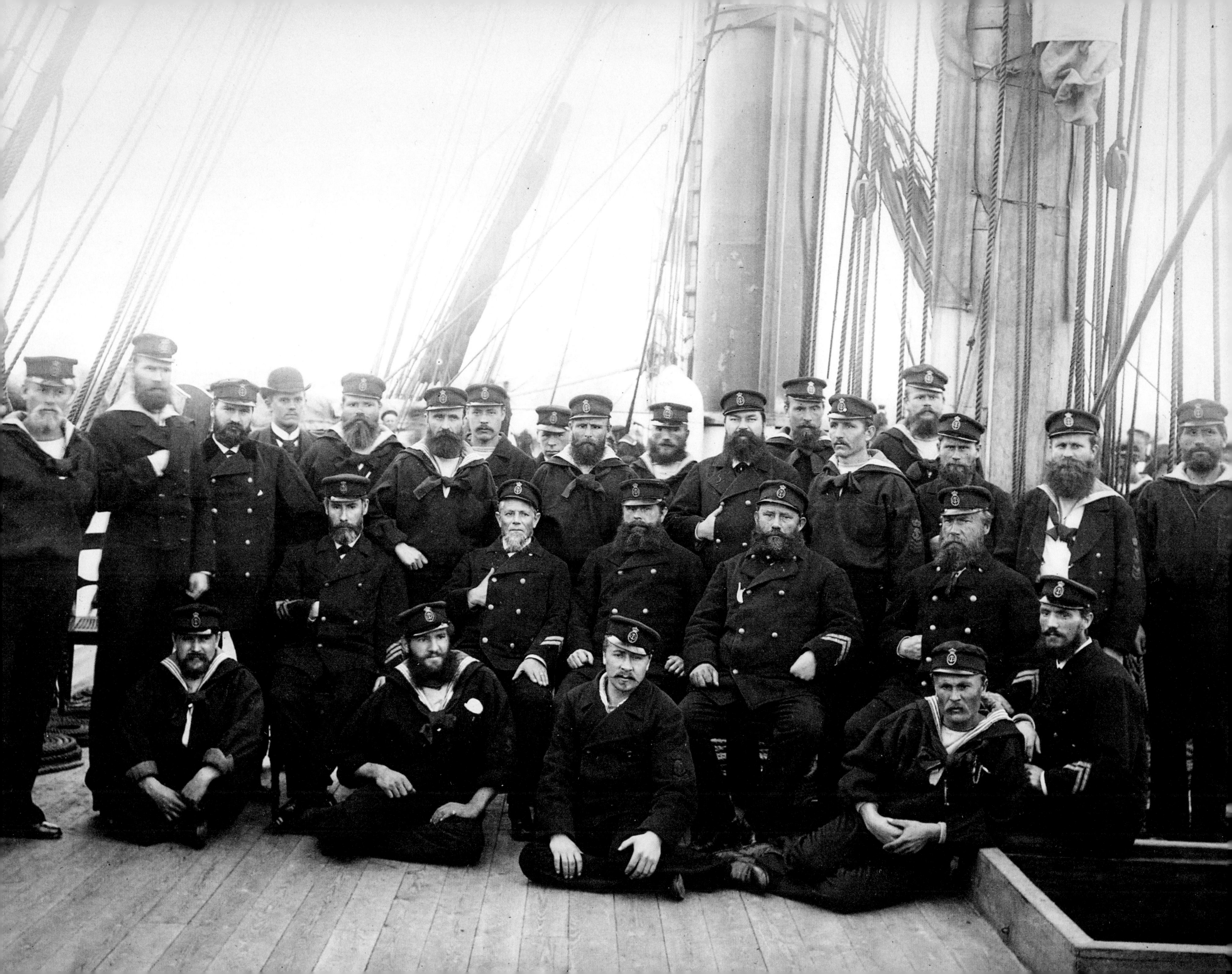

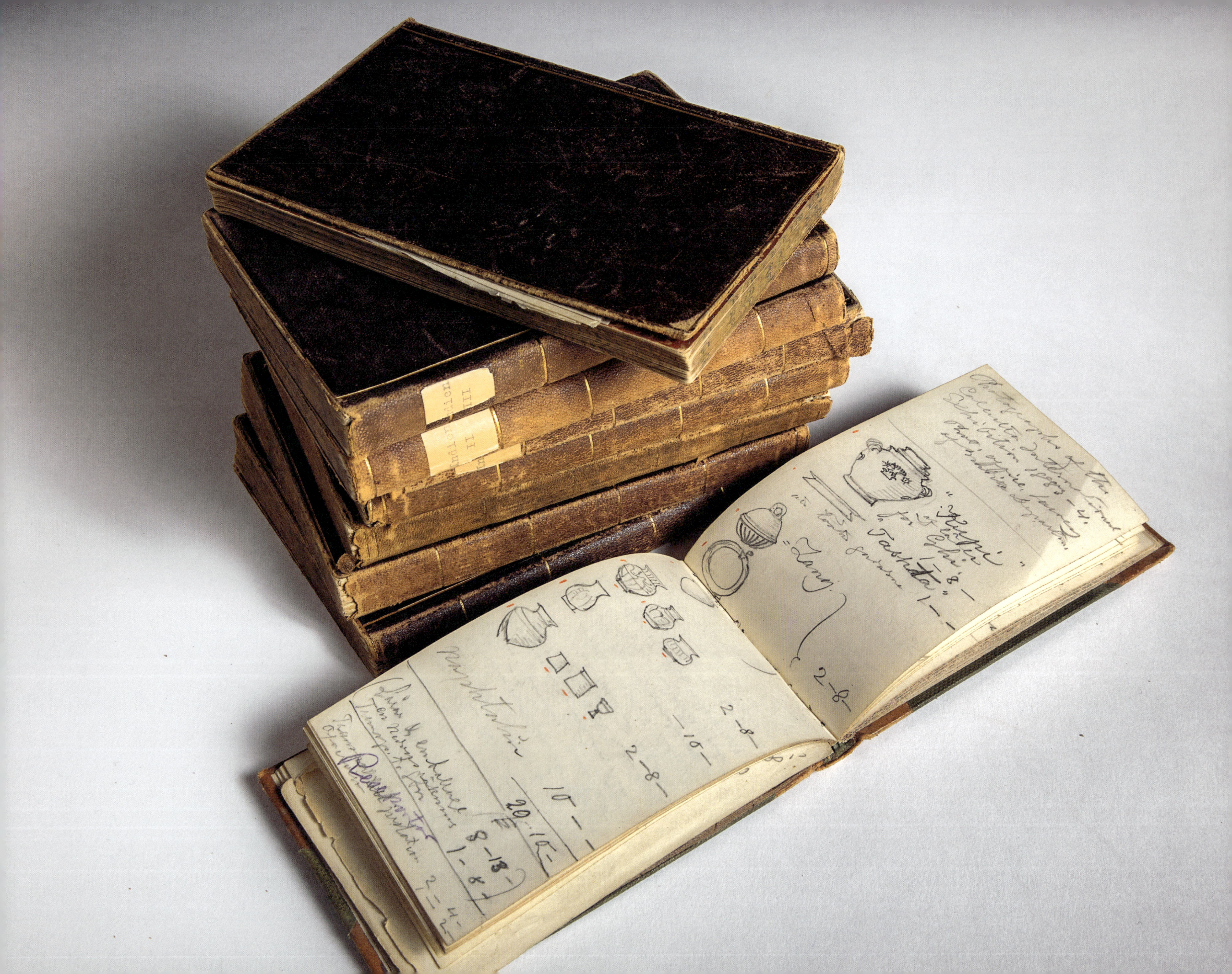

Stolpe always had a notebook in his pocket during his collecting and buying sorties. Of the total of ten books, however, No. 2 – the one about his activities in Chile and Peru – has been lost. The remaining ones are kept in the archives of the Museum of Ethnography, Stockholm.

The letter to Emmy was date-marked 1 July in Honolulu, arriving a month later after a long and circuitous journey across Sweden. But arrive it did – unlike many letters today.

After a successful career as an archaeologist, carrying out excavations in Birka and Vendel, Hjalmar Stolpe (1841–1905) became an ethnographer. Towards the end of his life he said: 'I'm an ethnographer through and through, though I have to steal time for this from my archaeological work, so I'm shaky in both camps.' He made his living from archaeology, but it was ethnography that he lived for, though the impression he made on posterity was that he had a firm footing in both disciplines, as he became a pioneer in both sciences.

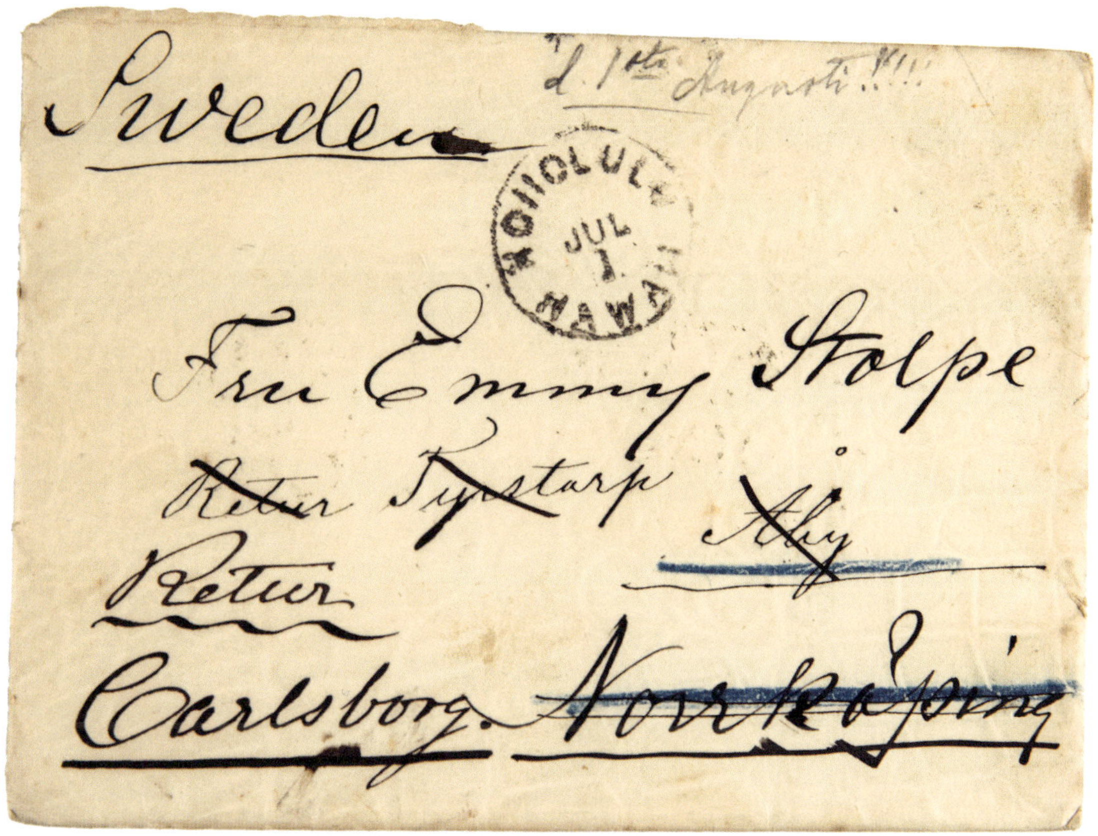

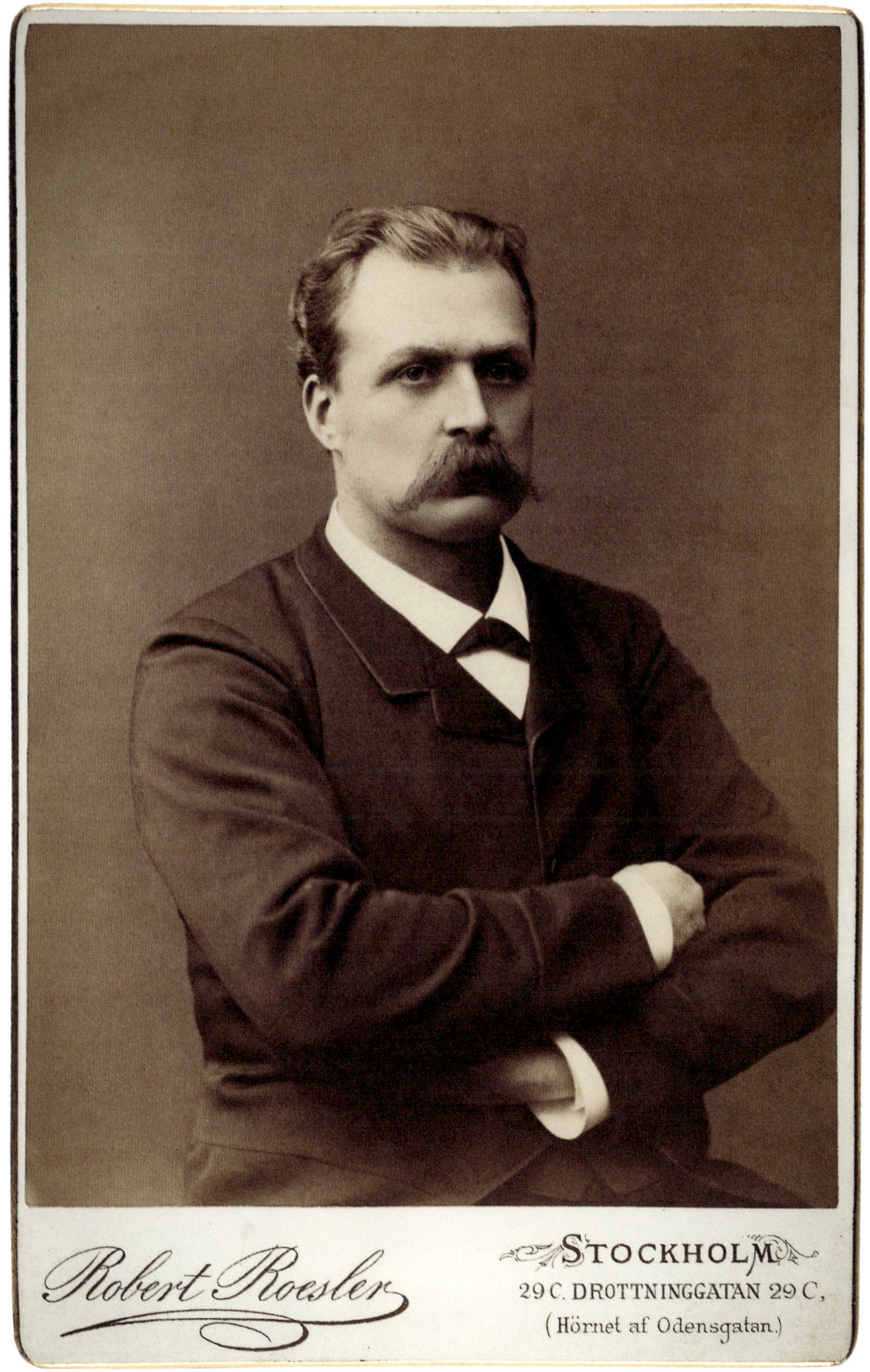

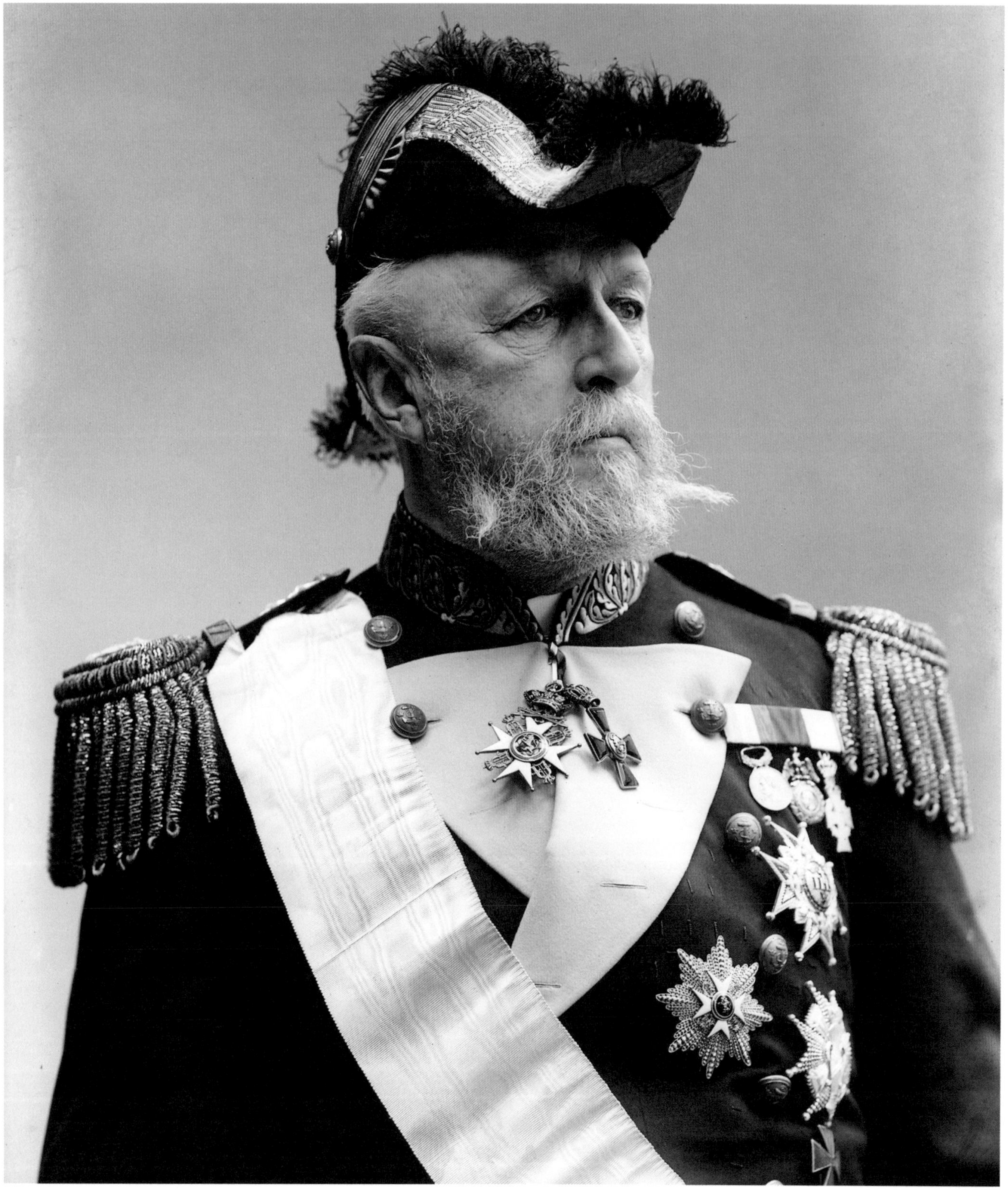

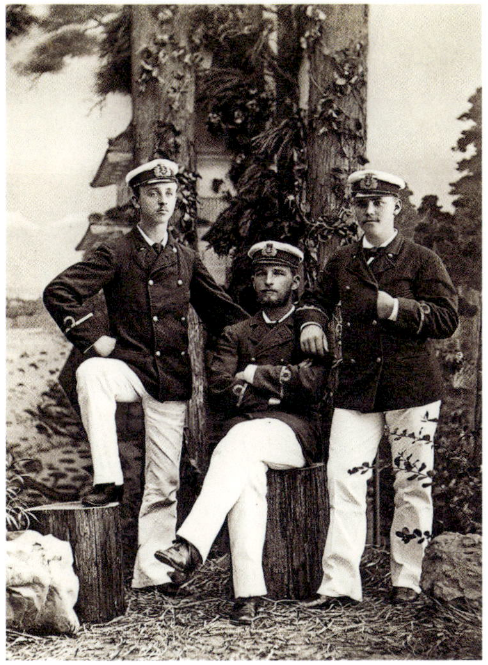

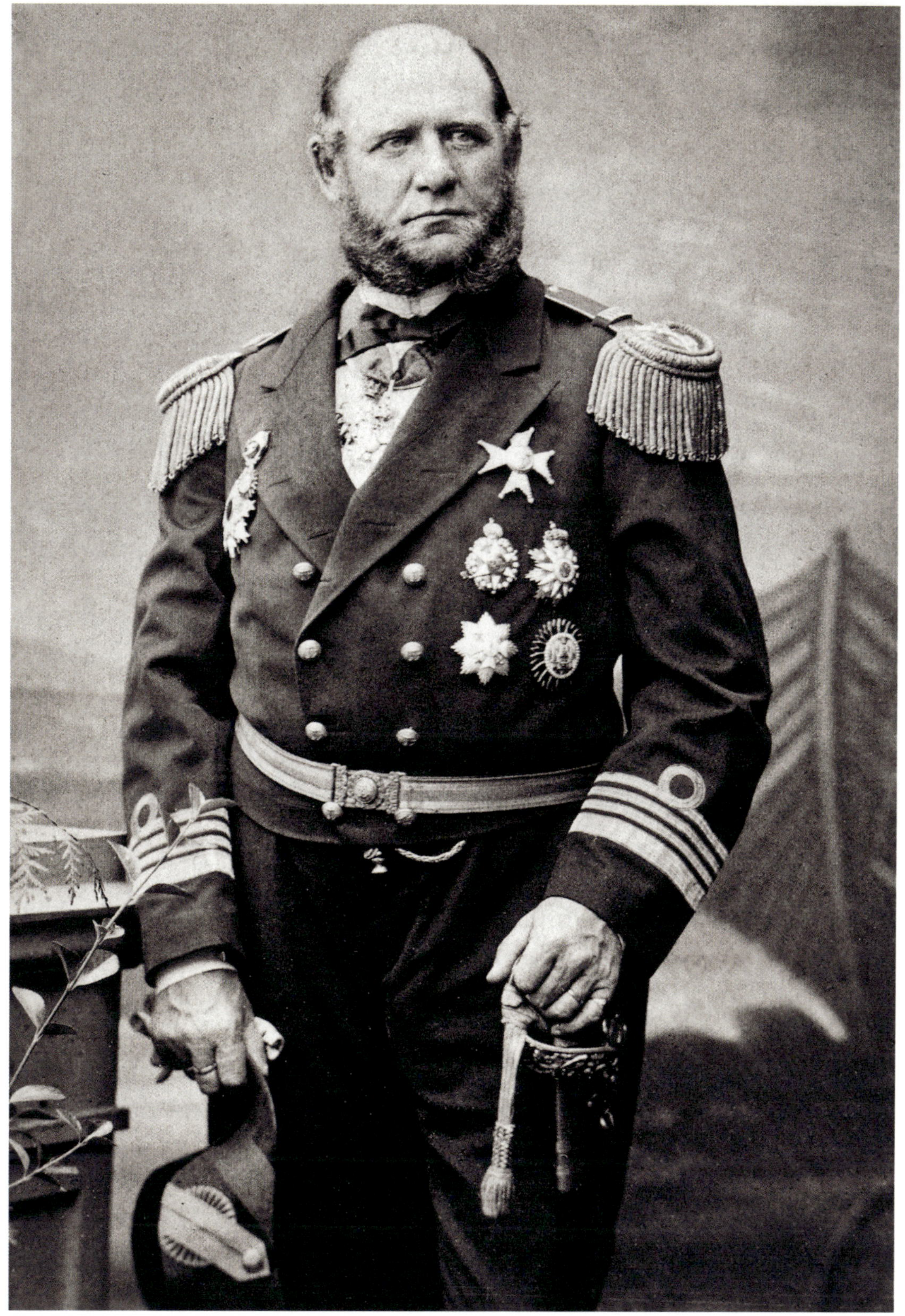

King Oscar II played an important part in the Vanadis expedition in several ways. Allowing his second son Prince Oscar to do his naval training on the vessel bestowed royal lustre on the expedition and drew attention to it and its tradepolicy remit. Ever since the Birka excavation in the 1870s, Hjalmar Stolpe had enjoyed a special relationship with the king and his sons, who visited the excavation site on Björkö and spent the night in his sailing boat. King Oscar had also directly intervened to secure Stolpe's accomodation on board the Vanadis as an ethnographer, and through a generous personal donation he later rescued Stolpe's collection work.

A trio of second lieutenants from the Vanadis in a Yokohama photographic studio, autumn 1884: Fredrik Peyron, Henning Krusenstierna and Arvid Lindman.

The commander-in-chief Otto Lagerberg (1827–1907) was in command of the Vanadis during the circumnavigation. He was a fierce disciplinarian, and it was said that on board the frigate there was only one will that meant anything – the Captain's. Lagerberg was an old seadog who had made a career in both the navy and the merchant fleet, and had crossed the Atlantic several times. He thought Prince Oscar bestowed a royal lustre on the journey across the world and the encounters with kings and emperors, but he did not give the prince preferential treatment. After the Vanadis voyage Lagerberg became a commandant in Karlskrona, and in 1889 he was made a rear admiral. He showed little understanding for the civilian Stolpe and his ethnographic assignment.

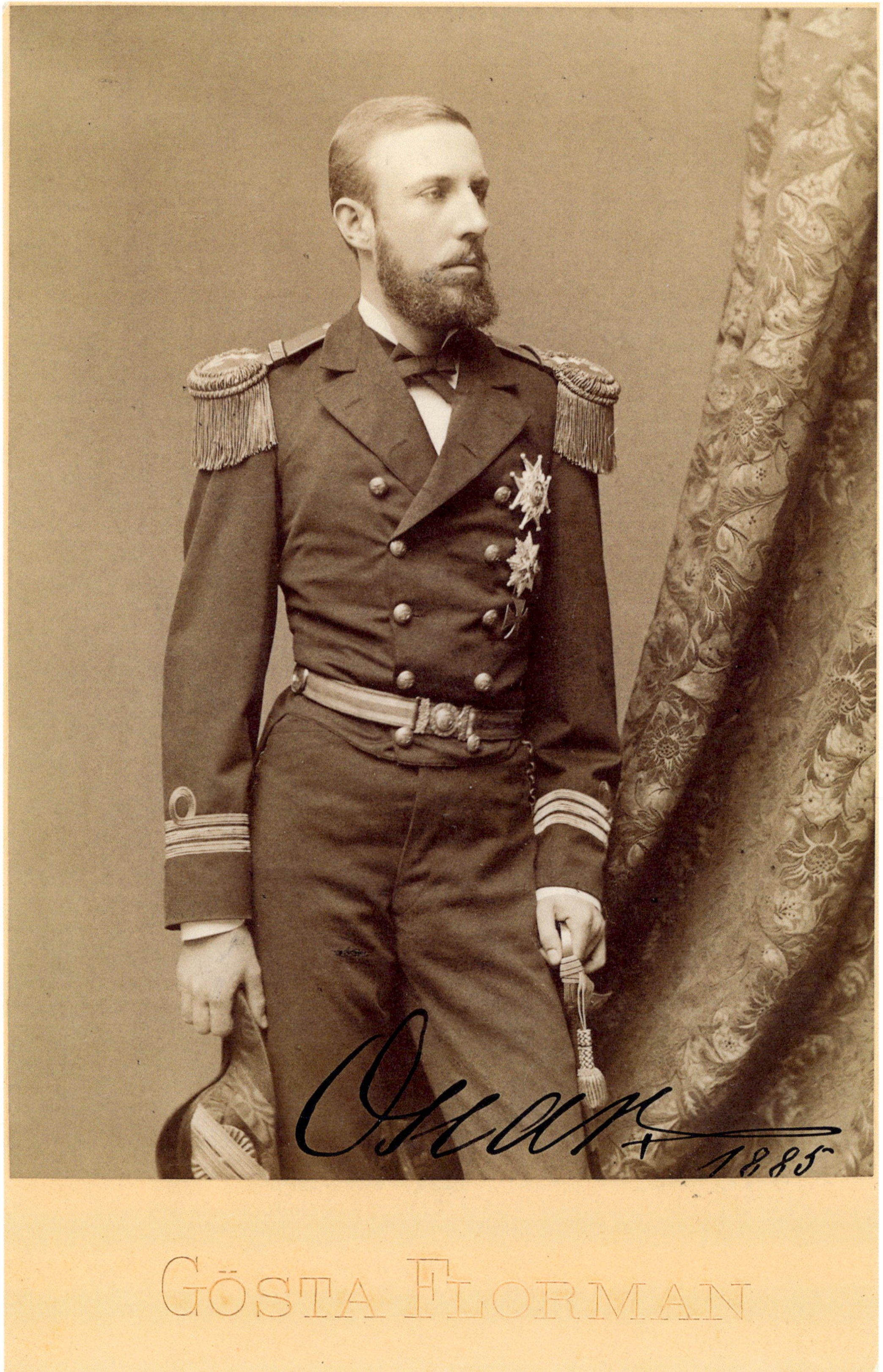

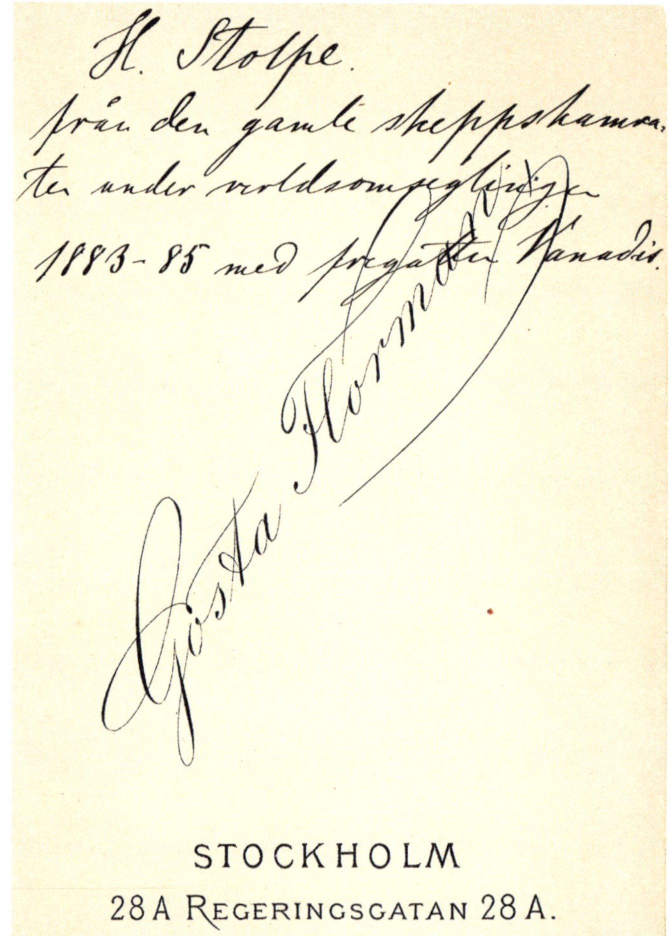

According to the logbook's list of ranks, Prince Oscar (1859–1953) was Officer No. 2 on board. He started as a lieutenant, and during the journey he was promoted to captain, entirely in keeping with the plans of his father, the king (Oscar II). The prince's small, elegant salon was just above Stolpe's much smaller cabin. The two of them had a friendly relationship, and the prince wrote on the back of this photograph: 'H. Stolpe, from his old ship companion during the 1883–85 circumnavigation in the frigate Vanadis'. The prince's status was also indicated by the fact that he had his own on-board valet with him – officer's valet Svante Segerman. The latter fell ill in Honolulu in 1884, and died in Chicago.

Opposite page: Interior of Prince Oscar's cabin on board the Vanadis. On the wall one can see his mother Queen Sophia, and King Oscar II.

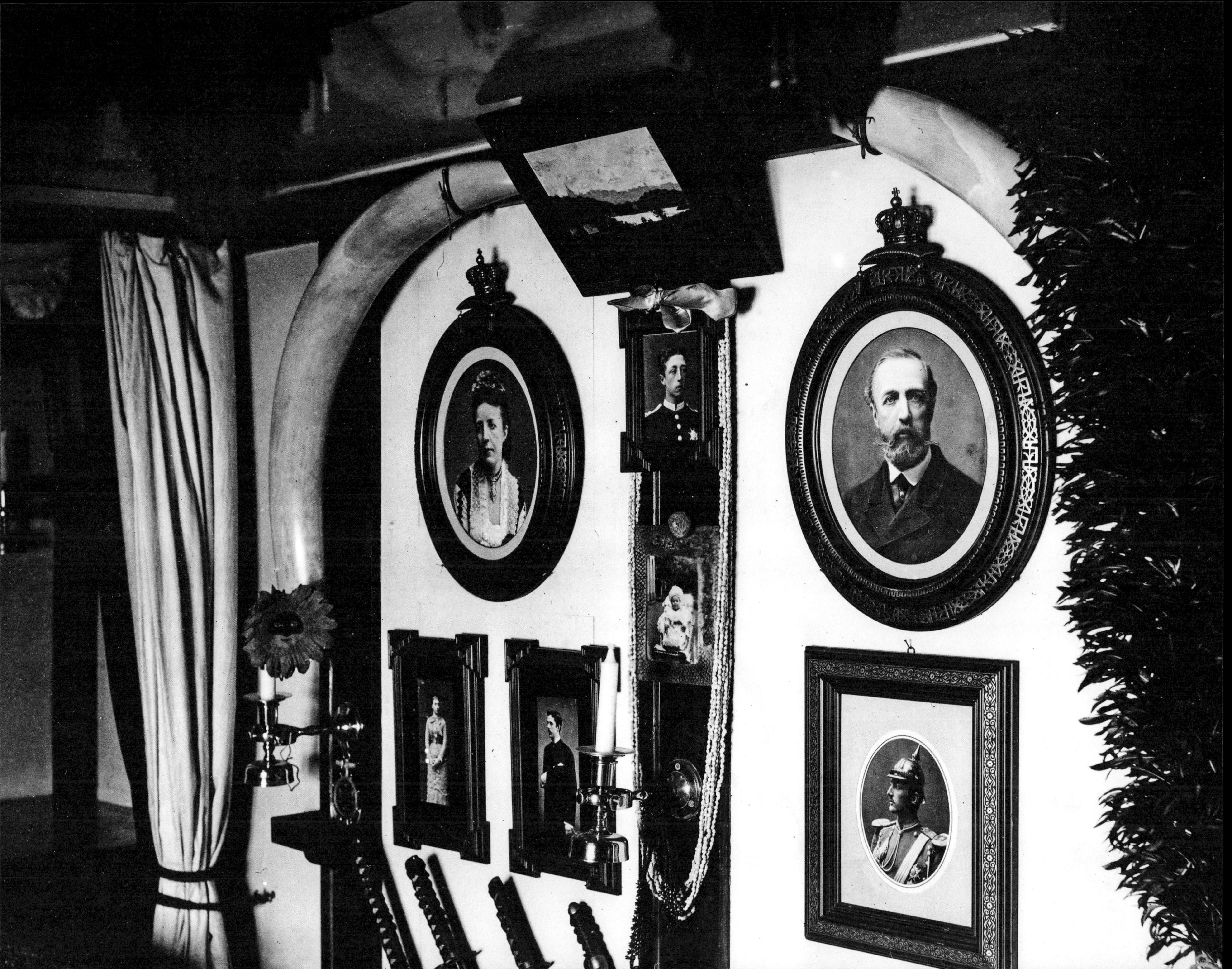

The Karlskrona shipyard was at the beating heart of the Swedish naval fleet for centuries, with more than 400 vessels launched from its docks. The old mast crane was built at the beginning of the 19th century, and is over 40 metres high. For the heaviest lifting nearly a hundred men were needed to operate the capstans. In 1844 the frigate Eugenie was launched at the shipyard. It took 100,000 man-days and eight years to build her. She was followed in 1862 by the frigate Vanadis. The shipyard has one of the world's oldest dry docks, Polhemsdockan, known as the eighth wonder of the world. In Dry Dock No.3 is Sweden's first monitor, the John Ericsson, completed in 1865 and named after her eminent designer. In the middle of the dry dock one can see the unique rotating gun turret. John Ericsson's armoured creation has been called the first modern warship. His first monitor was built a few years earlier for the Northern Fleet, and played an important role in the American Civil War. The hull of a frigate is to be seen by the mast crane. The old shipyard is now on UNESCO's World Heritage List. The photograph was taken c. 1870.

The picture on the opposite page dates from 1909 and shows Docks 1–3 in Karlskrona. In one of them is an old wooden warship. International developments in the field of naval technology left Sweden trailing some way behind. At the beginning of the 1860s the country's fleet was in fact partially unusable. It largely comprised wooden sailing and rowing vessels, whilst overseas, warships were being made of iron and steel and were driven by powerful steam engines. The Swedish shipbuilding industry faced major changes in terms of being able to meet the fleet's requirement for armed boats. The fresh thrust commenced with an order for four monitors, John Ericsson's revolutionary warships, but they were built at Motala Verkstad's shipyard in Norrköping. This marked the end of an era in Karlskrona.

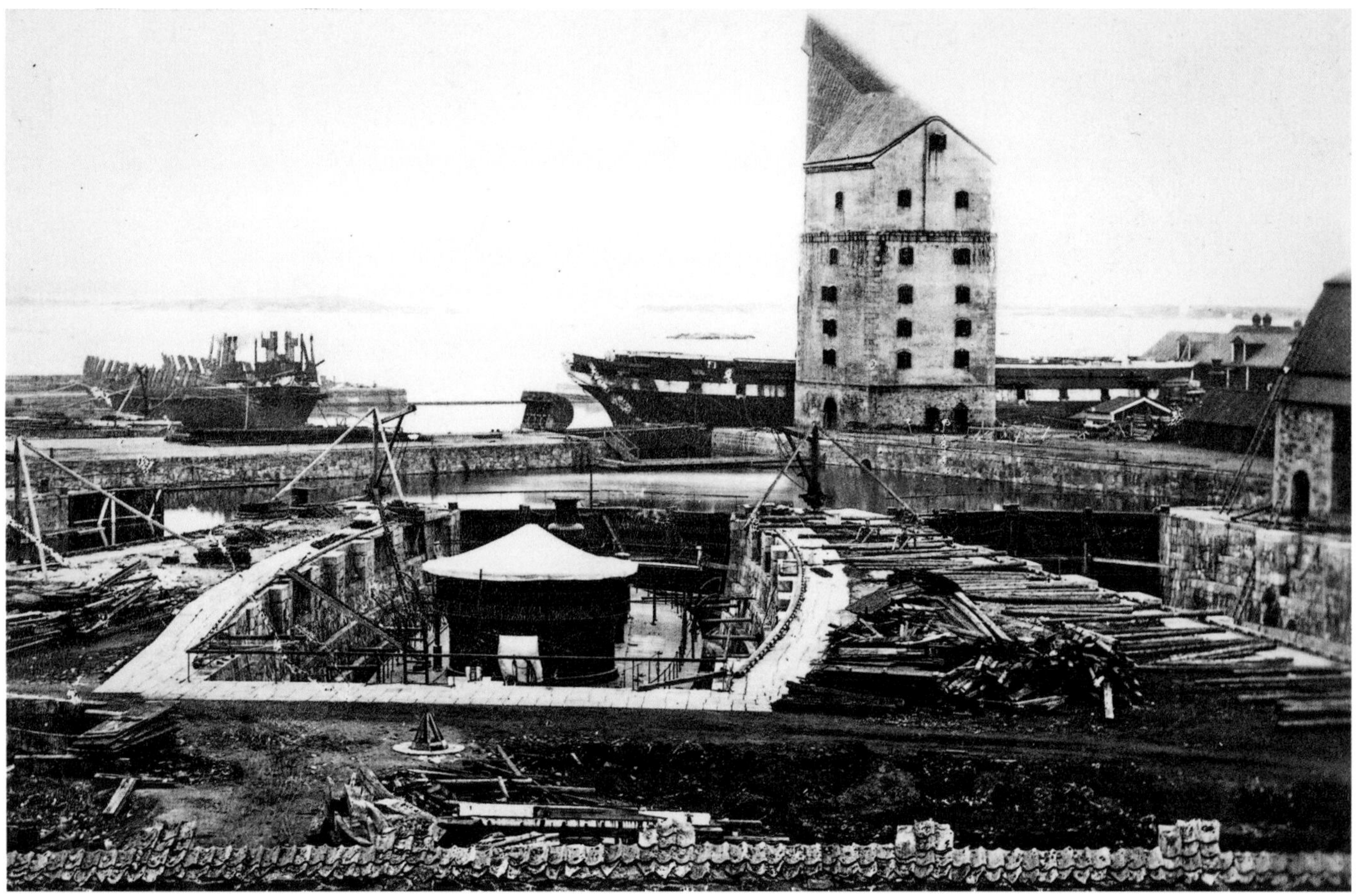

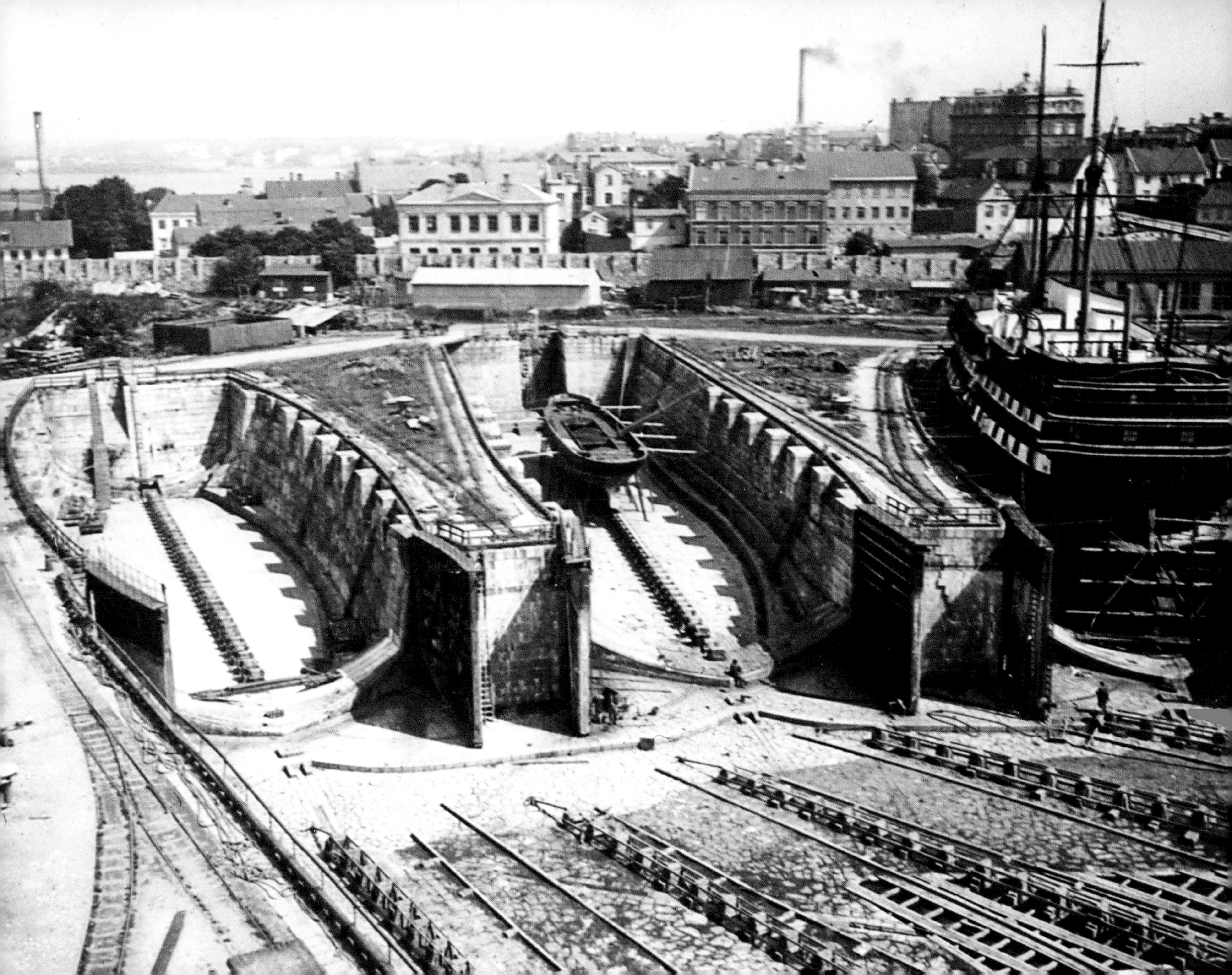

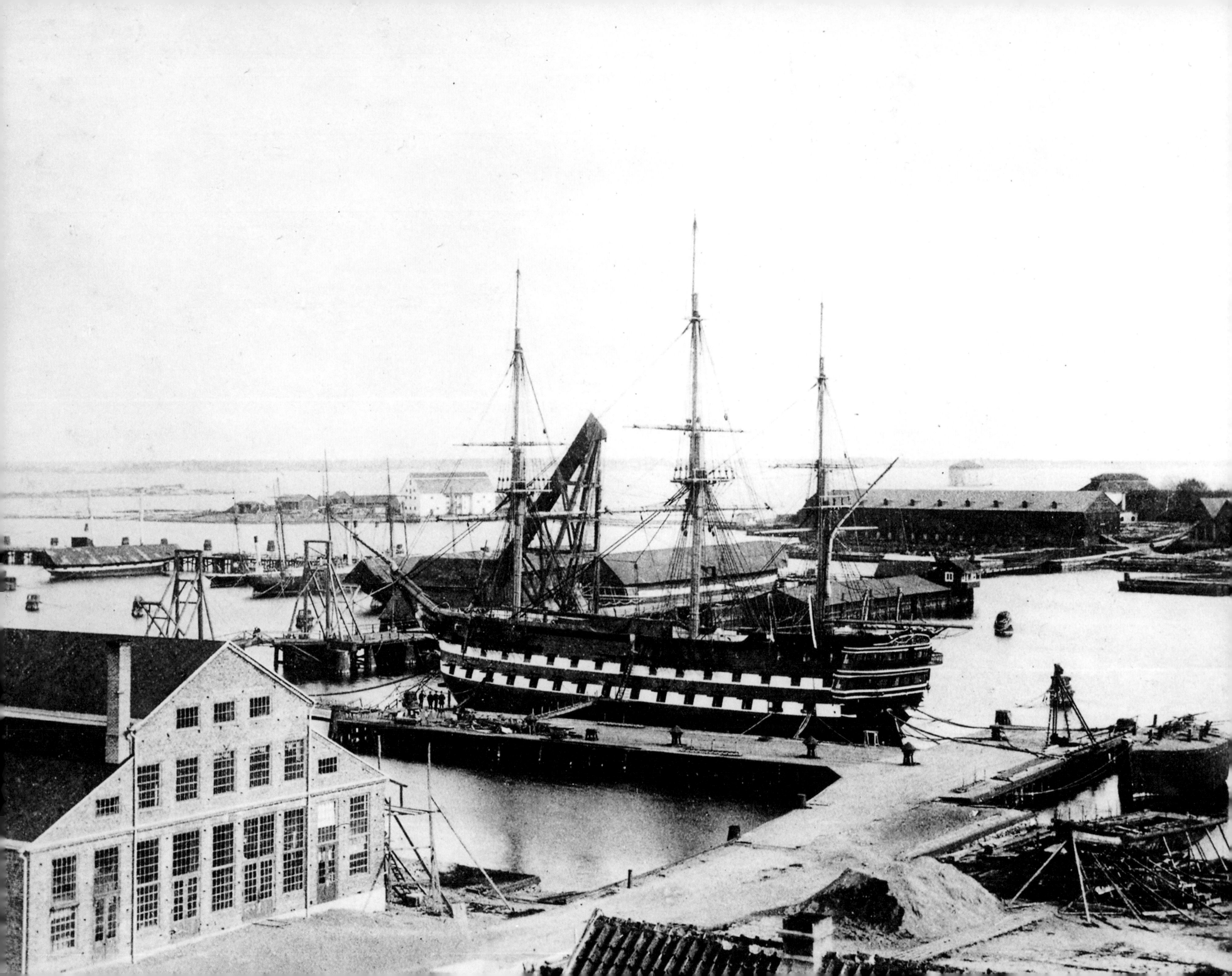

The keel of the ship of the line Stockholm was laid down in Karlskrona naval shipyard in 1832, but she was not ready to launch until a couple of decades later – in 1856. She was 57 metres long, and equipped with two battery decks with a total of 64 cannons, the cannon embrasures being within the white-painted areas. She was actually a steamship of the line, equipped with two steam engines and propellers. Her life as a warship was short, and as early as 1872 she was converted into a school ship. Ever since the 17th century the biggest sailing warships had been known as ships of the line. The unfortunate Vasa was such a ship.

The masts and sails of the two frigates Eugenie and Vanadis were similar, both of them being capable of carrying approx. 1,700 square metres of sail. Their mainmasts were nearly 50 metres tall, and had lightning conductors at the top. Gybing and furling the sails was tough work. The hemp mainsail alone weighed over a ton when dry, and twice that in rain and fog. Climbing up the rigging in a howling wind and high seas was a highly risky job. Boatswain Humbla on board the Vanadis described a stormy journey in the Mediterranean on Easter Monday 1885, when the sails were to be taken in: 'Up in the main topyard the 1st Captain of the Maintop, Seaman 301 Hultberg, was hard at work taking in the topsail when he slipped in the heavy seas and fell. He hit the deck at the front edge of the mainmast, and death was almost instantaneous.' He was buried in Gibraltar. The photograph probably dates from the 1880s, is the only picture from the Eugenie's rigging to be found, and was taken from the mainmast. High up on the front mast of the three (the foremast, 42 metres) are four seamen, and down on the foredeck a few further crew members are to be seen.

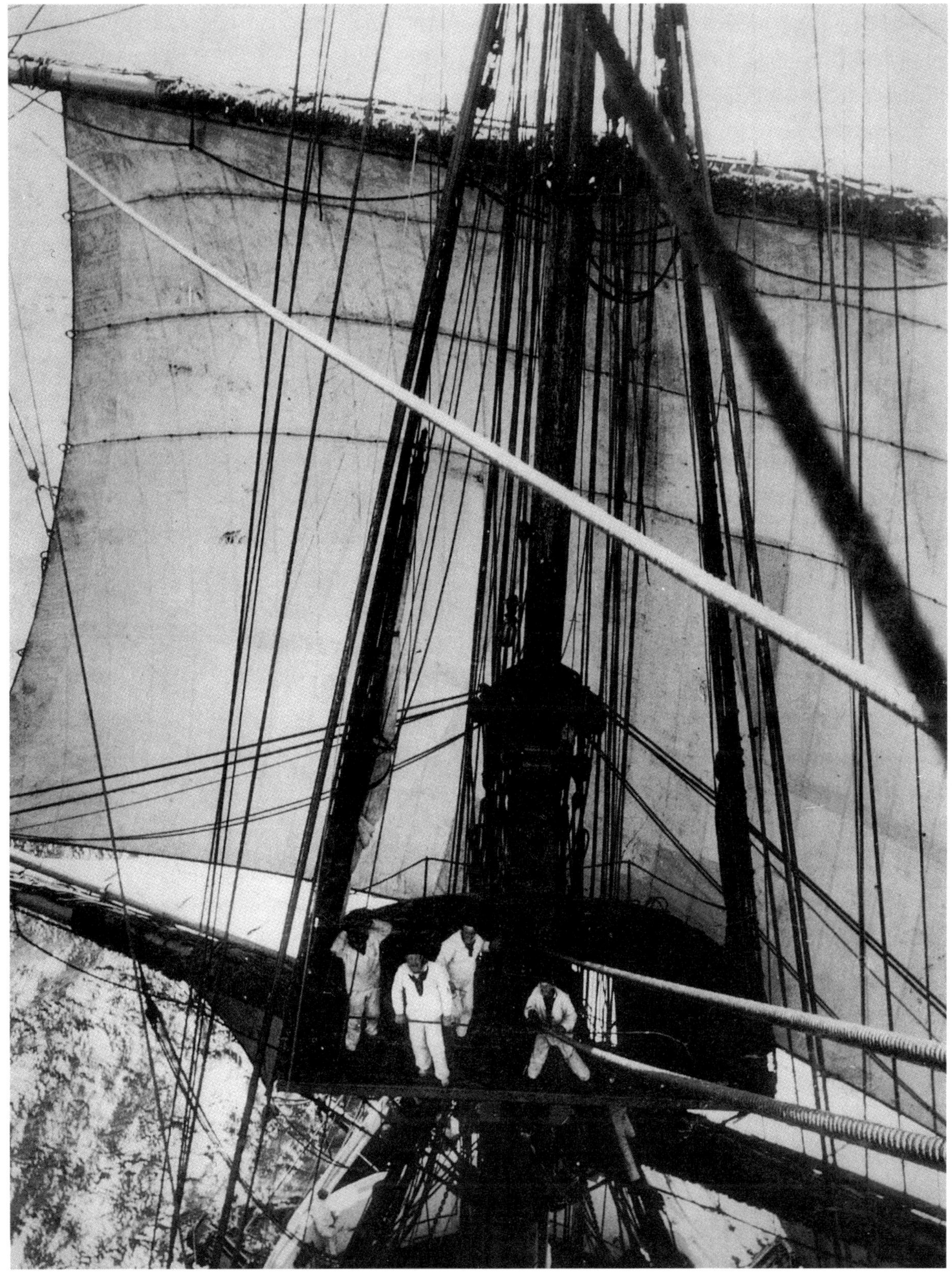

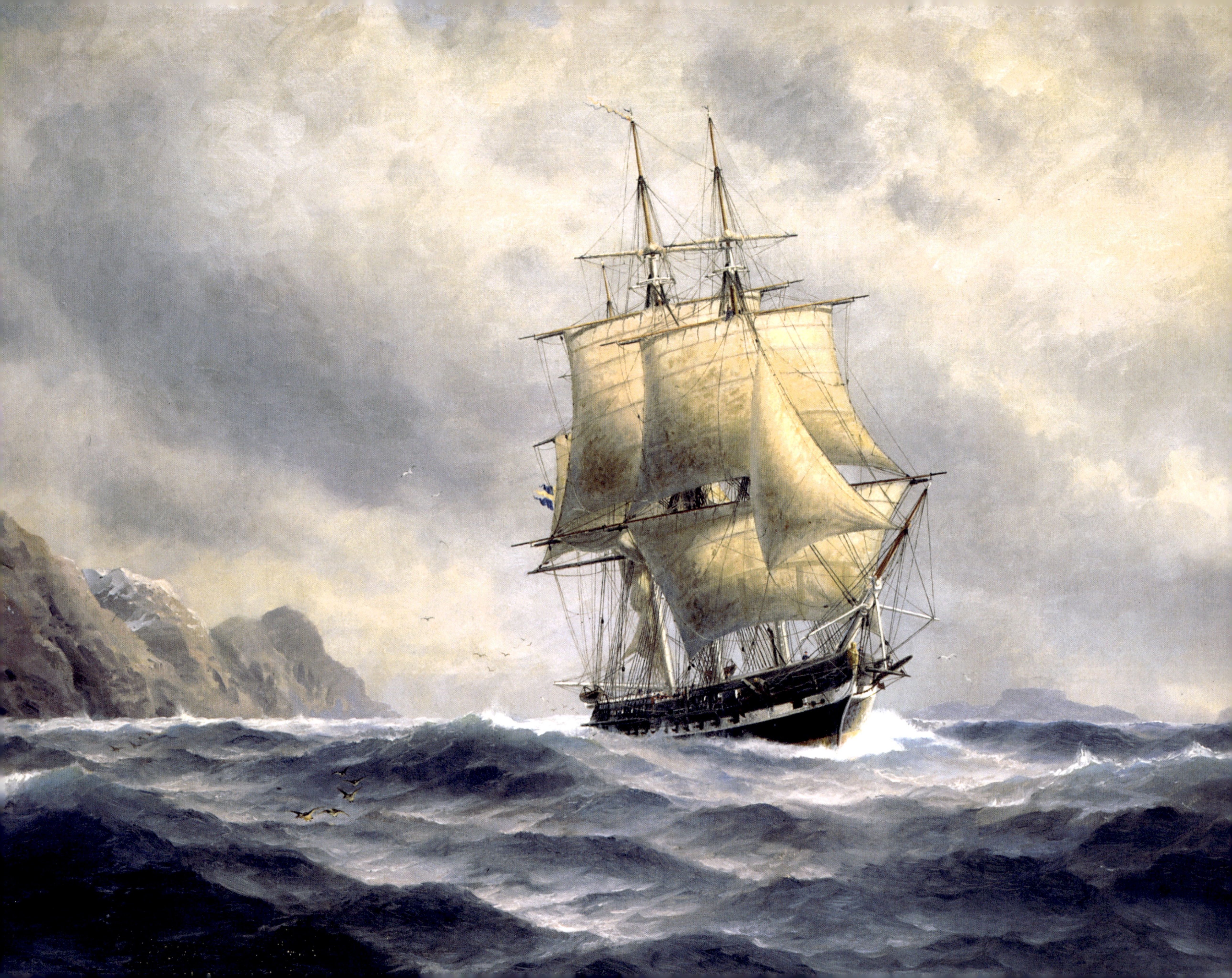

CIRCUMNAVIGATIONS

At the time of writing – the year 2020 – it is exactly 500 years since the Spanish ship Vittoria struggled around the southern tip of the Americas on its way from the Atlantic to the Pacific. The newly discovered route was named the Strait of Magellan. Two years later, in 1522, the Vittoria was back in Spain, thus becoming the first ship known to have circumnavigated the globe, though it returned without the captain Ferdinand Magellan, who was killed in battle on the homeward journey.

This first voyage around the globe provided the impetus for Europe's nations with maritime ambitions to repeat the achievement. A circumnavigation created national prestige, but the driving force was principally practical and political. Military strength on the world's oceans benefited trade and the establishment of colonies. These military and mercantile interests were gradually supplemented with purely scientific ones. James Cook's three circumnavigations during the period 1768–80 were shining examples of the new type of research expedition, which was a logical consequence of the scientific revolution in the fields of astronomy and physics.

On board Cook's ships were the Linnaeus disciples Daniel Solander and Anders Sparrman. And when, a century later, Hjalmar Stolpe boarded the Vanadis he became a member of that exclusive band of Swedish scientists who had circumnavigated the globe. He doubtless felt them to be kindred spirits (despite his travelling by train for the final section through Europe). He was moreover familiar with Sparrman's travelogues from the 1770s with the imaginative title *Resa till Goda Hopps-Udden, Södra Polkretsen och Omkring Jordklotet samt till Hottentott- och Caffer-Landen* (Journey to the Cape of Good Hope, the Antarctic Polar Circle and Around the Globe, and to the Hottentot and Kaffir Countries).

But Sparrman's work was just one of many that Stolpe studied in conjunction with the Vanadis voyage. His card index of both loaned books and those from his own collection – the lists of literature and loan slips from the Royal Library in Stockholm (500 of which still exist) – indicates great diligence. It of course included classics such as James Cook's travelogues and Charles Darwin's *A Naturalist's Voyage Round the World*, as well as Jules Dumont d'Urville's eight-volume work on voyages of discovery made in the three centuries following on from 1520. Ahead of the Vanadis journey Stolpe had made a special study of a dozen of these early circumnavigations.

The travelling library in his little Vanadis cabin contained 30 or so titles, including, in addition to those mentioned above, the standard ethnographic work *Die Völker des östlichen Asien* (The People of East Asia), written in the 1860s by the physician and research traveller Adolf Bastian. The latter's opinion that there were '*Elementargedanke*' or universal elementary thoughts was shared by Stolpe. All the earth's peoples were psychologically similar, and groups of people at the same developmental stage all over the world found similar solutions to problems. This was one of Stolpe's key concepts as an ethnographic researcher. He thus did not share the view that 'primitive' peoples were stupid by nature – a notion not uncommon amongst Europeans of the day.

The Nordic countries commenced their circumnavigations in around 1840. The Danish naval vessel Galathea carried out a successful scientific expedition, and in Stolpe's on-board library one finds the captain Steen Bille's description of the journey, *Beretning om Corvetten Galatheas Reise omkring Jorden 1845–47* (Report on the Corvette Galathea's Journey around the Earth 1845–47). Despite the ship's being only 42 metres long, it brought back over 110 large boxes of zoological, botanical and ethnographic material. A few years later the 46-metre frigate Eugenie was the first Swedish naval vessel to circumnavigate the globe. Like that of the Galathea, this journey went down in history as a spectacular research expedition, and the ship returned with impressive collections of 'specimens', which are described below.

But the first Swedish ship to circumnavigate the globe was a small two-masted merchant vessel that achieved this feat without any planning at all and ended up in the record books.

MARY ANN

The Mary Ann was a mere 22 metres long and was built in Luleå using pine. In August 1839 she left Stockholm, returning in May 1841. If the frigate Vanadis's voyage was a controlled adventure, the Mary Ann's circumnavigation was perceived as a daring exploit expertly performed by the captain Nils Werngren, who on setting out was a mere 24 years old.

His employer, the wholesaler Carl Fredrik Liljewalch, sent his little ship out as an initial attempt to open up direct trade between Sweden and Australia. She had a nine-strong crew, three of whom were inexperienced boys. The cargo, intended for the settlers, consisted of various building materials, timber, doors and windows. To prevent the cargo shifting at sea, empty spaces in the hold were filled in with oats.

Severe storms beset the ship while still in the English Channel, as Werngren relates:

In the 1850s the Eugenie became the first Swedish naval vessel to circumnavigate the globe (in 647 days). Here one sees the 46-metre frigate at Cape Pillar, having passed through the Strait of Magellan on her way into the Pacific Ocean.

> We in particular contended with a very severe southwesterly, which often filled the entire deck with torrents of water, smashed our water barrels to pieces, filled the galley and all our pots and destroyed the Christmas food, which had been prepared the Swedish way. We still ate some of it, despite its being mixed with salt. What wouldn't you do when you're overjoyed it's Christmas, even at sea and in a severe storm! The damage we'd suffered to our sails, rigging and water barrels should have prompted us to return to England, but we were too young, were enlivened by our audacious venture and were not to be that easily deterred by the first heavy seas – after all, so many more awaited us.

The Mary Ann rounded Africa's Cape of Good Hope, and after another storm they were able to replenish their food stocks with 25 albatrosses they had caught, which were stored beneath a boat on deck. Wernggren was well aware that the crew's health depended on fresh provisions, fruit and fresh water. In May 1840 theirs was the first Swedish ship ever to reach Sydney. The cargo was quickly sold, and the Mary Ann continued on to Chile laden with coal. On Christmas Eve she rounded Cape Horn with copper ore on board destined for England. A cargo of salt was loaded there, and was taken to Stockholm. The newspaper *Aftonbladet* reported this in May 1841 under the heading 'Trade Notices': 'With the skipper Werngren from Swansea, C.F. Liljewalch, 1050 tons of salt'. There were no ovations for what was the first Swedish circumnavigation of the globe.

The young Werngren comes over as being as skilled a businessman as he was a sea captain, and a teetotaller to boot. Alcohol consumption on board was prohibited. He assiduously wrote letters to his shipowner back in Stockholm, and was sent out on several voyages round the world. During the period 1841–43 he sailed in the brig Bull to Australia, China and the Philippines. As before, the first cargo, destined for the Australian settlers, chiefly comprised timber in measured lengths for houses, doors, windows and small boats, carriages and wheels. And, as previously, the empty spaces were filled in with oats. Werngren completed his second circumnavigation in the same vessel during the years 1843–1846, visiting China and the west coast of North America. He periodically spent time in completely unknown waters, and met groups of people in the South Sea Islands who had never seen a ship or a white man. The trading and provisioning seem for the most part to have gone peacefully, and nails and pieces of iron were often exchanged for provisions and sought-after goods, including sandalwood for transportation to China. Werngren describes a visit to one of the islands, where he sought sandalwood trees to fell:

> The chieftain came to meet us with his warriors, in large covered double canoes, though not with hostile intent, as he had children with him. Presents were handed out, noses pressed together and names exchanged. As a sign of friendliness tapa [cloth made of bark or bast fibre] was waved. A white lad of Scottish descent lived amongst the savages. He was the sole survivor from a brig from Tahiti, which had been burnt and whose crew had been devoured, all because of their inability to interact with the wild natives.

And on another island – Wea (Uea) in the South Pacific – the chieftains came on board the ship with their families and warriors. To divert them and show them respect, Captain Werngren displayed an unusual object:

> I took out and placed on the table a small music box, which with a little dexterity I got to play. The wonder and respect that this music achieved was extraordinary. Everyone lowered themselves to the ground, bent their legs and whistled loudly, until I noticed that the music from the box had slowed down, whereupon I touched it to make it stop. In their eyes this manoeuvre gave me a great power that their gods did not possess, and I can truthfully admit that it saved our lives on many an occasion, as these cannibals lusted for white meat. I always took my music box with me when I was walking with them on my own and unarmed in big forests, to show them sandalwood. Nobody dared to walk on the side where I keep my music box in my pocket, out of respect for this great marvel.

But Werngren was also prepared for a hostile reception, the crew being armed and the 30-metre brig equipped with six cannons. On one of the islands of New Caledonia, hostilities came about:

> When we at great risk found an anchorage, natives immediately came on board, some of them swimming, some in canoes, all of them armed warriors. Their appearance did not appeal to me, and did not bode well. They wanted to have everything but had brought nothing with them by way of exchange. Towards night they wanted to stay on board, but this could not be allowed. They finally had to be driven away rather violently. During the night large fires were lit, and they spread up the mountains in two places. The night was calm, but at daybreak a large number of big war canoes were discovered, all with decks full of armed warriors.

Violent combat broke out, and one seaman was killed, Werngren noting: 'We were assailed with spears, as the women swam behind the men and passed spears to them'. A sea breeze saved Bull and his 20-man crew from this superior force, which according to Werngren comprised '300 cannibals'.

The pioneer and circumnavigator Nils Werngren concluded his career by managing steamer traffic between Malmö, Lübeck and Copenhagen. The Mary Ann, Sweden's first circumnavigating vessel, ended its days as a breakwater. She was sunk in 1878, and the wreck lies in the Väddö Canal at Häverö, north of Norrtälje.

EUGENIE

Ten years after the voyage of the Mary Ann, the frigate Eugenie became the first Swedish naval vessel to complete a circumnavigation. She started off in Karlskrona in September 1851, and on Midsummer's Day 1853 anchored off Gothenburg. On board this 46-metre three-masted ship was a big crew of nearly 350 men – as compared with the Mary Ann's 22 metres and nine men. The Eugenie's remit was largely identical to that of the Vanadis 30 years later: to protect and foster Swedish/Norwegian trade, carry out scientific work and train the crew in the arts of warfare. The frigate was armed with 34 cannons.

The Eugenie was solely a sailing vessel, and took an extremely long time to build – she spent over eight years at the shipyard in Karlskrona. The stem, keel and ribs were made of oak, and the rest of pine that had been dried for over five years. The original intention to make the entire vessel of oak could not be fulfilled, as this would have required access to 1,500 fully-grown trees. The total number of man-days involved in the ship's construction is estimated at 100,000.

Ahead of the voyage around the world they had stocked up on water, herring, butter and grain for two months, peas, aquavit and bread for four months, and salt pork, meat and tobacco for six. According to the 1841 Food Regimen for the fleet's vessels, a piece of rock-hard bread, a knob of (often rancid) butter and a dram were served for breakfast. The midday meal comprised salt beef or salt pork on alternate days, plus peas in the form of soup thickened with pearl barley. The evening meal was always pearl-barley porridge with a knob of butter or fat. Plus a dram. There was sometimes a 'live larder' on board consisting of hens, sheep and oxen. When they were sailing in tropical regions the drinking water was filtered and freshened up with vinegar, lemon or aquavit.

This diet applied to the crew and the non-commissioned officers. The officers' menu was completely different, and the contrast became brutally clear the first Christmas on board. As usual the crew's evening meal comprised porridge and a dram, whilst in the officers' cabin pickled salmon, turtle soup, lamb cutlets, chicken, Christmas cake, oranges and bananas were served. Nineteenth-century society and its stratifications were illustrated with alarming clarity in the fleet's small floating communities. This applied not least to the approach to crime and punishment. An offence on the part of an officer led to a few days in the arrest cell. The infantry could of course also end up there. But the main punishment meted out was flogging. When the lookout, boatswain No. 46 Röd, had been inattentive while up on the mast, he was punished with 35 lashes of the blackjack. The punishment was conducted in front of the entire crew, after the ship's priest had first thanked providence for the vessel not having run aground on a coral reef.

In February 1852, the Eugenie was struggling westwards through the Strait of Magellan and the Patagonian Channels at the southern tip of the American continent. The journey was completed in 11 days – a 'world record' at the time for this type of sailing vessel. The achievement was acclaimed abroad, but was to be criticised back home in Sweden. The captain, Christian Adolf Virgin, was deemed to have unnecessarily put his crew and boat at risk in these hazardous waters, instead of going around the archipelago – around Cape Horn. Unlike the Vanadis, his vessel did not, after all, have a boiler and propeller.

The Eugenie ran aground, almost catastrophically, and it was only at the fourth attempt that they managed to pass the infamous Cape Froward. When tacking in the narrow strait, the big sailing vessel had to go about no less than 35 times, turning through the wind every 20 minutes. This achievement was considered by sailing connoisseurs to be well-nigh improbable, and the exhausted crew had to be kept going with the aid of alcohol. During the course of the day Virgin ordered three 'back-ups' of wine and aquavit – 12cl wine and 8cl aquavit per man on each occasion. He had been well warned before embarking on this audacious voyage, thousands of seamen having lost their lives in the Strait of Magellan. Virgin thus took a calculated risk, but despite the successful outcome he was met with royal disfavour when the Eugenie returned home from the circumnavigation.

From a scientific standpoint the expedition was particularly successful. Large collections of specimens were taken back to Stockholm – an estimated 10,000 objects from the mineral, plant and animal kingdoms. Of the latter, birds, reptiles and fish in their thousands had been preserved in alcohol, although sometimes, unfortunately, the solution was too weak. In addition there was a live menagerie in the form of guenon monkeys, parrots and tropical birds. Under the auspices of the Royal Swedish Academy of Sciences, 11 volumes of data on the collected objects and meteorological and physical observations were printed.

The collections were owned by the state, and were taken to the various departments of the Swedish Museum of Natural History, including the ethnographic department. But some of them ended up in private collections, such as at Örbyhus Castle, which in the final decades of the 19th century was owned by the baron and member of parliament Fredrik Barnekow. Not far from the castle, at Lake Vendelsjön, a unique archaeological discovery had just been made – an ancient burial site with remains of Vendel chieftains, generation after generation of whom had been buried there for 500 years, starting from the sixth century AD. The excavation was led by Hjalmar Stolpe, who considered the boat graves to be the most important archaeological find of the century. He once again emerged as a masterly field archaeologist, as he had done earlier, when exploring Birka.

But, after all, Stolpe was also Sweden's first great ethnographer, and it was in this capacity that he was asked to help Barnekow at Örbyhus. The castle's ethnographic collection from the Eugenie was in disarray, and Stolpe created clarity regarding a hundred or so clubs, spears, arrows and bows, and established their geographical origins. In his role as an amanuensis at the National Museum of Antiquities he also inspected the castle's famous prison, where King Eric XIV died in 1577, possibly from eating pea soup laced with arsenic. The prison 'has been wrecked for a long time', observed Stolpe, who became the castle owner's friend and hunting companion and sang Schubert by way of thanks for the excellent suppers he received there. Stolpe found Barnekow particularly congenial, and noted, self-effacingly, that the baron 'puts up with a poor amanuensis'. As a photography enthusiast, Stolpe had a camera and glass plates with him while at Vendel, and his photograph of the Barnekow family on the castle staircase is to be found in the Stolpe family archive. He perhaps also photographed the excavation of the famous boat cemetery, but no photographs have been found.

Captain Nils Werngren is one of the unknown big names in the history of Swedish navigation. At the tender age of 24 he became captain of the schooner Mary Ann, which in 1839–41 achieved the first Swedish circumnavigation. But the historic journey garnered no salutes or accolades. The only record of her return home in May 1841 was a notice in the Swedish daily Aftonbladet *under the heading Commercial Notices: 'With the skipper Werngren from Swansea, C.F. Liljewalch, 1050 tons of salt'. Liljewalch was a ship owner. Captain Werngren was a charismatic leader and entrepreneur who was equally skilled at sea and in business. He was furthermore a man of honour and an improbable survivor of dramatic adventures amongst the 'cannibals' on the South Sea Islands. A little later, in 1843–46, he completed his second circumnavigation.*

The small schooner Mary Ann was built in Luleå, and was 22 metres long. On her way home on New Year's Eve 1840 this small merchant ship rounded Cape Horn loaded with copper ore destined for England. She encountered a storm and heavy seas: 'With her heavy cargo of ore, our little Mary Ann rolled unusually steeply, but succeeded in rounding the Cape with no damage or injury apart from a seaman falling down onto the deck from the foreyard, smashing one eye and cracking his skull, but following stitching and bandaging he recovered, and after a few weeks regained his eyesight to some extent, so he was able to work. Young people heal so easily at sea in the fresh sea air,' declared the hardy young captain Nils Werngren.

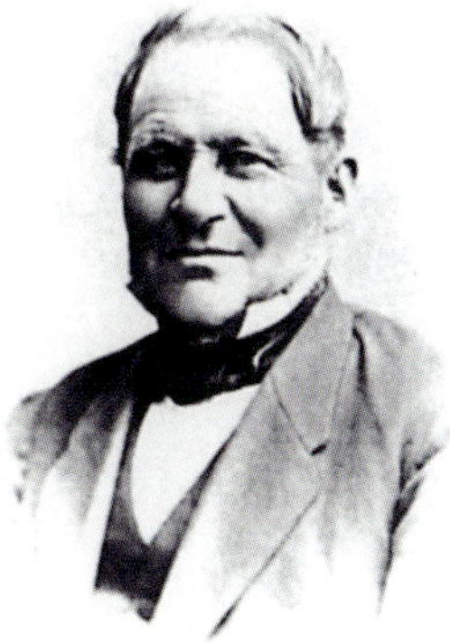

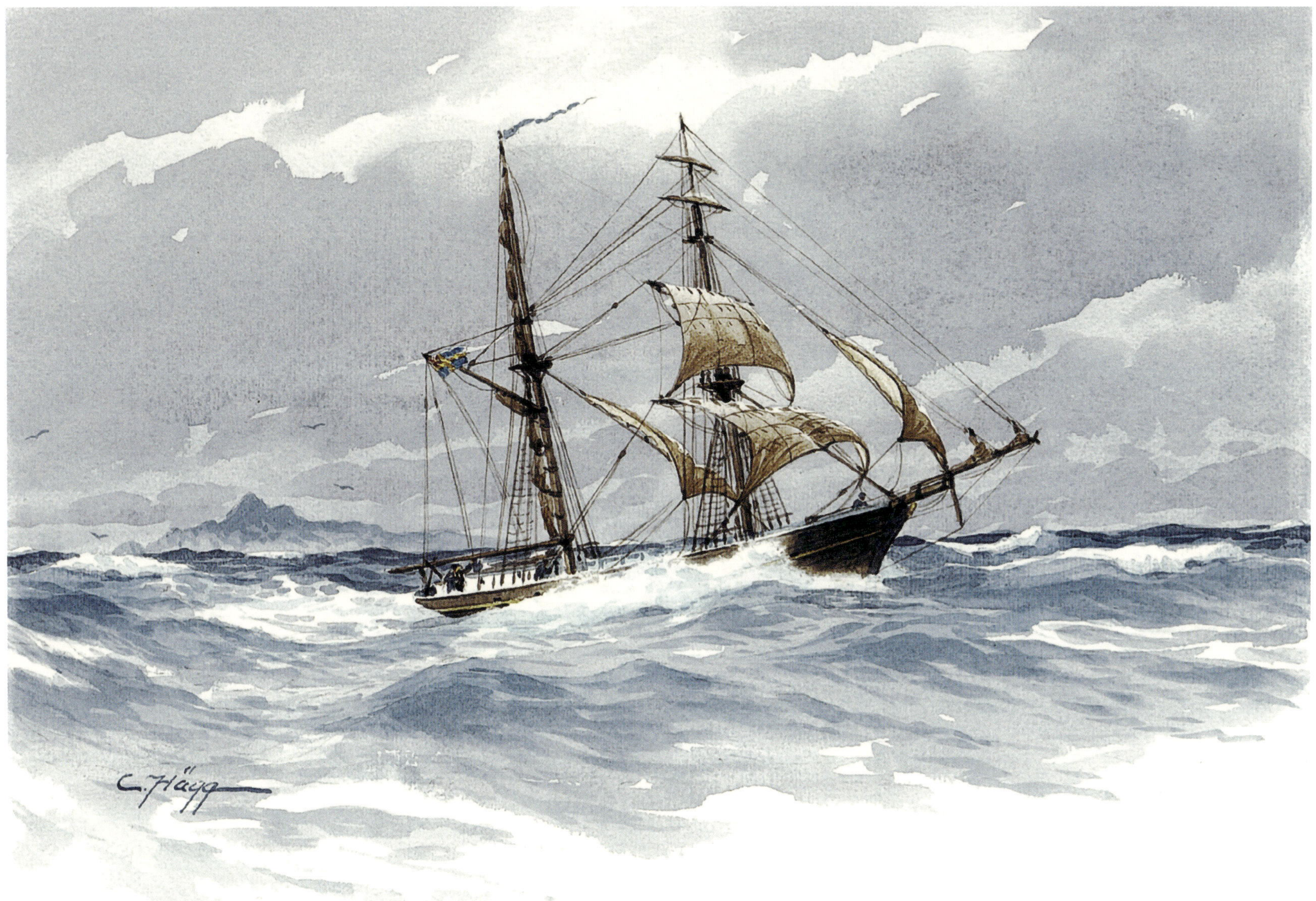
C. Hägg

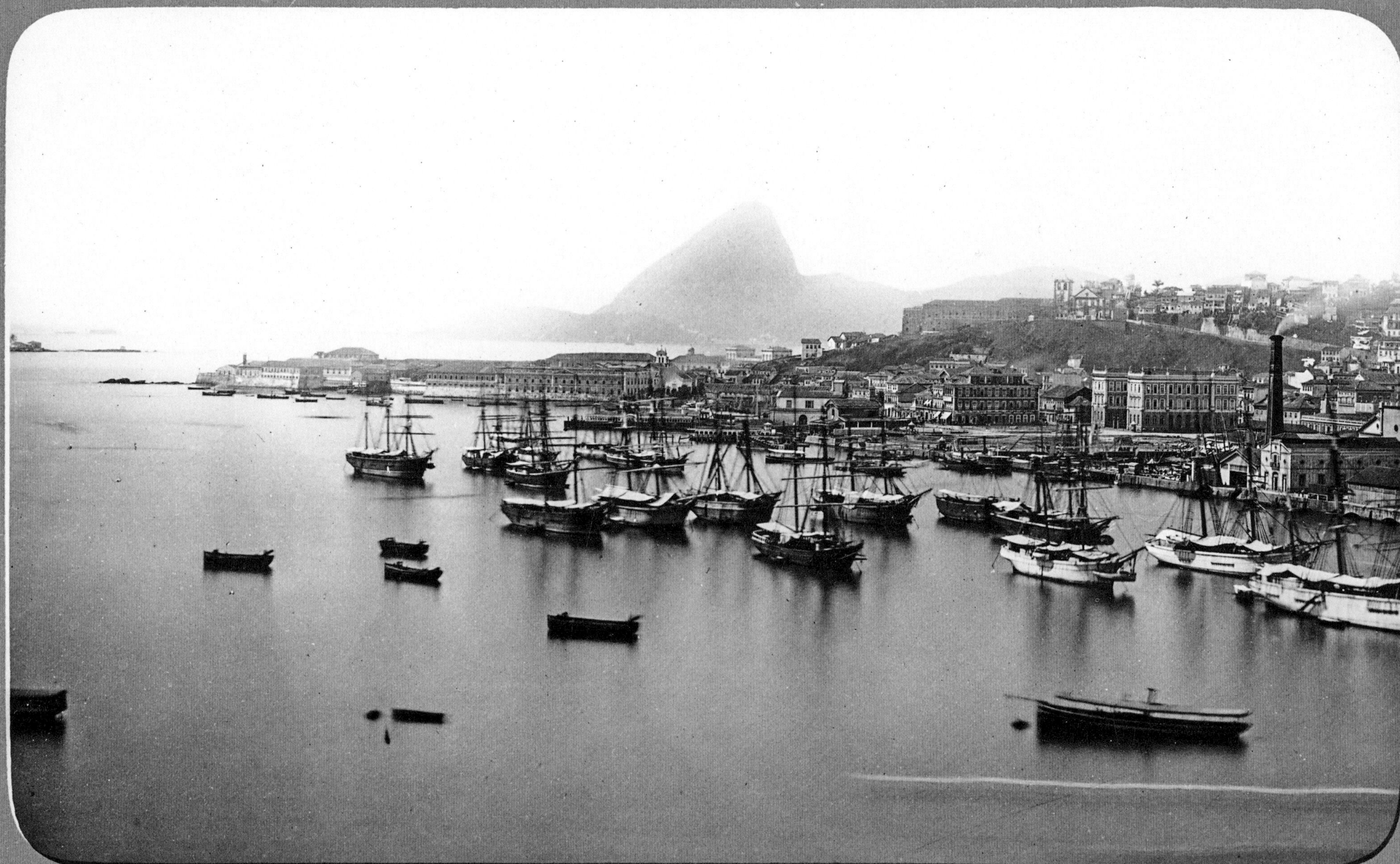

Entrée de Rio

SOUTH AMERICA

On 29 December 1883, the Vanadis left Lisbon and Europe, heading for Rio de Janeiro – a trip lasting a month, mostly under sail so as to conserve coal, and often at a low speed. Hjalmar Stolpe had been looking forward to this voyage across the world, and had great expectations. It was important for his ethnographic career and the Vanadis book he was planning to produce.

But his encounter with the reality of life at sea came as a shock. Stolpe never got used to his four-square-metre cabin. He was unable to cope with the constant movement of the ship, suffered from headaches, and found it hard to read and write. In the end he did not write at all on board, and did so only on terra firma, in his hotel room. 'The swaying troubles me as soon as I try to think of anything ...' he wrote. 'This is insufferable. And the table rocks in all directions and blood goes to my head'.

He was unprepared for such problems, and the enforced idleness had a big effect on this hard worker, leading to despondence and apathy. Witness to this is his very first notebook from the journey across the South Atlantic towards South America. He listlessly noted birds resting in the rigging and shot at empty bottles with his revolver, which otherwise hung on the wall by his bunk, together with his knife – in between hung a picture of his wife, Emmy. And, what is more, this unhappiness also affected his relationships with his travelling companions:

> I've withdrawn into a shell in which I've absolutely nobody to talk to. They're decent people by and large – well all of them, actually – but their educational level is so far removed from what *I* need ... It's simply not worth singing. Not one of them is musical. Not even jolly ditties take their fancy. Of an evening I thus stand on my own on the small steps by the port gunwale near the stern, gently humming a tune for Little Sparrow whilst beholding the firelight in the sea. I can stand there on my own for hours ... My thoughts constantly revolve around you, otherwise life would be insupportable. My, how it's rocking!

Hjalmar became increasingly irritated and critical during his 'life in a floating barracks', as he called his time on board, and he became furious with everyone around him in turn – barring the second-in-command, Hjalmar af Klintberg, and Prince Oscar. His mood of course fluctuated, and for the most part he held his tongue. But sometimes things snapped, and he turned his back on some of his cabin neighbours for a while. Especially in tropical waters the cabin turned into a veritable pressure cooker in two senses, and his tendency towards pessimism and lamentation was allowed free rein. His letters home to his wife Emmy became his safety valve, giving vent to his unhappiness in the form of criticism and condemnation. The person most affected by his disfavour was the ship's commander, Otto Lagerberg, and the antipathy was mutual, but the worst thing for Stolpe was the commander's complete lack of interest in the expedition's ethnographic remit and his tendency to rush past places where there might be finds. This ultimately had dramatic consequences, with Stolpe and his photographer Ekholm leaving the Vanadis prematurely, in India – something we shall be returning to.

This is the dark and unknown side of Ekholm and Stolpe's voyage across the world. But the bright side is so much more substantial and enduring. Barely had the Vanadis dropped anchor wherever it might be in the world before the duo rapidly disembarked, bursting with energy and eagerness to collect and photograph. Even their hauls from the brief one-day forays were impressive, whilst the results of the long periods in Japan and India were of an extraordinary scope and quality. Stolpe himself described his delight in collecting as verging on the manic.

On 1 February 1884, they arrived in Rio de Janeiro after a month at sea. Stolpe quickly fled the Vanadis and made for the Hôtel Vista Allegre up in the mountains, where he felt very much at home and had a view over the bay. The melancholy of the Atlantic journey was effectively blown away. He was 'healthy and well, feeling like a pearl in gold', and spent time walking, riding a tram drawn by hinnies, finding rare ant species, buying stuffed hummingbirds for his daughter, admiring the local nature, which, he wrote, 'surpasses everything I've seen'. And he bought 10 boxes of gypsum for anthropological castings.

By chance he met an old friend, Elie van Rijckevorsel, a physicist from Holland and a 'gentleman scientist' so typical of the age, blessed with both a fortune and a predilection for research. For years he had been making magnetic observations in Brazil and on Pacific islands, 'at his own expense, for the sheer love of science! What a fortunate man!' Hjalmar exclaimed, voicing his usual litany: 'We, on the other hand, must needs drive forward'. On one of the evenings in Rio there was singing down by the harbour, prompting this description:

On 1 February 1884 the Vanadis arrived in Rio de Janeiro.

> Some of our singers went out onto the terrace a little later, and broke into a number of songs well known to us. You should have heard Doctor Stolpe singing our national song *Du gamla, du friska, du fjellhöga Nord!* ['Thou ancient, thou free, thou alp-crowned North!']. Resplendent fireflies traversed the air, the night sky was a blaze of wondrous light, and the waves lapping at the foot of the terrace provided an accompaniment to the doctor's glorious voice, at one and the same time strange and heart-rendingly beautiful.

This description of Hjalmar Stolpe and his solo appearance in Rio was the work of Svante Natt och Dag, a second lieutenant on the Vanadis and, following a career in the fleet, a member of parliament. The following evening saw a charity concert by the Vanadis band, comprising five horn players and a drummer, who performed the reel '*Näckens polska*'and the folk song *Kristallen den fina*. Prince Oscar wrote as follows about the entertainment: 'We went to the Pasejo Publico park, where our music was played for a collection for the ransoming of slaves. Because all slaves born before 1870 are still serfs. The most burning issue in Brazil is whether or not *all* slaves should immediately be released'. It was estimated that the takings from the concert would suffice to ransom one slave. Just a few years after the Vanadis visit, both slavery and the empire had been abolished. Pedro II was forced to abdicate and flee to Paris.

After a few days of fuelling and provisioning in Rio de Janeiro, they sailed southwards. Oscar Ekholm related to his brother Fredrik:

> Our journey is very slow, for the wind is constantly against us, changing and gusty. I am writing this on the morning of 25/2. Three degrees north of the Falkland Islands. The frigate is bobbing up and down on high seas with close-reefed topsails. I have not in the least suffered from seasickness, and have seen no more than two men looking in a bad way, namely Dr Stolpe and Consul Ehrenborg. In such weather Stolpe is always to be found sitting *silently* on a chair in the battery, and Consul E lying on the deck – the latter, it should be added, is an excellent person ...
>
> I am longing to buy a few apes or parrots to take home, but written orders state that *no* live animals may be taken on board. But there is one thing they cannot stop me doing. The turtles (as you can put them in a box like potatoes and nail it shut, and they will live until I get home), because they are very cheap, costing around 2 kronor each.

A few days later they reached the Strait of Magellan. The crew and officers had already started to grumble about the commander's haste along the South American coast and the beaches of Tierra del Fuego, where smoke rose from the Indians' fires and sea lions basked on the islands. But nobody dared protest. Nobody, that is, apart from Prince Oscar, who was quietly mutinous, pointing out that his father, Oscar II, 'presumably had intentions for the expedition other than just fuelling and provisioning'. This had an effect, as nobody dared displease His Majesty. The prince had considered leaving the expedition. But Hjalmar Stolpe was the angriest: 'I'm in a constant state of fury at the thought of what I could achieve under another commander! Dashing through the Strait of Magellan in the way we did is inexcusable'.

The Strait of Magellan is the route separating the Tierra del Fuego archipelago from South America, connecting the Atlantic to the Pacific. Discovered back in 1520, the strait became a nightmare for sailing vessels because of treacherous shelves, strong winds, currents and fog. The Vanadis travelled downstream at the 'breakneck speed' of 16 knots, and at six o'clock in the evening of 29 February, the frigate anchored off the world's southernmost town, Punta Arenas, Chile's former penal colony. As Stolpe described to the daily paper *Stockholms Dagblad*:

> A few rows of single-storey wooden houses (at best provided with frontages à la Upsala), a beacon made of wood, and behind all this an immense tract of bare, grey tree trunks, witness to a raging forest fire – this in a few words is how Punta Arenas, the southernmost town on earth, appears in the eyes of newcomers. Fairly high, leafy mountains are to be seen in the background.
>
> As soon as was possible I made haste to disembark. It was here that my activities were actually to commence, and I was keen to trawl the sales premises before others did. Yes, Punta Arenas really does have shops, in pretty much every house, but it should be added that they to a large extent merely contain bottles, naturally filled with brandy, whisky, gin and other refinements of spirituous cereals, which I have heard called *potates*. After that, paraffin seemed to be the most important traded article. And travellers are lured by diverse 'curiosities' in the shape of model boats from Tierra del Fuego, more or less factory-made bows and arrows, as well as baskets from the same source, and, if you really want to push the boat out, guanaco or ostrich fleeces.
>
> The raid was soon completed. Amongst a lot of dross there were also many fine objects, mostly from the southern parts of Tierra del Fuego, and the museum achieved an excellent yield. But the price! If such curiosities are in general expensive when *one* traveller is speculating, it's even worse when a big warship with a number of collectors on board arrives, and in many cases they went up to far more than the things were actually worth. However, if the collection arrives home safely, we can now create a pretty good impression of Fuegians' life, in particular thanks to the contributions I received a few days later through direct bartering with them.

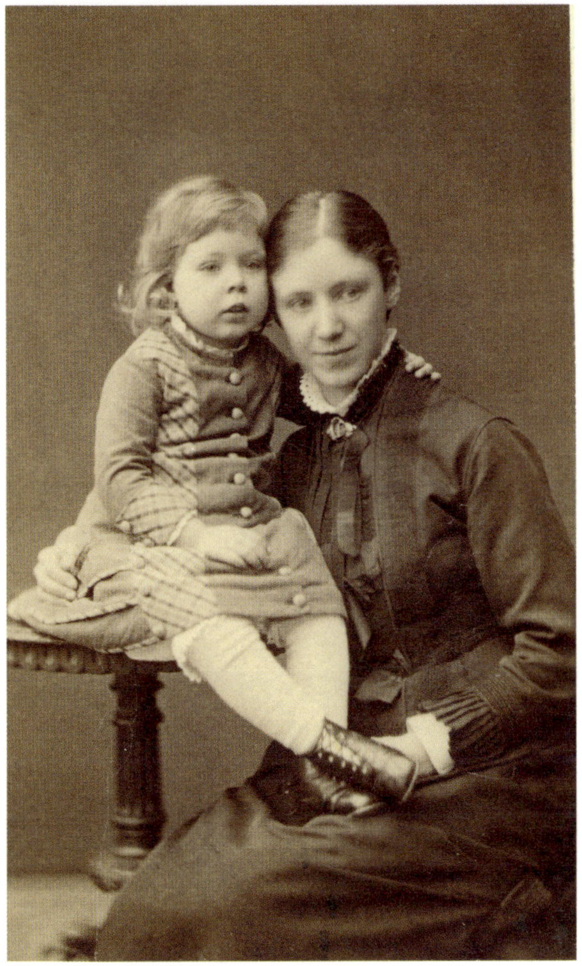

Emmy (here with their daughter Estrid) sent Hjalmar this letter, dated 13 February 1884. It reached him at the beginning of April, when the Vanadis was at anchor in the port city of Callao in Peru. Emmy could not reckon on getting a reply from him until some time in May. The post took a long time, but it was well organised. In the government gazette *Post- och Inrikes Tidningar* and at post offices, the ports the Vanadis was to call at and the planned days of arrival were listed. Emmy's letter from Norrköping went via Malmö and Copenhagen to Southampton, after which it continued to Callao via Panama or the Strait of Magellan. As one can see, she used four 20 öre stamps – approx. SEK 40 or £3.50 in today's money. For some obscure reason there are sometimes no stamps on Hjalmar's letters: 'This really irritates me. I'd really have liked to have a complete collection of my stamps from all over the world.'

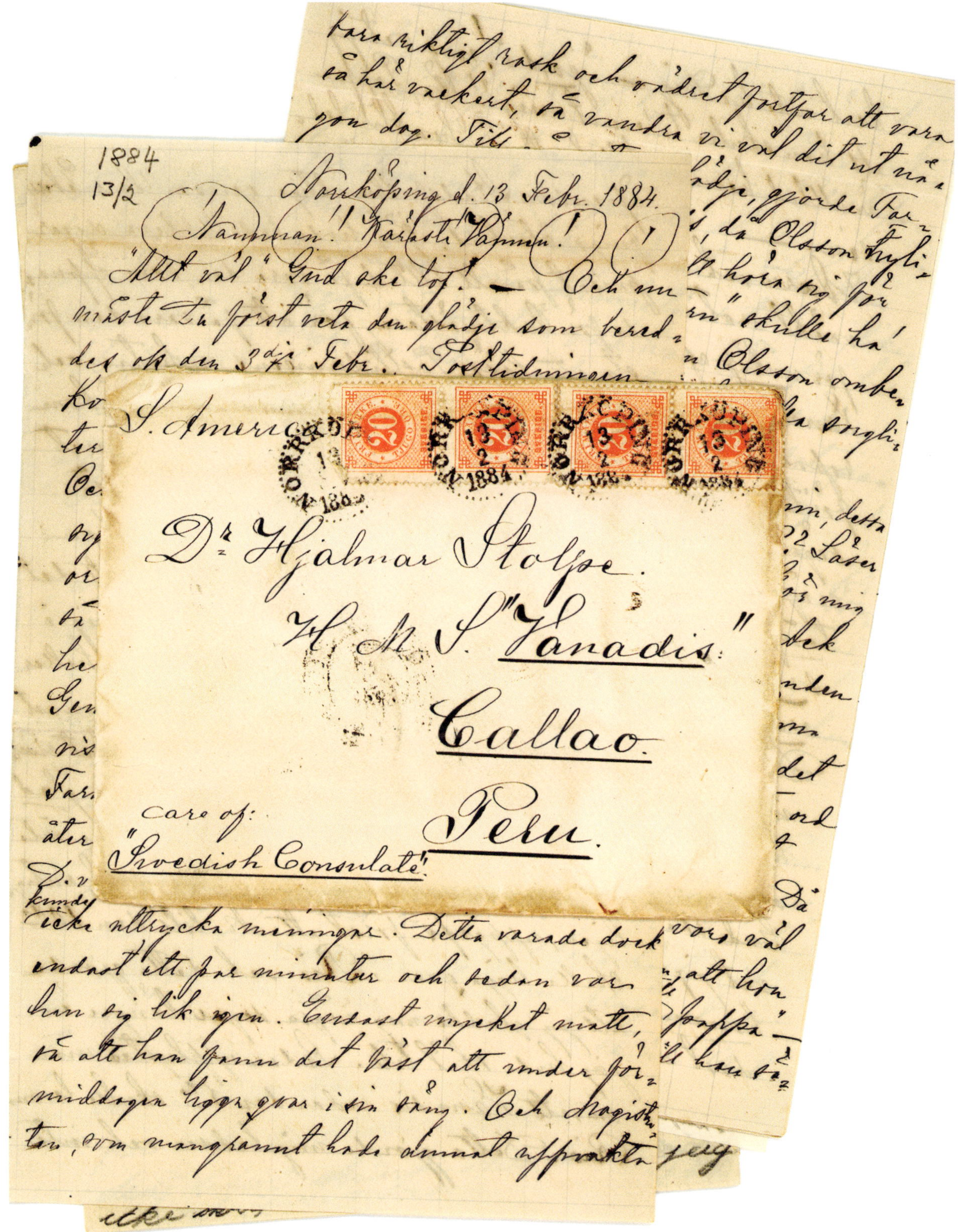

When the dealings were over and done with, the party gathered to await boats at the so-called Union Hôtel, where I met two adventurers, gold-washers, guanaco hunters and heaven knows what, one of whom turned out to be from a good English family, and provided an abundance of useful information. This included the fact that a few weeks prior a troop of around 150 Patagonians had been in Punta Arenas. When they came riding into town – something that happens from time to time – there was a dreadful hullabaloo, in which I conjecture the town's ill-tempered dogs played no small part. The Patagonians brought with them guanaco and ostrich fleeces, and it was unfortunately during the bartering process that the above-mentioned bottles played a major role.

The Vanadis only stayed for 24 hours, to stock up on coal, and Stolpe and Ekholm rushed ashore with photographic equipment and anthropological measuring devices to document two Patagonian women – Maria and her daughter Marguerita, representatives of the indigenous population. It cost a pound to photograph the latter:

Took 2 bust pictures of her outside the cabin. Then came the worst part – the measuring. The tables involved around 80 measurements. The girl – a stocky lass stated as being 15, but probably older – protested ... But in the end she got rid of her mother, and I was on my own with the beauty! She now became more acquiescent, and meekly took off her dirty attire (cloth boots with *high* heels, and otherwise European clothes) to reveal a *dirty* reddish-brown body, which I at once set about taking very precise measurements of.

You should have seen me kneeling before the beauty, my forehead perspiring, and eager – I can affirm that I was nothing short of a scientist honing his skills. The badly made equipment was a great hindrance. The girl was troublesome, and would not stand still, moving at the very moment I was to read off the measurement. I was enraged. And then, when the job was 2/3 done, she said 'No quiero' and started getting dressed. What was to be done? Over an hour had been lost. I gave her £½ extra and that did the trick.

Your faithful Namman

In due course this report from the world of science got through to his wife. Her enthusiasm was moderate. And she was unaware that the report was censored. For her spouse had also plucked 'hair from the mons Veneris' – the mound of Venus. After completing the job our physical anthropologist and his photographer trekked out into the Magellan countryside, where Hjalmar tried to shoot vipers with his revolver. He also encountered gold-washers – something that was documented.

With the help of Punta Arenas' only pilot, the English sea captain Pritchard, they travelled southwards into the Strait of Magellan, in the dark, against a headwind and in ice-cold rain. Pritchard was an old seal hunter, and with his assistance Stolpe drew up an ethnographic map of the various species' dissemination on Tierra del Fuego and the archipelago on the western side of Patagonia. When the Vanadis had passed the southernmost tip of the American continent – Cape Froward – they encountered the coastal inhabitants in canoes, Ekholm noting:

We had a pilot, and when in the morning we were close to Port Gallant we saw smoke rising on the northern shore, and the pilot said it was an Indian signalling fire. We sailed landwards, and a canoe soon appeared, carrying 4 people (2 men, 2 women). I will never forget their shouts and the commotion in the flimsily sewn-together bark boat! They were half-naked. One of the men was light yellow and tall, clad in a uniform cap and a jacket. The others were red. One young woman looked really good.

It was absolute chaos on board. Everyone wanted to barter! With difficulty I reached a gunport and got 2 paddles, a bailer and a few other trifles. I was beside myself! A man and a woman (the older one) came on board. I measured their height and photographed the man naked, but without success. He would not stand completely still. The woman was wearing a dirty sheepskin around her body, and was allowed to keep it on. The photograph was a complete failure. Every time I gave a sign that she was to keep still she nodded! All in all I got just 1 hour, and that was largely interrupted by others ...

She had been injured on her right temple by a falling aquavit bottle, and she bled and snivelled. Dr Holmer tended to the wound. The woman was measured (149 cm), and a hair sample was taken. Upon our attempts to get a hair with a root she proved to be completely insensitive. Her sole adornment was a fine sinew plait wound tightly around each ankle and cutting into her flesh. These ankle bands were taken charge of.

The Vanadis continued its journey, and at the little Rupert Island another canoe appeared. And this time, too, the commander allowed only a brief meeting – barely half an hour. But Stolpe gained sole rights to the bartering.

This boat held two men, two women, two older people and two small children, as well as a black-and-white dog and a small puppy. One of the men had pale-yellow skin, and his hair was cut short, sticking straight up on his crown, whilst at the sides it hung freely down his neck. Around his forehead he wore a kerchief. All the others had copper-red skin. The men's clothing was largely European, and the women had hides around their loins and shoulders. The small children sat naked on the bottom of the boat, protected on the windward side merely by a small piece of hide draped over their shoulders.

> The bartering was to take the shortest possible time. In their nervousness they offered me everything possible that I did not want to have. An accordion was deemed to be an article on which I should place value, and it was with clear signs of surprise that they found I preferred a tattered rush basket holding a pile of live mussels intended for a meal. They tore the hide off the backs of the poor children, and when I refused to take it they offered the puppy.
>
> The yield from this boat was of great interest, as it included nearly everything a few such families take with them on a boat trip. It comprised a very long oar, the blade of which was attached to a long pole using whalebone (there was another such oar in the boat), a short paddle and a piece of baleen, from which bristles are taken for various purposes.
>
> But the most important thing was a round wooden box holding a host of small items such as a leather bag, containing cut-down quills from a bird, which according to the pilot are used in sorcery, some bone harpoon tips, pieces of glass for making arrow tips and a few pieces of bone probably used for the same purpose, a grooved sandstone for polishing arrow shafts, an iron axe and an iron chisel, a necklace made of threaded marine-worm casings and finally a small toy – a boat – carved from the trunk of a big willow. Even in these inhospitable climes the children delight in toys!
>
> Some rush baskets and raw materials to make them, a few necklaces like the aforementioned, a few hides from a kind of deer – presumably from the rare [south Andean] huemul, no less – made up the remainder of the barter, but unfortunately I saw several other nice items leave with the boat, when after a 25-minute stop it had to put off from the frigate.

The journey continued along the Patagonian Channels, sparsely populated by groups who moved around constantly in their flimsy canoes. Seemingly unafraid, they had paddled out to the Vanadis to barter, and even positioned themselves in front of Ekholm's camera. The canoes consisted of sewn-together pieces of bark, and on the bottom of them a fire was kept going on a bed of tussocks, sand and stones. It was the job of the womenfolk to keep the embers going and paddle the canoe, which had to be continuously bailed out. The visitors came from the American mainland, and appeared accustomed to contact with passing vessels, while the natives on the other side of the strait, in Tierra del Fuego, were already doomed to destruction when the Vanadis passed.

Fifty years previously, Charles Darwin had visited Tierra del Fuego. He stated his opinion about the indigenous population, the underdeveloped Yaghan people, and suspected that in them he had found evolution's so-called missing link. Hjalmar, too, considered that they were a long way down on civilisation's development curve. But he looked on them with respect:

> In each small trodden path one reads a small piece of the history of a people who are engaged in an arduous struggle with nature – a people whose standing is not in the slightest lower than once that of our forefathers, but who will not be granted the opportunity to rise to a higher position like our race, and are instead doomed to be annihilated by the fiery breath of civilisation.

This civilisation's burning breath had already affected Tierra del Fuego's native peoples. As from around 1880, more and more colonisers arrived from Chile and Argentina to set up sheep farms and prospect for gold. Merciless pursuit of the population commenced, ending in genocide, with a small number of them seeking refuge at the island's mission at Beagle Channel. When the Vanadis sailed along Tierra del Fuego's coast, Stolpe stood at the gunwale and gazed out at the mountains, where he saw smoke rising from a fire. He probably knew what was going on in the interior of the island, but despite this he wanted to go up into the mountains. It would be another ten years before the Swedish geologist and geographer Otto Nordenskjöld arrived at Tierra del Fuego with his expedition to study the indigenous population.

Ekholm took lamentably few photographs of the Strait of Magellan and its island world, though he was one of the first photographers to go there. But the commander only allowed a few brief stops, and the time pressure ruined the chance of unique documentation of the coastal Indians' life and culture. Mere remnants of this indigenous culture now remain.

A couple of decades before the Vanadis, in January 1866, another Swedish vessel – the steam corvette Gefle with its 160-man crew – had passed through the Strait of Magellan. Several times the ship had been driven back by storms, despite its being the first Swedish naval vessel equipped with a propeller drive. The Gefle had no remit to gather natural-history specimens or ethnographic objects, but the ship's doctor Georg Westergren gave a lively and detailed account of his encounters with the Fuegians. Unlike the Vanadis, the Gefle anchored alongside one of their settlements, Wigwam Island, where bartering took place. Knives and clothes were exchanged for everyday objects of a type similar to those later collected by Stolpe: woven baskets, harpoons and arrows, necklaces and hides. Their chief was offered aquavit and was on the verge of death, according to Westergren, who upon examining his inebriated patient made a remarkable discovery: 'Unexpectedly and to my great surprise I found in the old man's mouth a large piece of glass, which had lacerated his mucosa in several places. I did not at the time know what I in due course came to realise that afternoon, namely that all Fuegians, be they men or women, children or adults, carry in their mouth a piece of glass the size of a hazelnut'.

Westergren believed that the pieces of glass were fetishes that were in some way connected with the Great Spirit, and he also found that the Fuegians liked music and danced with the ship's boys on board the Gefle. He noticed in the canoes 'poles and pieces of hide that in favourable circumstances were set up as masts and sails'. Georg Westergren's description from the Strait of Magellan was published seven years before Stolpe's journey through the strait, and it is likely the latter had read it.

CHILE

The state reception of the Vanadis in Valparaíso, Chile, demonstrated the value of having a prince on board. At sea Oscar was just a lieutenant who did dog watches like everyone else. But when the heads of state were to be visited he relinquished his incognito role and became His Royal Highness. The ship's priest wrote as follows about the official reception:

> When Prince Oscar paid his respects to the Chilean president [Domingo Santa Maria] he was greeted with the customary salute, and all the while parading troops were lined up to receive him along the entire route between the place of embarkation and the residence. The following day the president paid a return visit to the Vanadis, whereupon he was welcomed with the customary expressions of respect. On the frigate the men manned the yards to the sound of a thunderous salute, which was really magnificent, as all the naval vessels lying at anchor in the roadstead were courteous enough to take part in it.

From Valparaíso Ekholm and Stolpe took the train to the capital, Santiago, and Hjalmar offered his wife a description of the landscape. He rarely wrote like this on other occasions, and had even urged her to refrain in her letters from reflections on nature, as he found them uninteresting. But these lines were probably intended for the planned Vanadis book:

> One sees little in terms of buildings from the railway. The landscape is otherwise characterised by black-poplar hedges, which divide the fields into squares. Here and there are leaf huts – sheds in which the countryfolk live. There is dirt, and children with eyes like balls of fire and mops of horse hair. Otherwise fields, grazing herds and large vine plantations. And as a backdrop to all this the crumbling red mountains with small bushes that from a distance are tuft-like, or without any trace of vegetation. A haze surrounds them with a violet veil. The air trembles, everything distant becomes unclear, and in the distance the odd snow-covered mountaintop. This in concentrated form is the entire landscape.
>
> The journey was comfortable though dusty. I spoke to a couple of Germans, who were friendly and helpful. An old Chilean with a beautiful daughter (her teeth generously adorned with gold inlays!) was keen to learn about my interest in Santiago, and referred me to a Dr Lefevbre, who was said to own a collection of Peruvian vessels. Everything went via German, as the father only understood Spanish. The beautiful daughter mostly slept, and smiled a golden smile in her sleep. Put up at Hotel Oddo, the best one in town, but very shabby ... You will receive photographs from here. Unfortunately I cannot take any myself, as the plates we have brought with us are needed for the museum.

Hjalmar notes with regard to the ethnographic museum: 'I've worked like a slave all day; you know how it is in museums. Rubbed [copied ornaments] and photographed. The museum building stands magnificently in a beautiful botanical garden. The collection of ethnographic material is not big, but everything is of importance to me. Photographed a number of objects from Easter Island, amongst other things'.

After a trip into the mountains Stolpe returned to the hotel:

> It was nearly 10 when I got back, and still no word from Ekholm, who should have been at dinner at 6! I was worried, as was the proprietor. I was also angry, and determined to really teach him a lesson if I ever got hold of him again. I was afraid he had got lost, topographically or morally, and had strayed into the outskirts of the town, where knives are readily brandished. I went and stuck my nose in his dark room a few times, but in vain. It was nearly 11 o'clock when the proprietress saw someone coming from his ground-floor room facing the courtyard, and – unused to being able to sleep undisturbed he had lain asleep the whole time! Laughter all round!

PERU

The Vanadis stopped in Chile for ten days, continuing on 25 March to Callao in Peru, a country that had just concluded a humiliating war with Chile – the so-called Saltpeter War of 1879–83. The conflict had been about the rich saltpeter deposits in the Atacama Desert in the borderland between Peru, Bolivia and Chile. Following bloody combat Chile was now the victor, while Peru had lost its southern provinces. Chile quickly continued its expansion, suppressing the indigenous peoples in the south – the Araucanians. In one fell swoop Chile had doubled its area and laid the basis for a role as a major power on South America's Pacific coast. When the

Vanadis anchored in Callao, there were still a few Chilean military units scattered around the conquered country.

In the capital, Lima, Oscar and Hjalmar documented some private ethnographic collections, including a hundred or so clay vessels made by the indigenous Peruvian population. Large quantities of clothing and weapons were also bought from the native tribes in the Peruvian interior. Some of Stolpe's best finds were those from a French collector/cobbler who had gone into insolvency as a result of the war with Chile:

> Just listen to what I've achieved. I've bought a *large* and attractive collection of Peruvian clay vessels (over 200), as well as a number of other grave finds. It's cost quite a bit, but people say the prices are unusually good, and connoisseurs congratulate me. I could fill a room as big as my study at the museum just with Peruvian objects! Plus many excellent modern Indian objects. I'm more than content! Everything is already packed up and has been sent home via Hamburg, addressed to myself at the National Museum. It's got to stay untouched until I get back. Little Sparrow will be there to help unpack! It's been a hell of a job. Yesterday I sat all day long in a cobbler's shop, cataloguing pots!

After a few hectic days in Lima, Ekholm and Stolpe took the train for the 20-mile trip north to Ancón, where a colossal graveyard extends across the windblown sandy soil. For thousands of years the inhabitants had buried their dead in the easily worked ground, which by the same token unfortunately made things easy for tomb raiders, who increased in number when a new stretch of railway was built through the graveyard in around 1870. In a small village they put up at a ramshackle hotel: 'This is actually a seaside resort formerly frequented by Lima's rich families, but most of it has been shot to pieces by the Chileans. In my wall there's a gaping hole left by a grenade. Will photograph the destruction'.

Stolpe was to spend four days in Ancón, devoting himself to a cross between archaeological excavation and tomb-raiding – an unusual pursuit for this prim scientist. He brought in a professional tomb raider and the latter's five-strong digging team. They amassed a total of 80 human skulls and a dozen mummies. He described his time there to his wife:

> Unbearably hot. The burning sand is whirling around, penetrating my clothing and pricking me like red-hot needles. My eyes feel scorched, and I will doubtless get new skin. Work is thus difficult. The valley appears to be about 3 miles wide and 20 or so miles long. Consisting solely of graves, largely destroyed by treasure hunters. Parts of mummies, shrouds, clay vessels etc. are lying everywhere. Just by picking things up I've gathered a large number of objects. Eighteen (18) packing cases will be leaving here in a few days! I've bought a lot – *over* 1,000 kronor spent on clay vessels alone, plus a lot of other things. The Ancón finds will fill 10 packing cases.

The excavations were performed with the aid of hired workforce:

> The workers are provided with a long, thick iron wire that they poke into the sand. If it goes down easily and deep, then there's a grave there. We first hit on a layer of adobe bricks (air-dried bricks made of clay mixed with sand and shaped with water). In the pit below is a mummy, surrounded by pots etc. I've of course not been able to carry out the investigations the way I'm used to at home. I have scant idea about the nature of the graves, but at least that's something. Neither are the finds there valuable, but from the many thousands of graves (it seems to have been a general burial site for the mountain peoples, who found it easier to dig here in the sand, and furthermore needed all the arable land in the mountains) finding a lavishly endowed one wasn't easy. One mummy that I took out was clothed in a blue-and-yellow feather cloak with yellow feather adornments on the hood. I'm afraid that everything will fall apart when it's removed, and that insects will have been at work beforehand. But I'm proud of the vessels. Just imagine, at least 210 clay vessels. It will be a lovely collection! And such textiles!

In Ancón the routine was:

> to rise with the sun and then work until 11, when we dragged ourselves back for breakfast. I then bathed in the bay – excellent bathing, without which it would have been unbearable. A large seal kept me company one day, and we swam a mere few fathoms from each other. It looked surprised. After breakfast back out into the field until 6pm, then bathing, so that taking the desert hike into account I never got my dinner until close on 8 o'clock. We then sat for a while and hung out on the beach in the moonlight, after which we turned in, exhausted.

In Ancón a group photograph was taken that is one of the most frequently published pictures of the Vanadis journey. In front of five workers furnished with spades, Stolpe is sitting attired in a safari helmet, surrounded by skulls. In one hand he appears to be holding a measuring tool, in the other a clay vessel. It is a good picture with a powerful symbolic effect that has consolidated the perception of Hjalmar Stolpe as a physical anthropologist. But it is a deceptive picture. It shows Stolpe posing as something he never was: a physical anthropologist and skull measurer. He had admittedly been given an anthropological remit from the Royal Swedish

Academy of Sciences, and had been provided with equipment and a measuring plan, but he did the work reluctantly, and soon refused to carry out the assignment, pleading lack of time. Stolpe was never convinced of the scientific importance of skull and body measurements.

On his right-hand side sits a bearded man, hired by Stolpe, who was most definitely not interested in anthropology – the professional tomb raider George Kiefer, who was in charge of the five workers. For several days this group dug down into a dozen graves in the scorching sun and emptied them of their contents. Hjalmar describes them as 'circular pits in the sand of varying depths, 5–10 feet or more wide and 5–6 feet in diameter. In these pits the corpse has been placed in a sitting position with its knees under its chin, and wrapped in several kinds of clothing and rugs. All around are the deceased's belongings, such as *huacos* [clay vessels] and weapons (wooden spears)'.

Much later Stolpe had the mummies from Ancón examined using the new X-ray technology, resulting in a well-shaped skull, spine and ribs belonging to a young woman being visible on the plate without any damage to the shroud – a sensation at the time. The pictures of the other mummies were unsuccessful, as the mummy packages had been impregnated by the windblown sandy soil from Ancón. The X-ray images were taken in 1900 in collaboration with Thor Stenbeck's Roentgen Institute in Stockholm, which had been set up the year before. In conjunction with the publication Stolpe stated that the mummies from Ancón could probably be dated to the 16th century. This was probably the first time that X-ray images were used by a Swedish archaeologist and ethnographer.

The Vanadis stopped in Peru for a good ten days, and then, on 14 April, it put out into the Pacific Ocean, heading for the Marquesas Islands in French Polynesia.

One of Oscar Ekholm's first preserved pictures from the journey round the world shows the traditional line-crossing ceremony for all crew members who have not previously crossed the equator (21.01.1884, 12.15 pm). In the so-called 'baptism on the line' or 'equatorial baptism' they are prepared and cleaned ahead of their entry into the other hemisphere. The ritual is led by King Neptune (Senior Master Nilsson in the centre of the picture), whose courtiers heckle the newcomers. The ship's band is playing and the crew are drinking alcohol. Prince Oscar: 'I was the first to be called forward. I was daubed with whitewash and was shaved and washed, then my hair was cut – all using wooden instruments. ... I was let off being thrown into a sail full of salt-water, like all the others, in consideration of my eye condition (catarrh).' Stolpe said to his wife: 'I had no wish to make a fool of myself, and instead watched the prank from the bridge together with several other prudent men, such as [the ship's doctor] Rudberg, the purser and the priest. You know I find it hard to take part in games.'

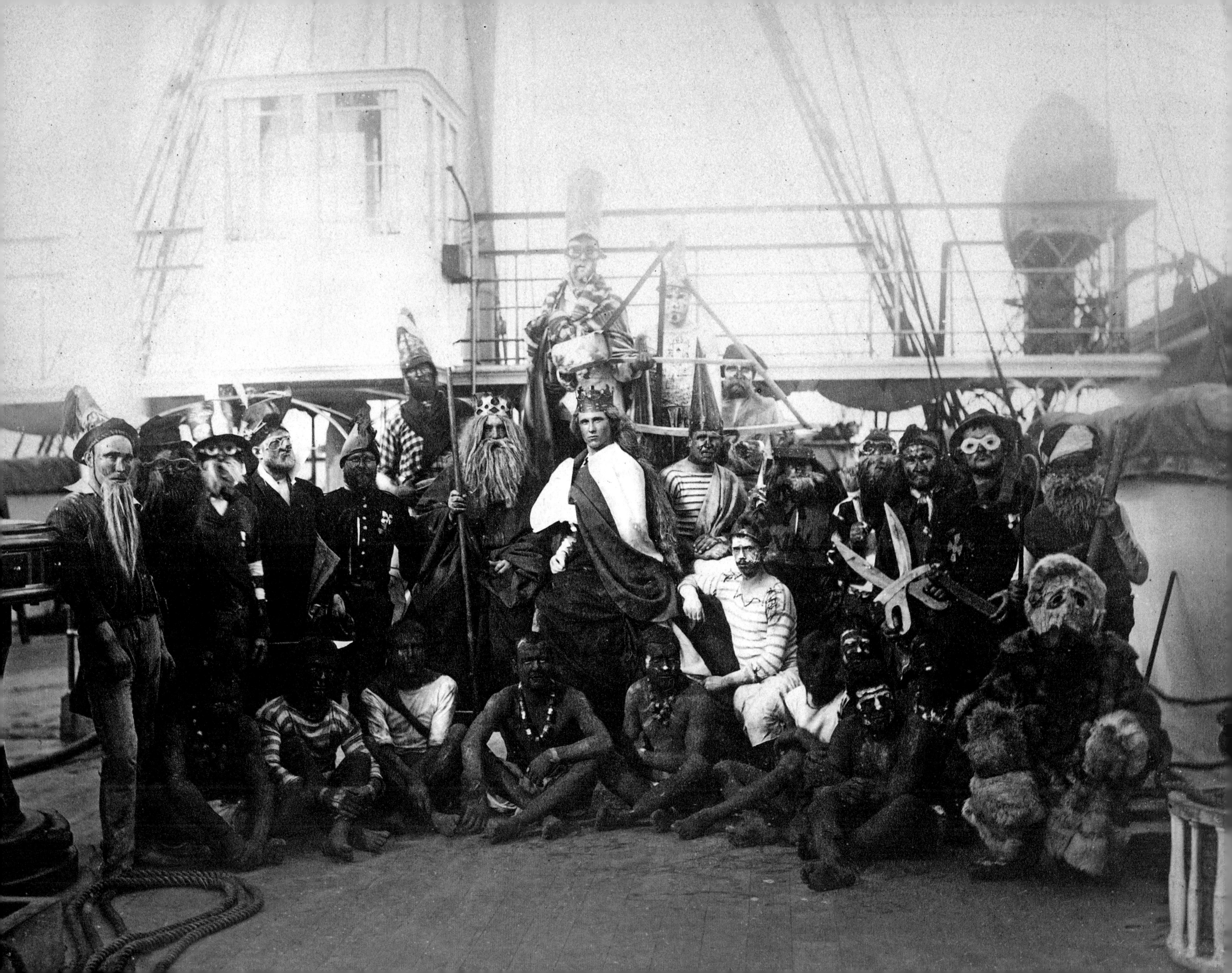

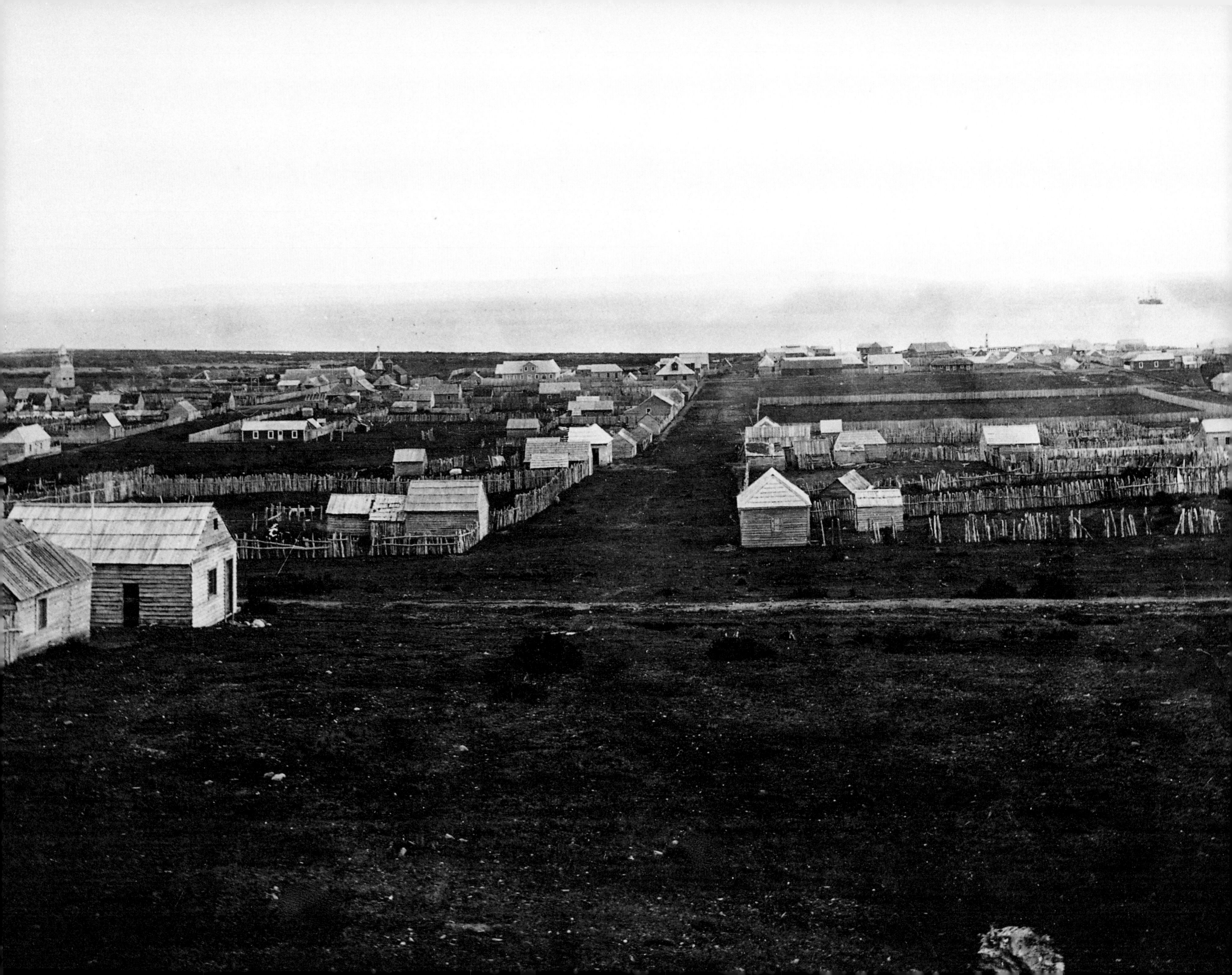

Opposite page: In the evening of the last day of February 1884, the Vanadis dropped anchor off Punta Arenas on the Strait of Magellan, the world's southernmost city and Chile's former penal colony. It accommodated around a thousand inhabitants, and was described by Stolpe as follows: 'A few rows of single-storey wooden houses (at best provided with frontages à la Upsala), a beacon made of wood, and behind all this an immense tract of bare, grey tree trunks, witness to a raging forest fire.' The vessel only stayed for 24 hours, to replenish her stocks of coal.

In a river valley in the forests near Punta Arenas, Stolpe and Ekholm happened on gold washers, one of them an 'Indian' who allowed himself to be photographed after having been bribed with a cigar: 'Positioned himself for a photograph. I got a family picture, with 3 ruffians and myself to keep the Indian still. Saw the gold they'd got – small, insignificant slivers.' The journey home for Stolpe – armed with a revolver – was hazardous: 'Waded over the river with the dangerous bridge. Broke into the anthem Thou Ancient, Thou Free *by way of a signal. Pitch black. Pack of dogs upon entering the city.' In the picture Stolpe is to be seen alongside the 'Indian' in a peaked cap and with a full beard.*

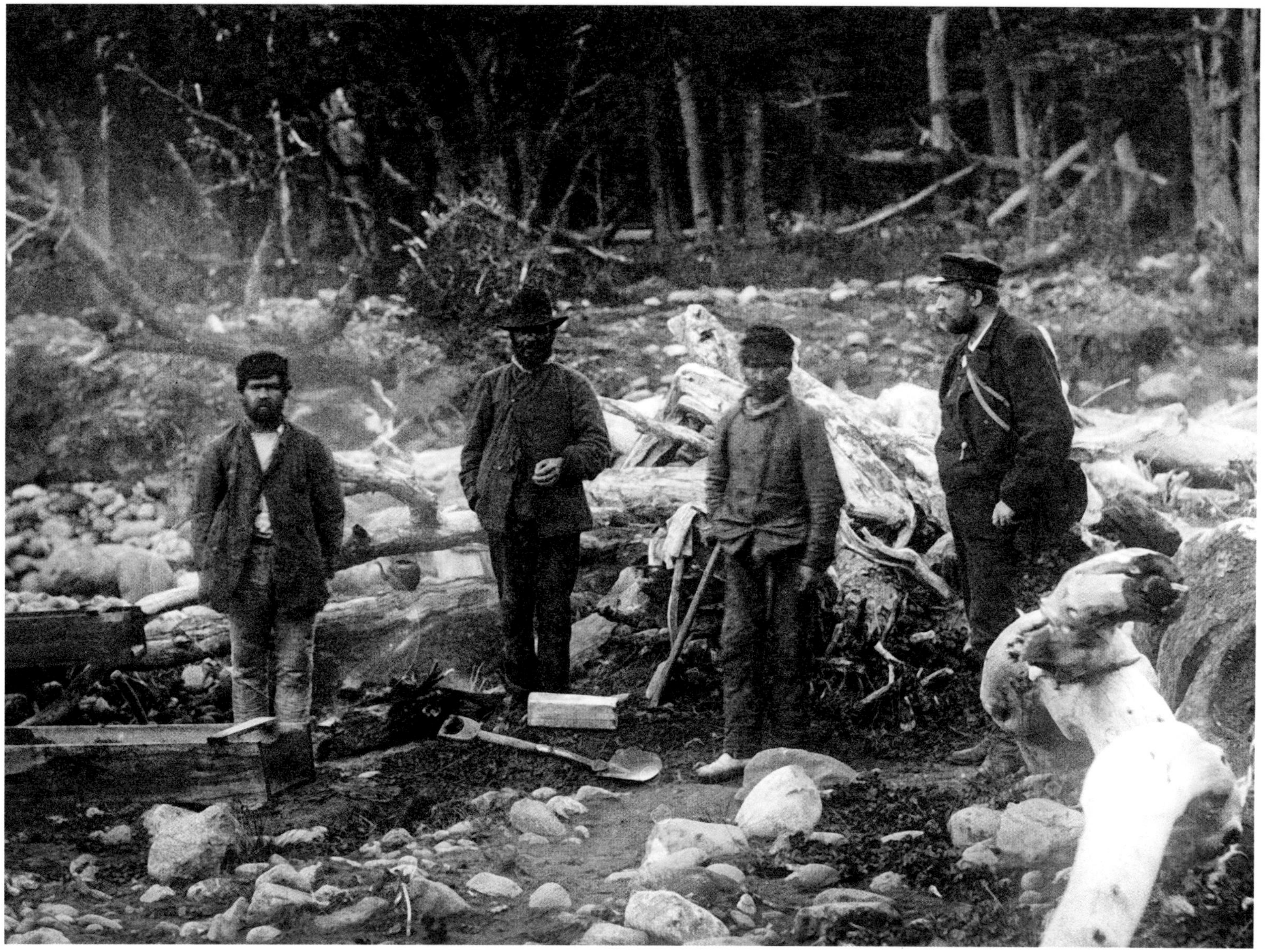

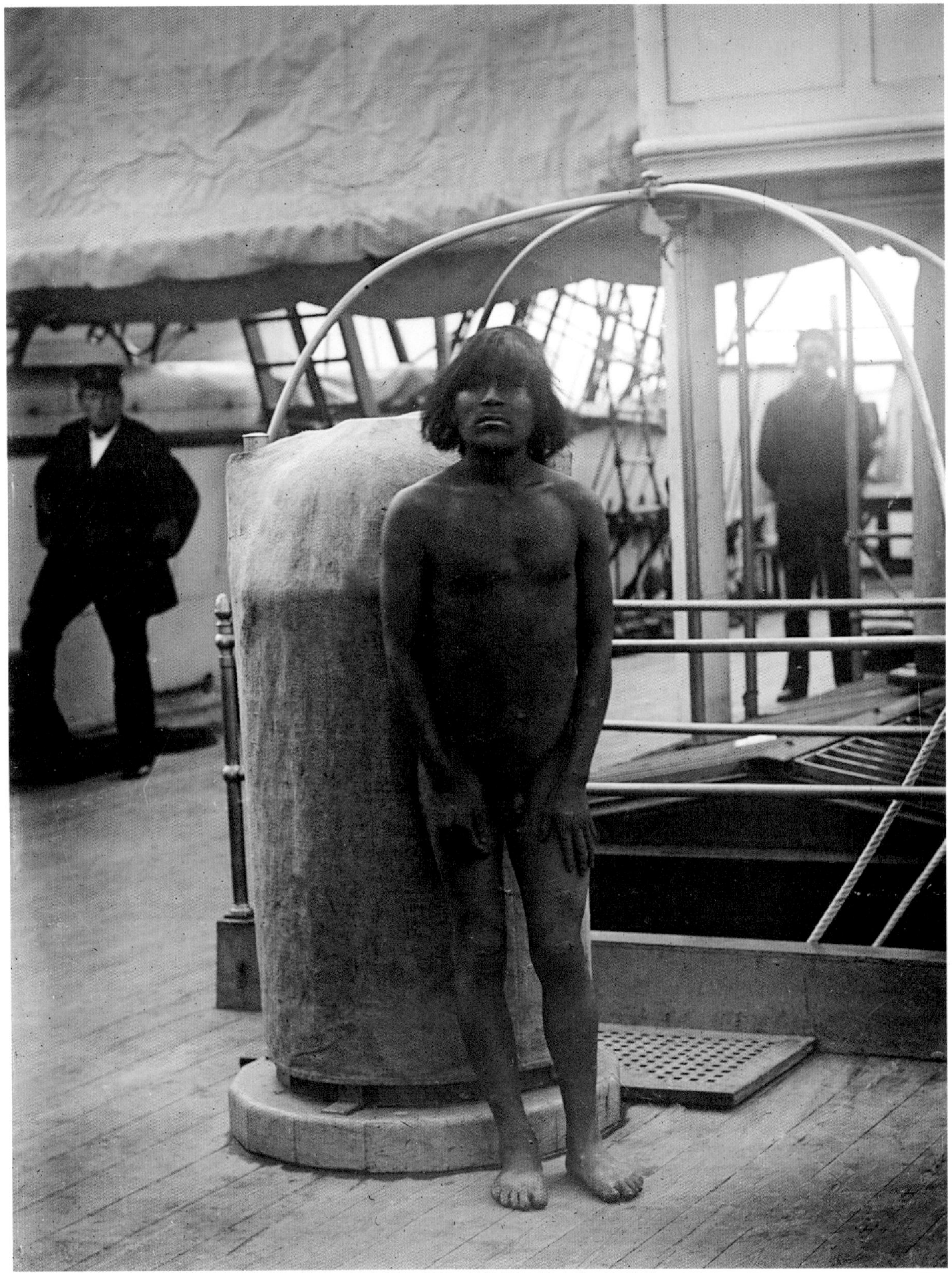

In the strait between the Tierra Del Fuego archipelago and South America's southernmost tip, Cape Froward, they came upon canoes bearing members of the Alakuluf tribe (now known as Kawésqar). 'I will never forget their shouts and the commotion in the flimsily sewn-together bark boat! The final one to come up was a red man, who was photographed naked and measured (H: 156 cm)', wrote Stolpe. After some bartering, the visitors disappeared with the Eskilstuna knives, tobacco and aquavit, whilst Stolpe got the contents of one of the canoes – an ethnographic rarity. Fifty years previously Charles Darwin had thought this indigenous people to be evolution's so-called missing link – the link between animals and humans. For Stolpe they were on the contrary 'a people whose standing is not in the slightest lower than once that of our forefathers'.

Opposite page: With Tierra Del Fuego's mountainous silhouette as a backdrop, Ekholm photographed the Vanadis, anchored off Borja Bay at South America's southernmost tip on 3 March 1884. In all, it took a week to pass through the Strait of Magellan and the Patagonian Channels, to get to the Pacific Ocean. The Vanadis became the third Swedish naval vessel to make this journey. In 1852 the frigate Eugenie had beaten the 'world record' by negotiating the infamous channel in 11 days using sails alone – lacking an engine. She was followed 14 years later by the steam corvette Gefle, the first Swedish propeller-powered naval vessel.

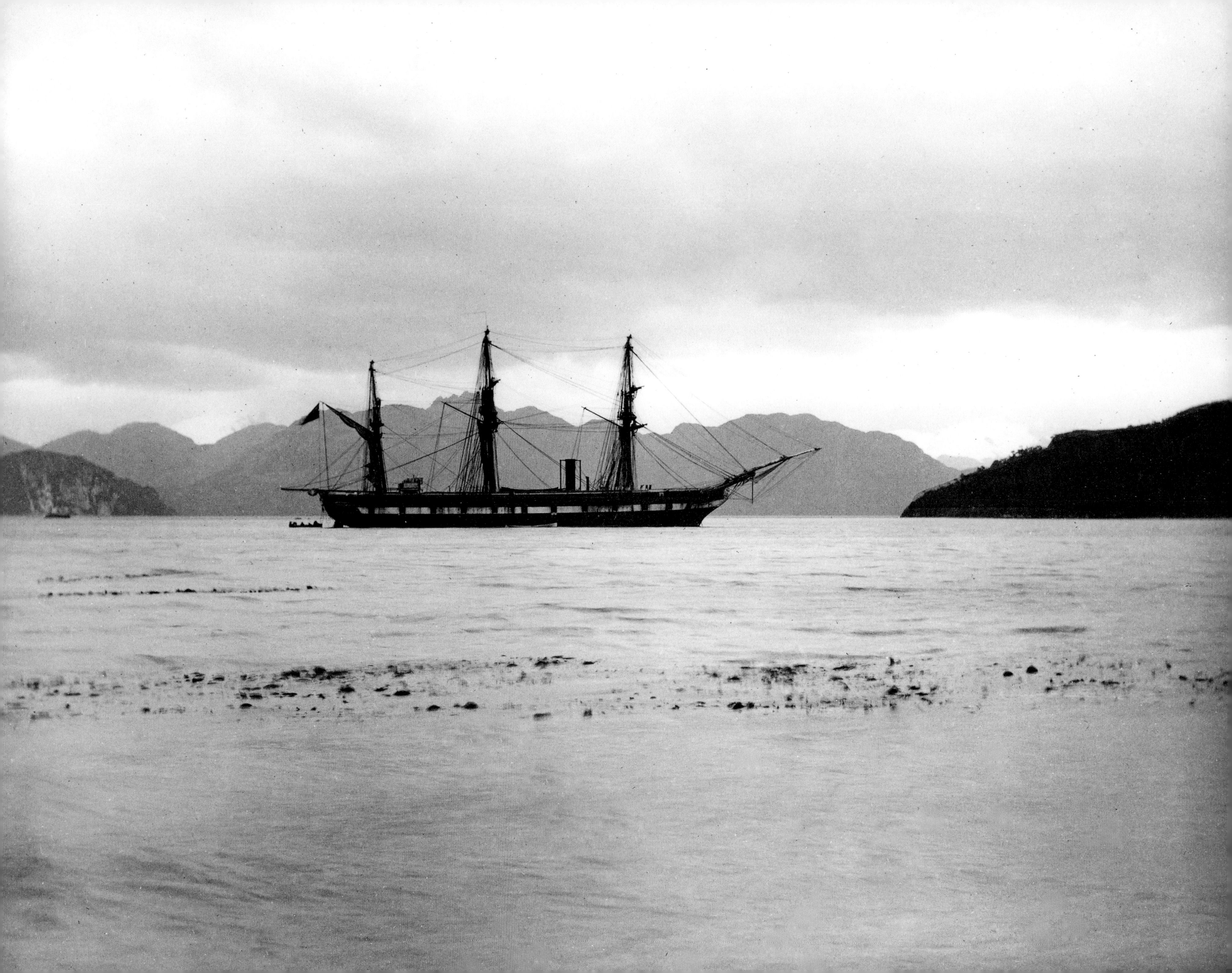

Opposite page: This picture of a marketplace in Chile was taken by the local photographer Spencer in Santiago. The year before the Vanadis came to Chile he had printed an extensive illustrated catalogue, from which Stolpe chose to buy photographs. This picture bears the number 483, and is described thus: 'Plaza de Abastos in Chillán during the fair.'

This page: On the back of the photo is the following, in Hjalmar's neat handwriting: 'Indians' huts and mussel shells (waste) Borja Bay.' He had chanced on the canoe-borne coastal Indians' settlement – two brushwood huts that had been covered with hides but were now abandoned: 'They were made of thin beech branches pressed into the ground to form a circle, their bent-together tops having been tied together with rushes to make a beehive-shaped hut. All around lay large quantities of mussels and other seashells, sealbones, whalebones etc. – in other words a complete kitchen midden, providing detailed information on the inhabitants' diet. Broken baskets lay everywhere, and here a child's most treasured toy was also to be found: a little rigged boat, made of sewn-together bark and even equipped with a deck and a mast – a clear imitation of a European sloop. ... These Alakuluf Indians never leave the coast, where they spend their lives in canoes. They never go up into the mountains. The spirits live there. I sometimes sank down to my shoulders between tree trunks, and had difficulty struggling to my feet. But it was great fun!'

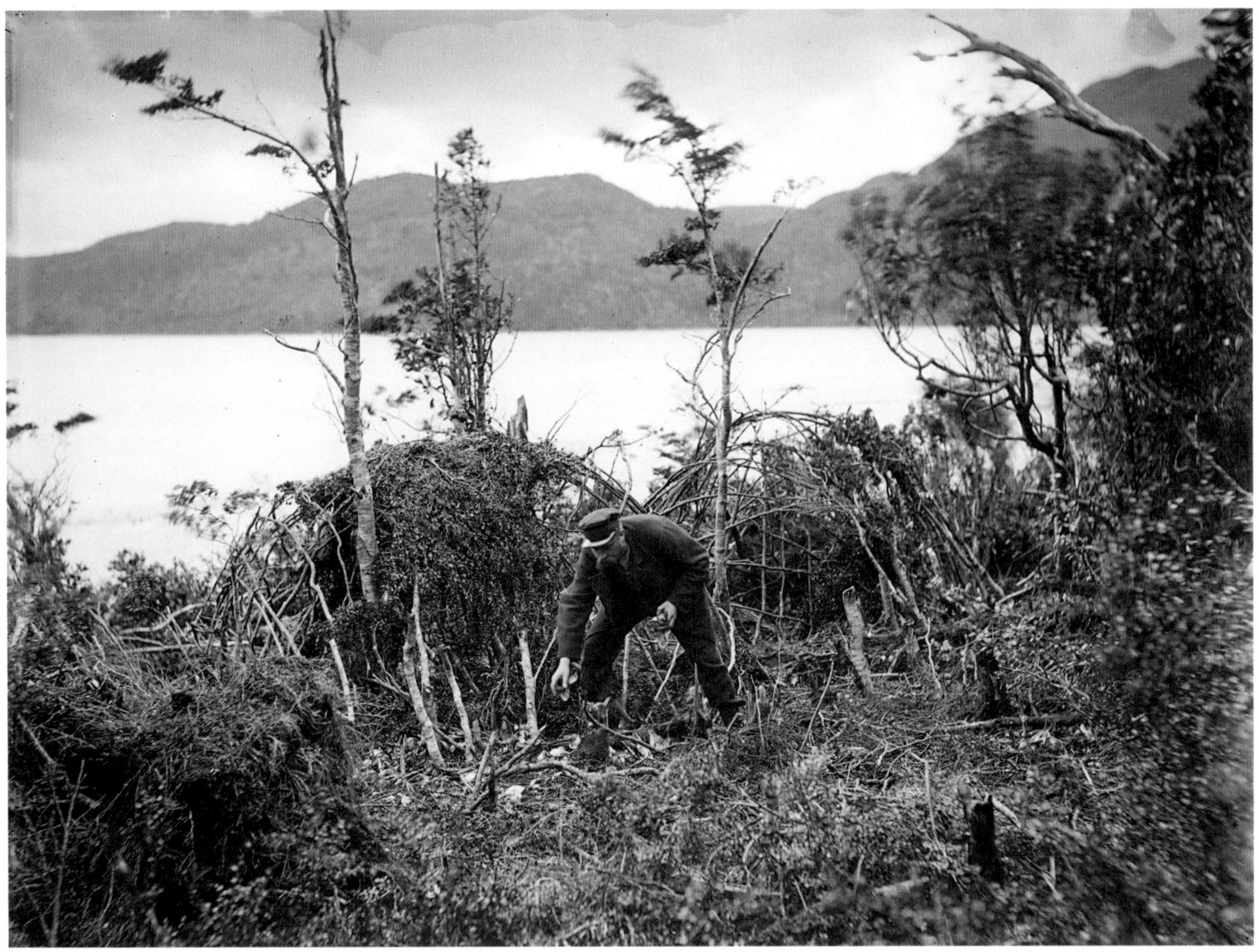

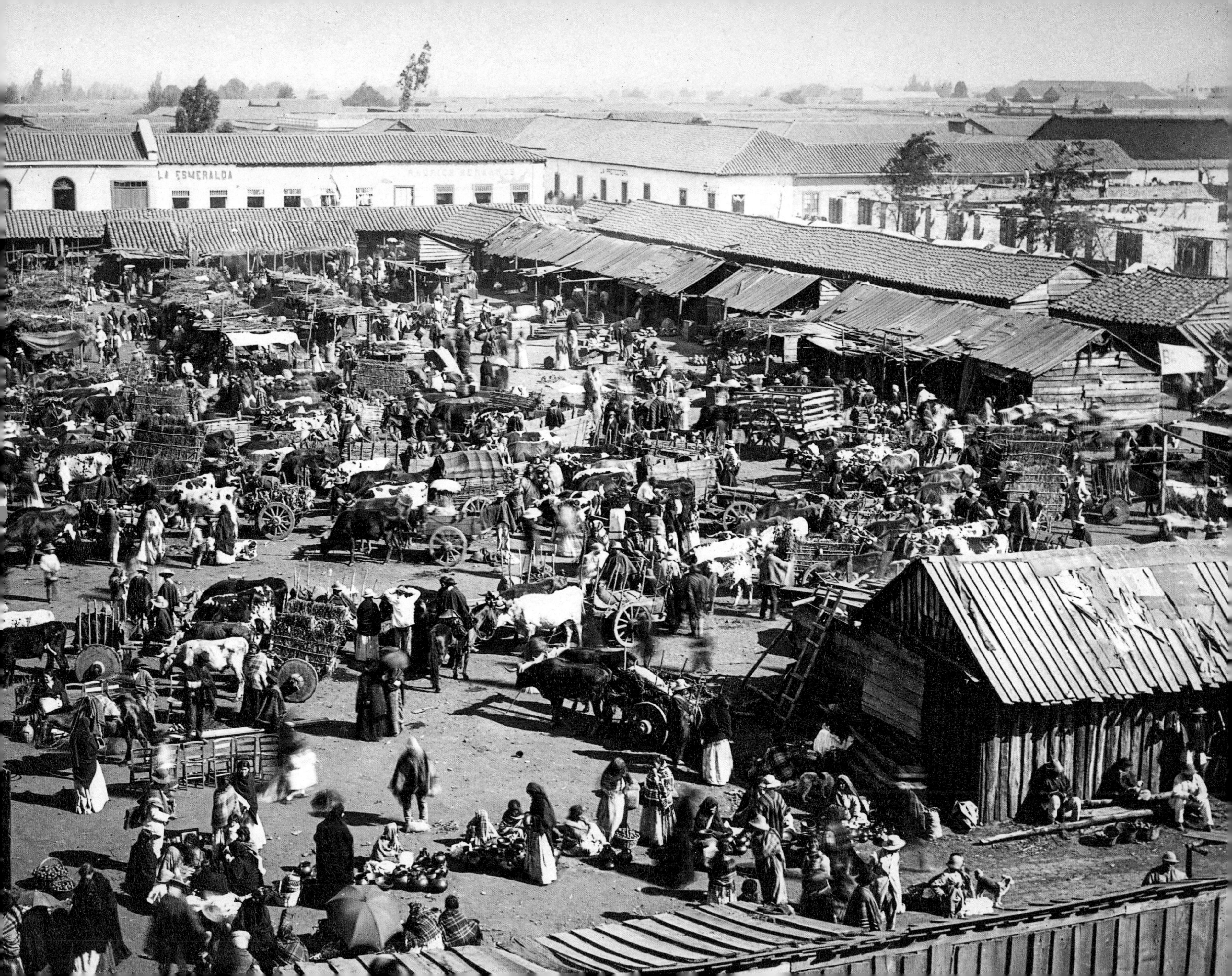
LA ESMERALDA

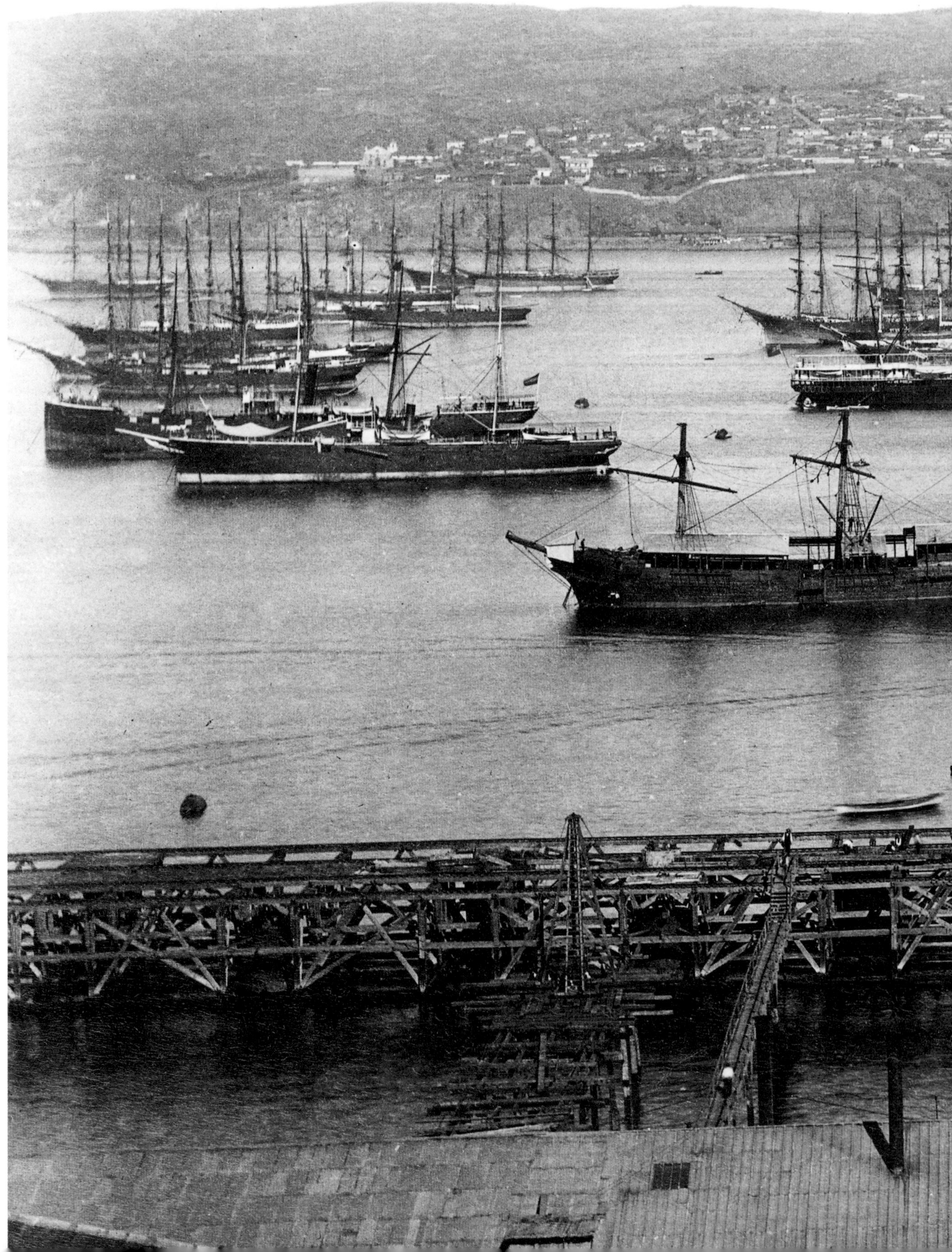

In mid-March 1884, the Vanadis arrived at Chile's port city Valparaíso, where they stayed for ten days.

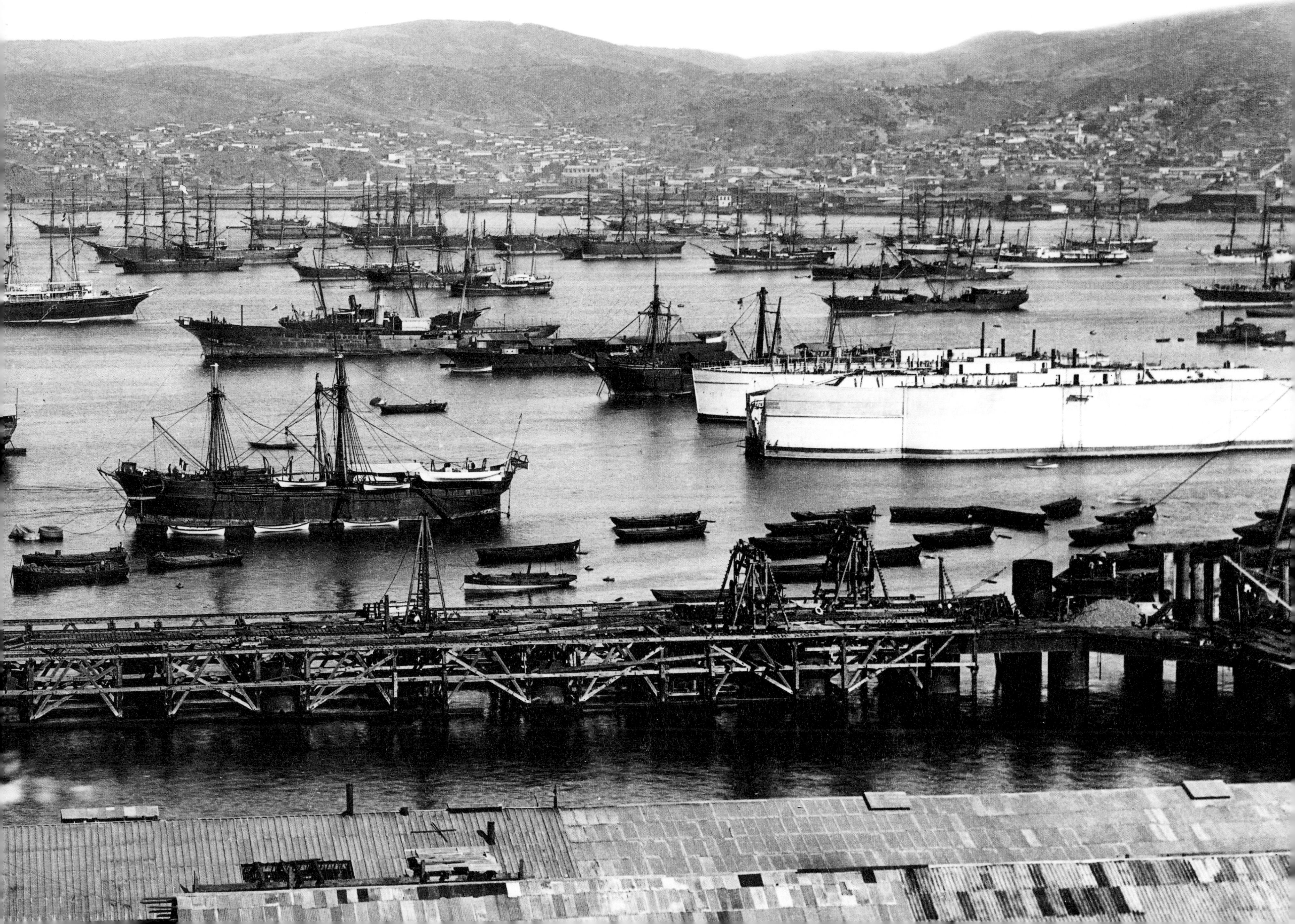

This page: The earth of the desert at Ancón was easy to work and had been used for burials for thousands of years. Stolpe recounts in his travel report to the daily Stockholms Dagblad*: 'Wherever you dig you hit upon a grave – or at least remains thereof, as the field has been most heinously plundered. ... The entire field is consequently now covered with fragments of pots, whole and damaged mummies, scraps of clothing – offering a rich variety fabric samples – and separate skulls and parts of skeletons. Seeing this destruction is heartbreaking!' Stolpe collected a total of 80 skulls, ten mummified humans and four dogs here and sent them to Stockholm.*

Opposite page: For four days in April 1884 Stolpe led a survey of the Indians' old grave field at Ancón in Peru, 20 miles or so from Lima. This excellent picture is one of the most frequently published photographs from the Vanadis voyage. It has sometimes been used to present Hjalmar Stolpe as something he never was: a race researcher. Putting it simply, 19th-century physical anthropologists tried to divide humanity into different races with different levels of ability, the 'primitive' peoples being at the bottom of the list. Stolpe's opinion was the very opposite. They were 'people whose standing is not in the slightest lower than once that of our forefathers.' The reason why he collected skulls and other body parts in Peru and at further locations during the Vanadis voyage was his instructions from the Royal Swedish Academy of Sciences. Ahead of the journey he had been provided with anthropological measuring equipment. But after just a few weeks visiting the islands of the Pacific Ocean he had refused to carry out this assignment, citing lack of time. And for the remainder of his professional life he was never to return to this 'science'. Sitting on Stolpe's right-hand side is the hired foreman and professional grave robber George Kiefer, and on his left-hand side is his German friend, the businessman Wilhelm Gretzer. The men standing up are the five Peruvian workers who shovelled away the windblown sandy soil so as to get down to the grave. Stolpe is holding a clay vessel and a measuring stick. The foreman Kiefer 'kept his books on the diggers by tying a knot in a length of string for each working day. The workers kept track in the same way.'

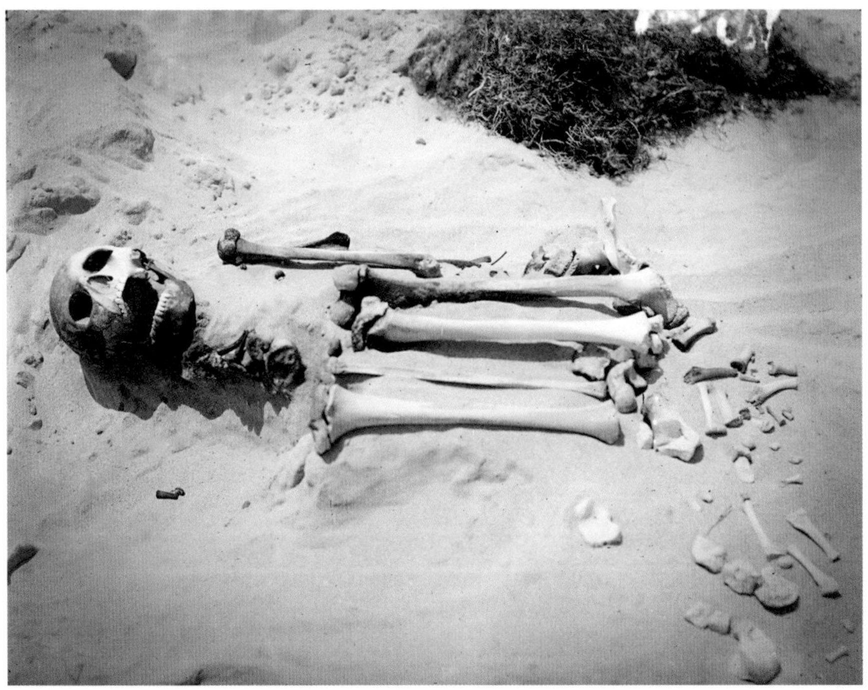

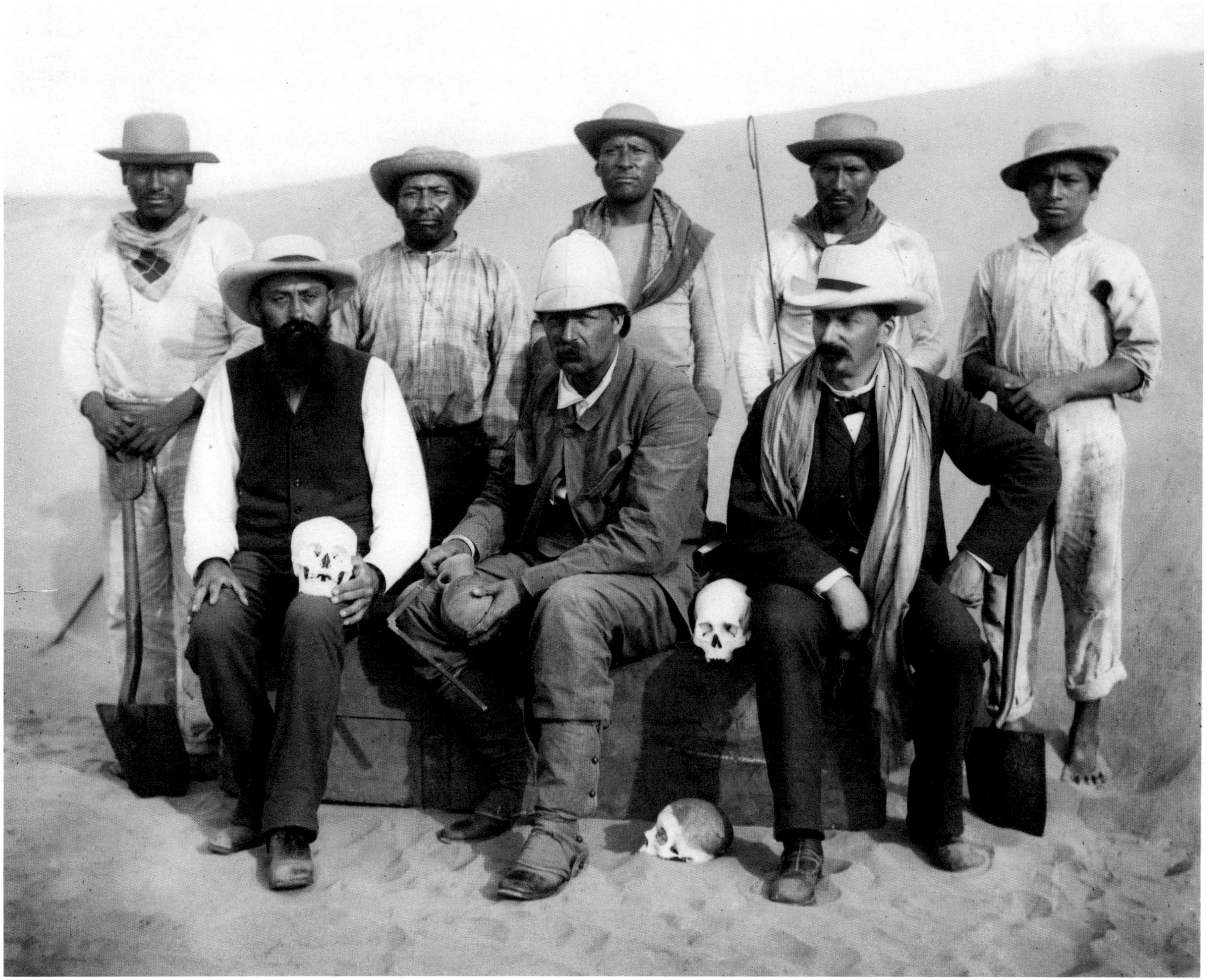

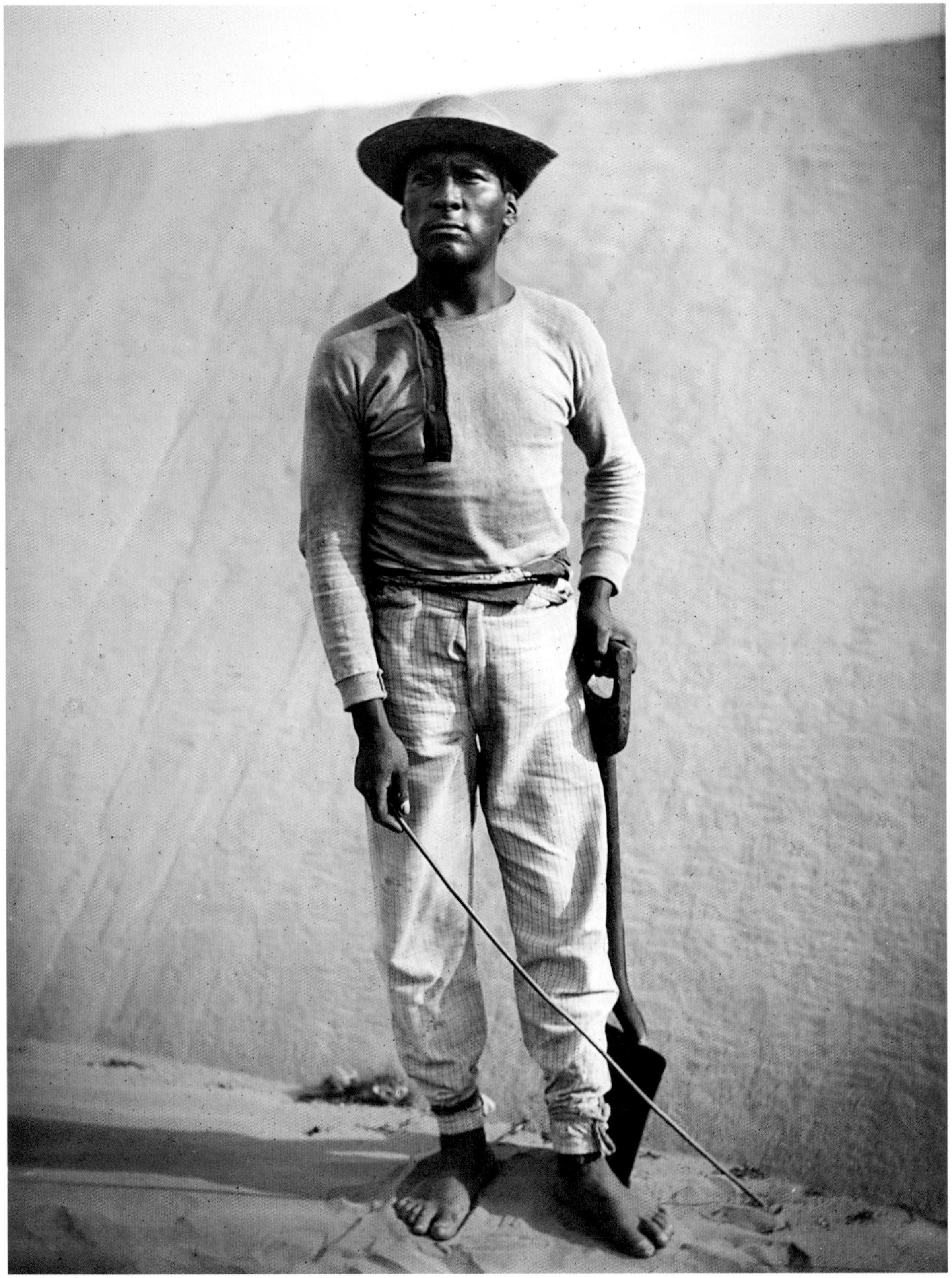

This page: The gravedigger Faustino Rojas wields the simple tools used on the Ancón grave field: a spade, and a long piece of coarse iron wire that was pushed down into the windblown soil to locate the graves. When the wire 'goes down more than 2 feet [60 cm] without any resistance, there will usually be a grave there.'

Opposite page: 'Ancón, group of buildings at the outer eastern edge', Stolpe wrote in one of his lists of pictures. The settlement is thus adjacent to the grave field he excavated.

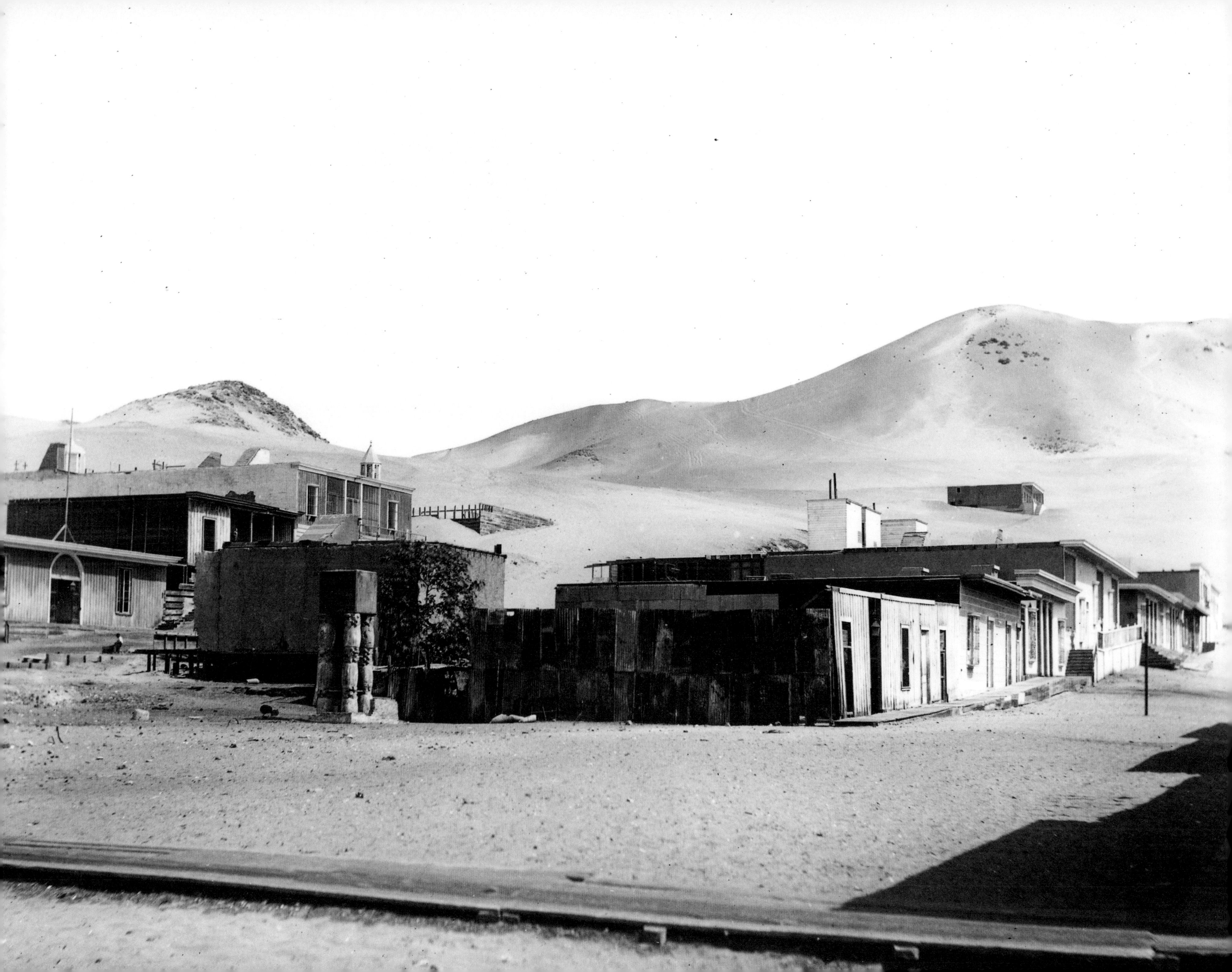

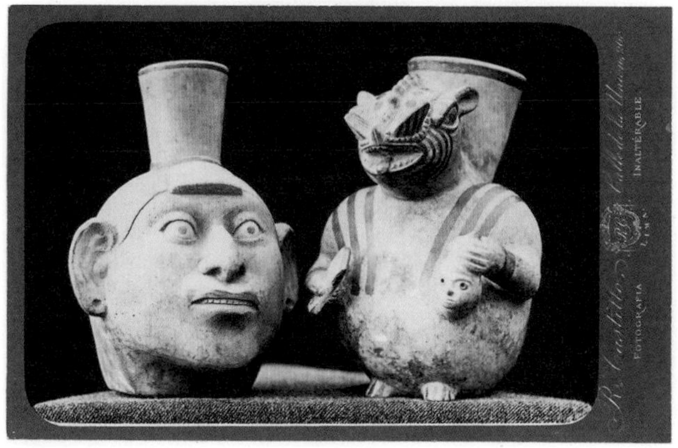

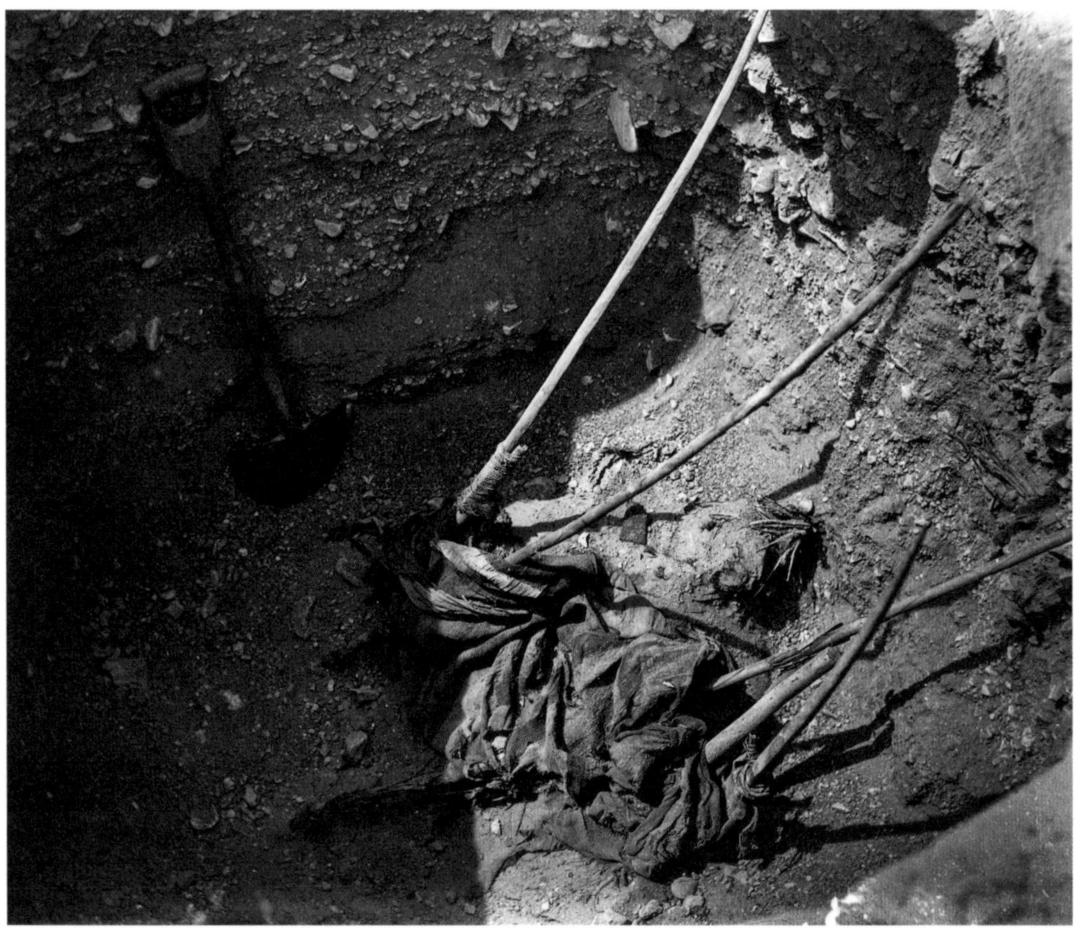

This page, top: In Lima, Peru, Stolpe studied a collection of old ceramic objects belonging to the surgeon José Marino Marcedo. The collection had been documented by a local photographer, and in 1884 Hjalmar bought these photographs, which Stolpe's grandson Björn Fabricius in turn sold to the Museum of Ethnography in Stockholm in 1932.

This page, bottom: Ekholm documented an excavated grave in Ancón's huge desert area, and Stolpe wrote: 'The graves are round pits in the sand of varying depths: 5–10 feet deep or far more, and 5–6 feet in diameter. In this pit the corpse has been placed in a sedentary position with its knees under its chin, and it has been wrapped in several kinds of clothing and rugs. All around are the deceased's possessions, such as huacos [clay vessels] and weapons (wooden spears). Everything is covered with so-called adobes – air-dried clay-and-water bricks.' Later, in Stockholm, Stolpe was able to date one of the graves to the 16th century using the new X-ray technology. In his list of pictures Stolpe states the following about this photograph: 'Huaca No. 1 with mummy'.

'Grave opening at Ancón', Hjalmar Stolpe wrote on the back of this picture, in which he is measuring the depth of one of the excavated graves. In his list of pictures he also noted the following: 'Huaca with roof made of adobe. I examined 11 such graves. They did not contain any precious finds, but some of the objects were excellent. The most important thing was becoming acquainted with the mode of burial. A number of boxes were filled with mummies, a lot of skulls, pots etc., and the boxes were padded out with all kinds of cloth found lying around on the soil, providing shrouds for damaged mummies.'

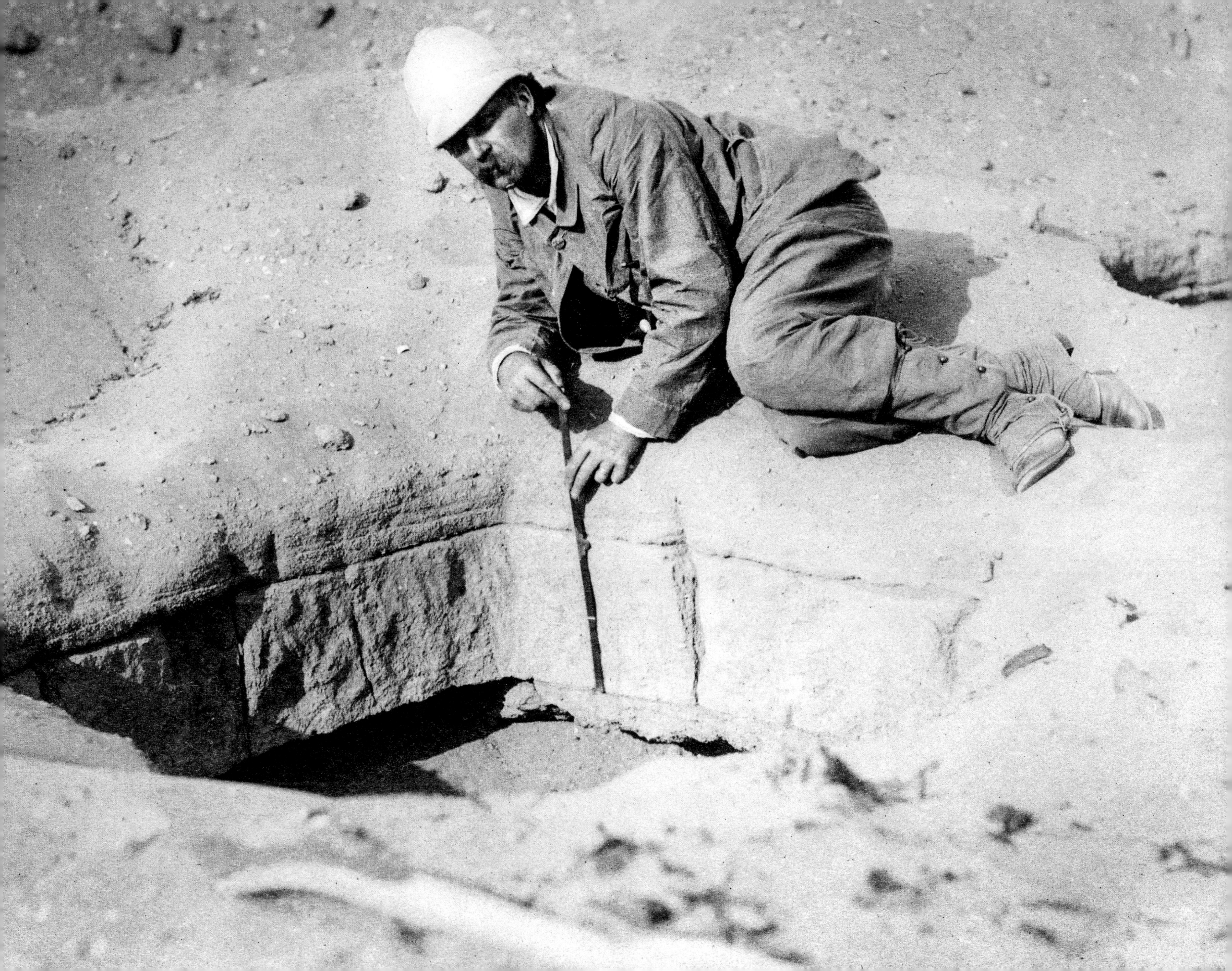

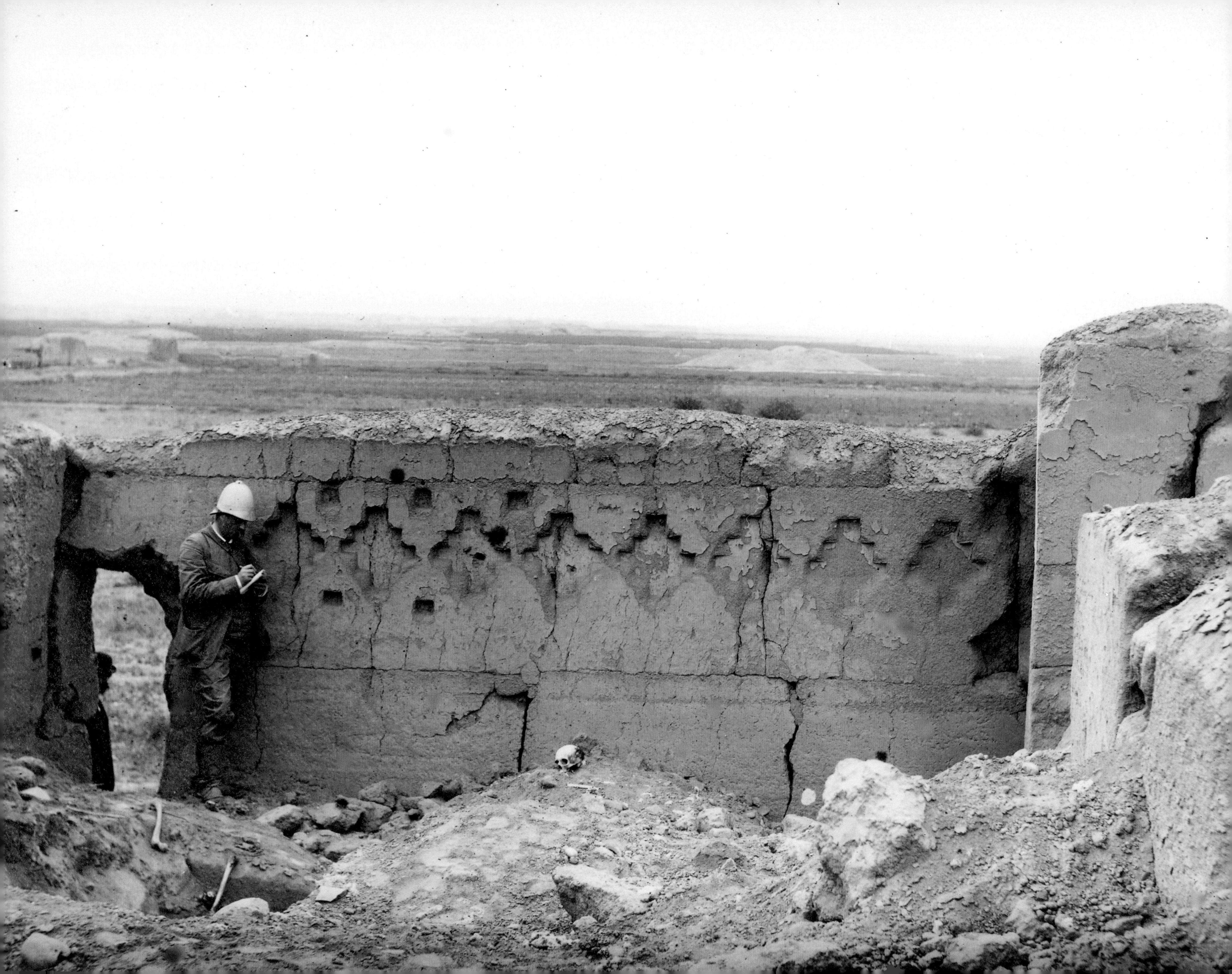

Opposite page: Stolpe is writing, posed in front of Ekholm's camera in Magdalena del Mar in Peru. Behind his back, his friend Wilhelm Gretzer is peeping through the opening in the wall, thinking he cannot be seen. Hjalmar left a total of ten notebooks about his journey round the world. The particular one he's writing in in this picture – No. 2 on Chile and Peru – has unfortunately been lost.

'On the beach, cavalry barracks, boats (I'm sitting in one!)' – the text on the back of the photograph Hjalmar sent home to his wife from Ancón. He said of the visit to the desert area that 'my face [burns] when I think about the heat of Ancón's fields of windblown sand, where it was impossible to stand still for half a minute without the hot sand burning through the soles of my shoes. And the sand whirled up into my face! And yet, how I would have loved to devote at least a month to surveying Ancón's desert!'

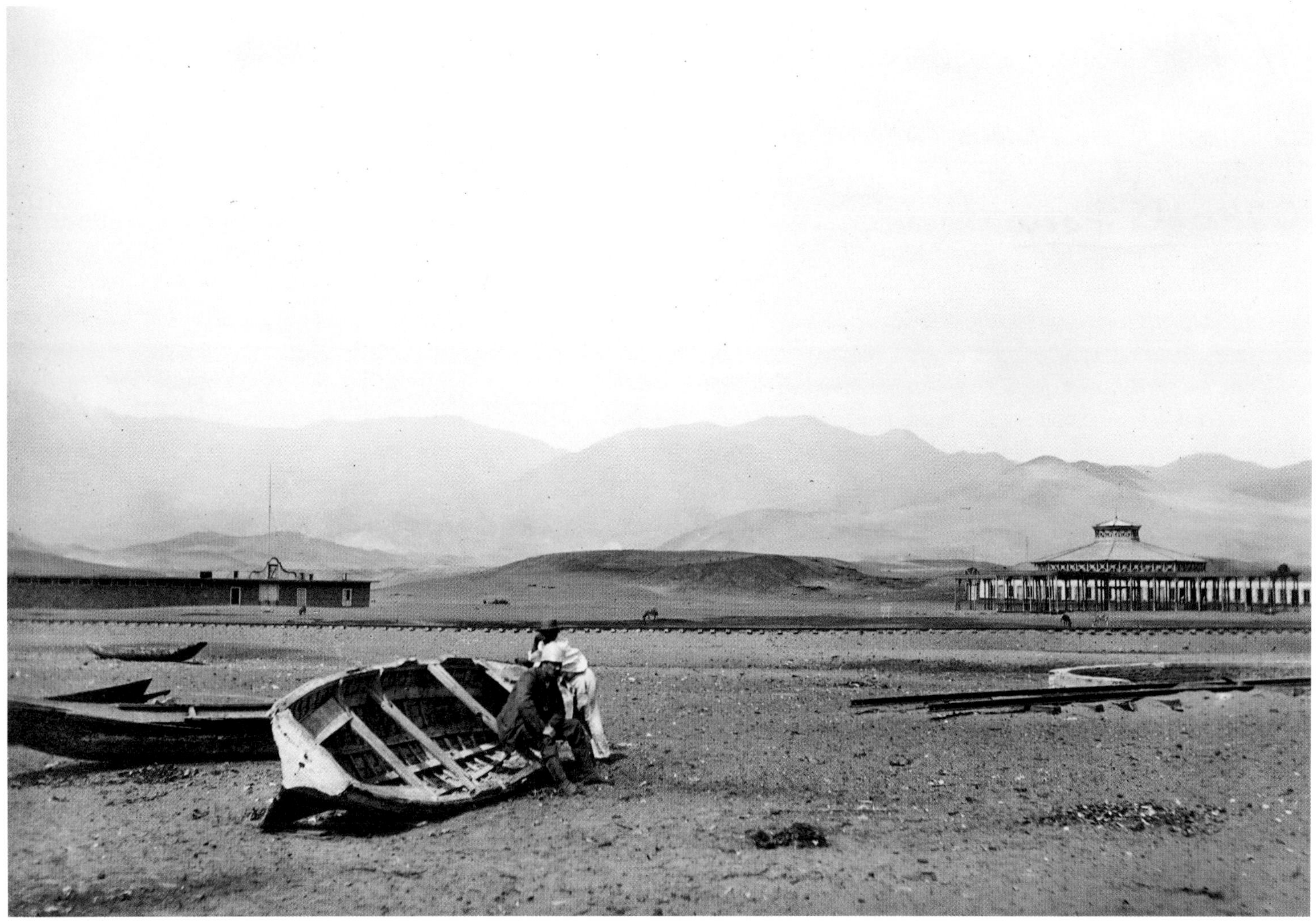

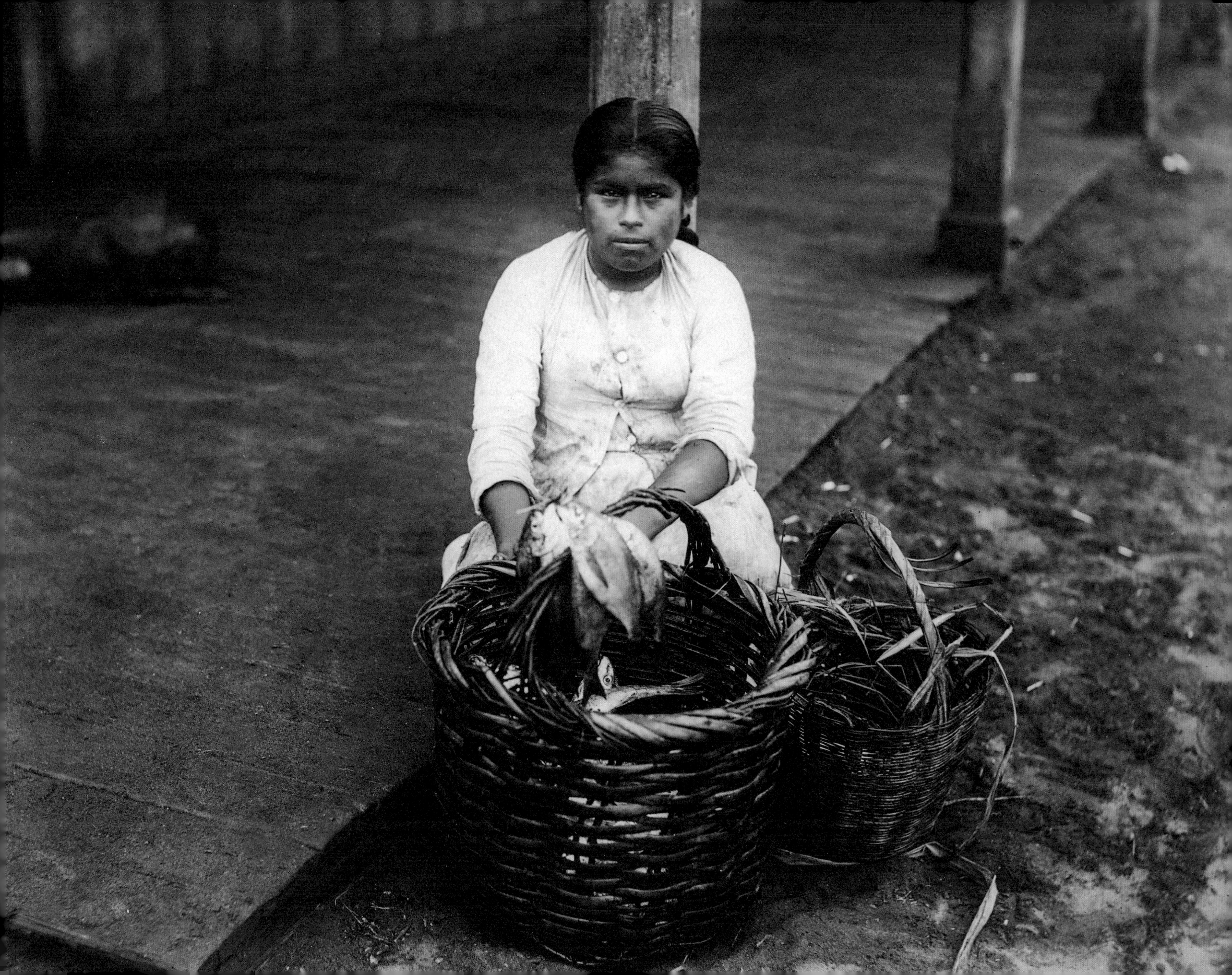

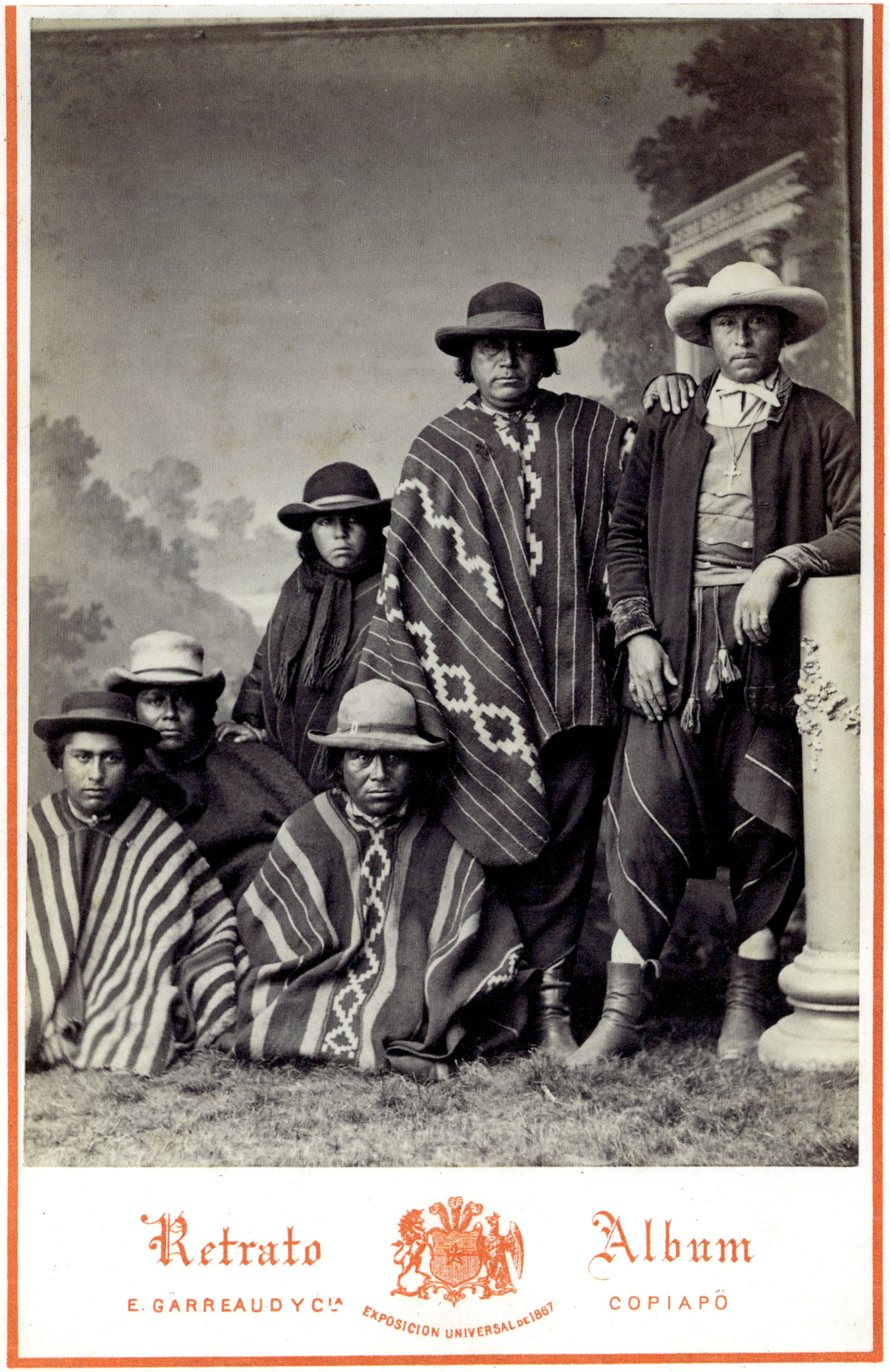

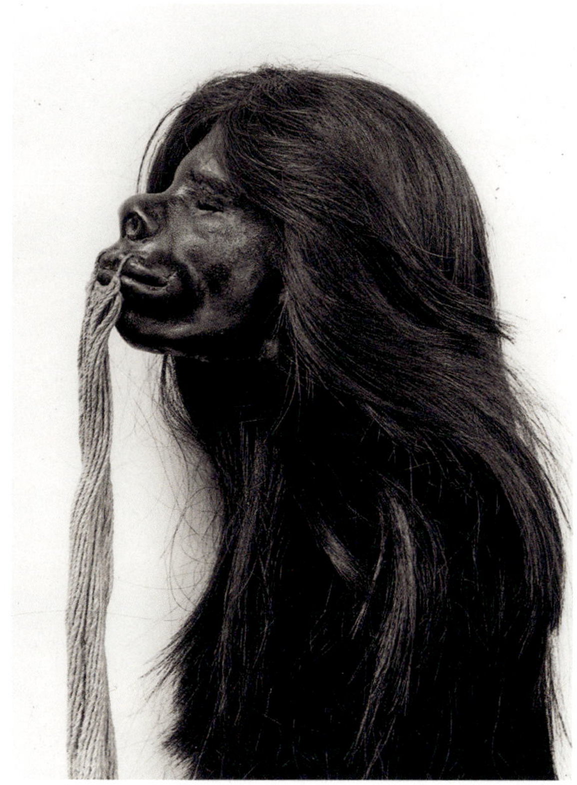

Top: Hjalmar wrote to Emmy as follows about the visit to Valparaíso in Chile: 'I came across a collection of Araucanian silver objects and a prepared Indian head from the Jivaro tribe in the upper Amazon. The heads of slain Ipurinas are prepared by removing the bone structure and drying the soft tissues with hot stones until the whole thing has shrunk to a diameter of a few inches, whilst retaining its shape. They're precious objects.'

Left: As with other contemporary ethnographers and anthropologists, Stolpe zealously collected 'ethnic types'. This photo from 1867 shows Araucanians, one of Chile's indigenous peoples. On the back Hjalmar noted that they were 'envoys' at some important meeting. The picture formed part of the collection of ethnographic objects that was given to Stolpe in March 1884 by Luiz Grosch, the German director of the Astronomical Observatory of Chile.

Opposite page: 'Fish seller Chola, aged 13', Stolpe wrote in his list of photographs about this Ekholm picture from Peru.

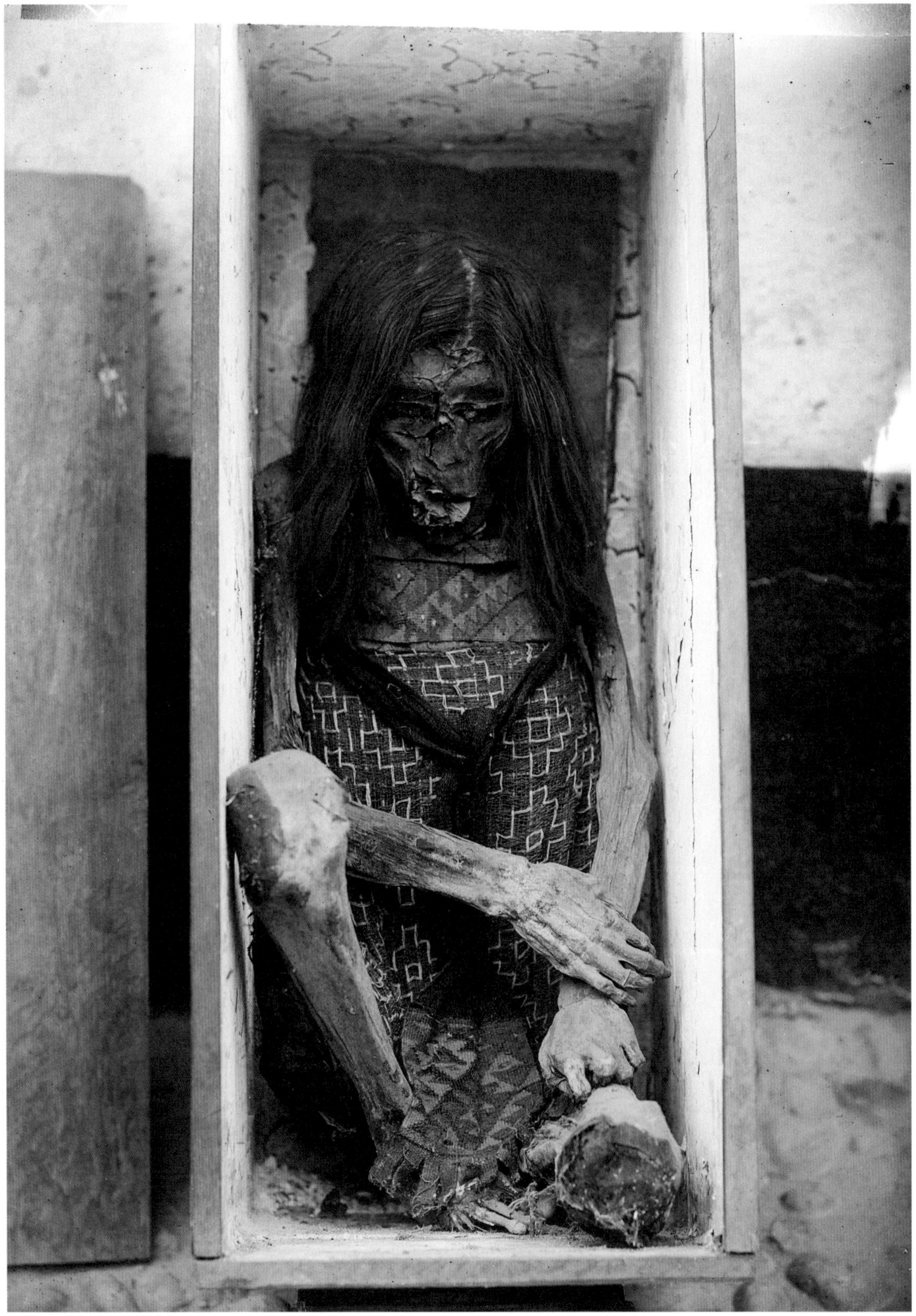

Opposite page: the picture is stated as showing a group of 'Patagonians' from the southernmost tip of Argentina, Patagonia, whence the people made their way to Punta Arenas, to carry out bartering.

Left: One of the mummies Hjalmar Stolpe took back to the Museum of Ethnography from Peru. The mummies would later be displayed at the Vanadis exhibition in Stockholm, and in conjunction he mentions a peculiar Peruvian burial tradition that involves collecting 'crystalline lenses from octopuses' and inserting them 'into the mummies' eyeholes, so they appear to glow with a gruesome lustre'.

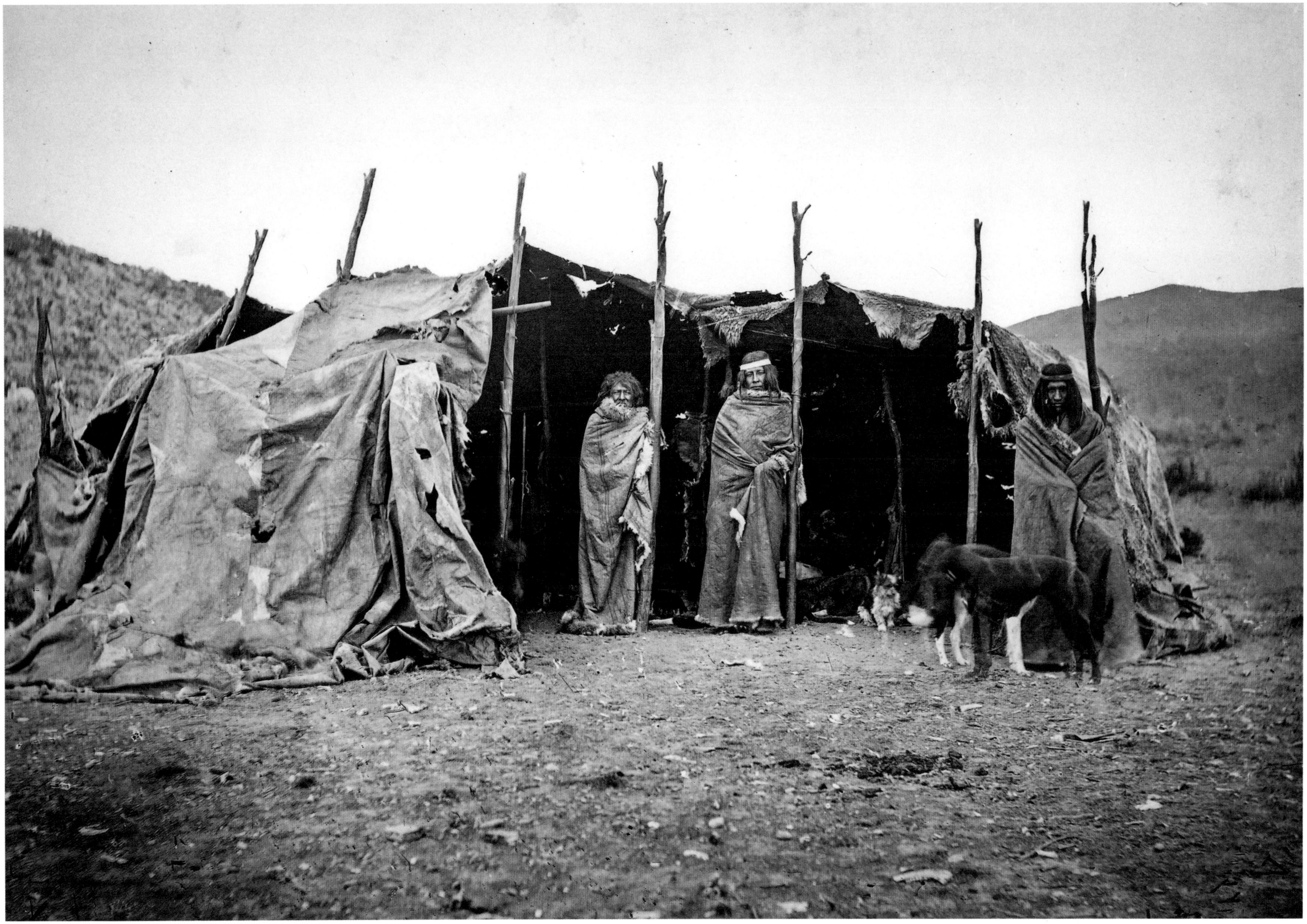

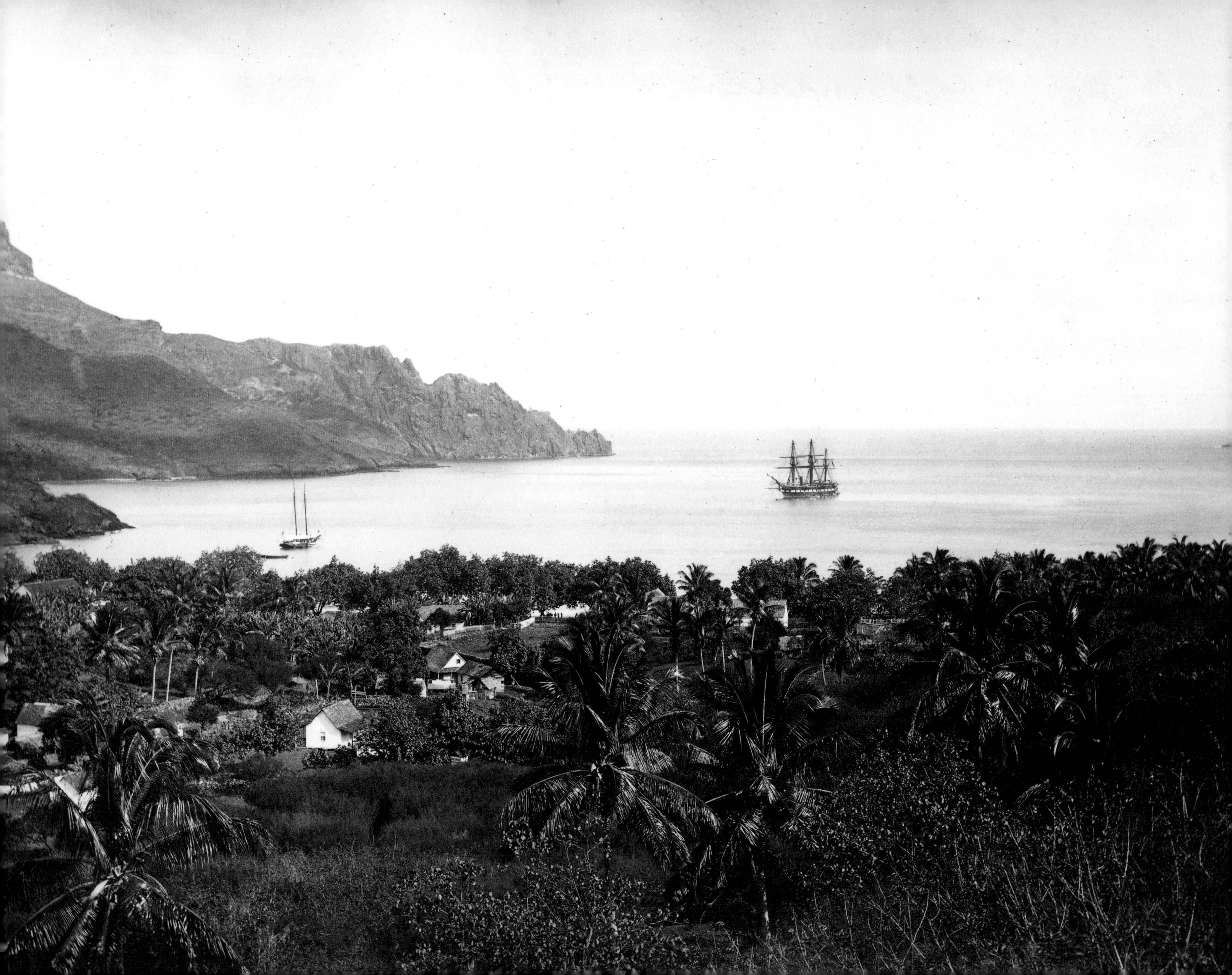

THE PACIFIC OCEAN

In mid-April 1884, the Vanadis left South America, and after the second-longest leg of its journey – which lasted 24 days – reached the mountainous volcanic island of Nukuhiva, the biggest of the Marquesas Islands in French Polynesia. The archipelago had become a French protectorate in 1842, and the governor resided on Nukuhiva, with its 1,200 inhabitants – collectively affirmed by everyone on the Vanadis as being the most beautiful place they had visited thus far.

Hjalmar Stolpe had as usual suffered as a result of his 'life in a floating barracks', and he continued his litany of complaints about the captain's dash round the globe: 'We rush from country to country, from people to people, and before one's had time to organise one's impressions and check one's notes, we're in the midst of a new world'. He had also got into a row with the ship's doctor, Karl Rudberg, and the gentlemen didn't talk to each other for a month. In one of Oscar Ekholm's many excellent photographs of this new world one sees the steam frigate at anchor in Nukuhiva's protected harbour lagoon. In his report to the Swedish daily *Stockholms Dagblad,* Stolpe wrote:

> The Marquesas Islands have always been famed for their wild beauty, and their inhabitants used to be reputed for being Polynesia's most beautiful people, just as they were incontrovertibly one of its wildest peoples ... The island, with its steep, cliff-lined shores and its narrow valleys separated by well-nigh inaccessible mountain faces, is the same, but how different the people are! One was previously received by a flotilla of canoes full of wild, naked warriors, tattooed all over in a style that had long been held to be the most beautiful in Polynesia. But now, on 8th May 1884, having arrived at Taiohae's harbour surrounded by steep mountains, we were met by a white-painted boat of European design, with a French non-commissioned officer and rowers dressed in the European manner, who did not at first blush appear suited to upholding the island's old reputation for beauty.

Oscar and Hjalmar were quickly ashore, and sought out John Hart & Company, who had long delivered ethnographic objects to the Godeffroy trading house and museum in Hamburg. The firm helped them 'photograph some Marquesan women and view the backside of an old biddy who was magnificently tattooed all over her legs!' As an ornament researcher, Stolpe was constantly hunting for tattoo patterns, which could not, however, be documented using Oscar's camera. Hjalmar now hit on the idea of 'attempting to cast in plaster of Paris a young beauty, onto whose (plaster) body I then intended to transfer the old biddy's tattoos'. There now followed a surreal sitting, which Hjalmar recreated for Emmy:

> It was hard to get anyone to consider the matter, and I had to make do with Akata, a by no means attractive girl but the only one on offer. There are not many people in Taiohae. I first tried a hand, but it was impossible to get her to hold her fingers still until the mould had hardened. My attempts failed time after time. I then twice sought to do her breast, but as that endeavour was also a failure (caused by my lack of practice mixing the plaster) I gave up on the whole thing for ever. It takes too much time, and I never even sufficiently manage to carry out the minimum necessary searches for ethnographic objects. In future I'm thus abandoning both casts and measurements, in other words everything purely anthropological, and will concentrate on museum-related matters.

Emmy commented as follows on this news:

> I cannot deny that I'm rather glad you've dropped the idea of measurements and plaster casts. I've not really wanted to admit it for my part, but I have to say that this – your work in the interests of science – has distressed me a little. So I have to tell you that you're – a little rascal! Why not also measure males? Is that maybe of less interest? I mean to science, of course! Well really! When you come home and are held to account by your little wife, I'm going to be very tough!!!

Stolpe had, after all, had problems with the anthropological measuring equipment back in Punta Arenas. As of now he was thus finished with plaster casts – one of physical anthropology's primary documentation methods at the time. Things went better when photographing and measuring Akata (168.6 cm) naked and without any plaster, as well as other natives in old festive costume. Those who ended up in front of the camera included the 'giant' Kabuphiau (178.8 cm) and Ovai (175.3 cm) – 'two fellows who had been interned on Nukuhiva for having been involved in beating to death and, according to some, eating an American a couple of years ago'.

Photographs were also taken of the interpreter and guide, Harone, clad in plumes and other festive attire. But one needed to watch out, as he was addicted to alcohol. On one of his sober days they went on a collecting trip in the valleys between the mountains, up to a thousand metres high. The expedition comprised five men:

After a 24-day voyage from Peru, on 8 May 1884 the Vanadis dropped anchor in Port Anna Maria on Nukuhiva, the biggest of the French Marquesas Islands. The visit lasted four days.

Stolpe, Harone, Ekholm and two men to carry the photographic equipment, contained in two substantial wooden boxes suspended from a carrying rod.

In one of the photographs a couple of stone ovens are to be seen in which food from the breadfruit tree was prepared; Hjalmar is looking at them, and on the back of the copy of the photograph he has written 'poi ovens'. The crushed fruit was mixed with coconut milk, and constituted this Pacific people's poi-poi dish. Some of Ekholm's other plates show how copra, i.e. the coconut's contents, was dried by hanging the endosperm high up in lopped trees. The yield of ethnographic objects on Nukuhiva was decent, and it included utility items, hair adornments, tattooing instruments, an idol and five human crania from grave holes up in the mountains. The total cost of the purchases was nearly 500 kronor, excluding the cost of 'photography of live, naked people'.

On Nukuhiva the captain and the officers organised a hunt for the island's wild goats, with 40 of the crew acting as beaters. During the hunt the seaman Otto Larsson fell from an overhanging cliff and died. He was buried in the French churchyard with military honours, in the presence of the Vanadis's crew in parade dress and the island's royalty, Prince Stanislaus with family. In total a dozen goats were shot, but two were caught alive. They were named Markus and Lukas, and sailed with them through to Stockholm, where they were portrayed together with the crew. One evening the officers were dancing on the beach with the Marquesas women of the island to music played by the Vanadis band. Stolpe noted: 'I went off along the beach in the moonlight and enjoyed myself singing songs just for Little Sparrow, where no-one else could hear me. I needed to exercise my lungs a little'.

Stolpe was highly critical of the driven hunt for goats. He was an experienced hunter himself – chiefly in the Kolmården area – and described the goat hunt as 'one of those wretched battues that poor hunters are so keen on'. In relation to this he wrote to his wife:

> Then and on many later occasions I've had reason to inwardly exclaim: in what company are you travelling around the world! I'm at times ashamed and at times disgusted. Whilst they're pretty decent – it would be a pity to say otherwise, some even very decent – they're without exception dreadfully undeveloped. The naval academy has gone down substantially in my esteem. It ought to be more improving, more educating.

But he was otherwise very satisfied with his few days on Nukuhiva:

> I'd happily have stayed on Nukuhiva! I found to my surprise that one would be able to create excellent collections there if one only had the time. On this occasion I just got relatively insignificant stuff at unreasonable prices. Competition from the general rabble naturally raised the expectations of the natives, who by the way don't have the slightest idea about the value of money, simply striking a deal at random.

FAKARAVA

During the short trip from Nukuhiva to Fakarava – one of around 80 atolls going under the collective name Tuamotu Islands – the Swedish Queen Sophia's name day was celebrated in line with the regulations, with the crew in parade dress and a 21-gun salute. Plus two inches of rum per head. Stimulating activities had to be found for the crew – and of course to pay respects to Prince Oscar's mother. In the daily newspaper *Stockholms Dagblad* Hjalmar Stolpe wrote of the inhabitants of the Tuamotu Islands (also known as Paumotu, the Low Islands):

> [They] of old had the same reputation for wildness as the people of the Marquesas Islands, but are now some of the meekest people in Polynesia. They are most closely related to the Tahitians, in terms of language, customs and appearance, and have constantly been in regular contact with the latter. Overall, when you start looking you find traces of vastly more lively communication between the different island groups – even the most remote ones – than was formerly believed possible. So much stranger is then the striking dissimilarity between the separate island groups in terms of ornamentation, tattooing and a number of social conditions.
>
> I was lucky enough to find an almost completely tattooed old man, and there is not the slightest similarity between his tattoos and the pattern for the Marquesas Islands ... I made an attempt to copy the complicated pattern, but time was so short that I was far from finishing the work. May the old boy Tairi, the fortunate owner of a large two-cornered hat, tailcoat and trousers with galloons (yellow strips) – the *admiral*, as we called him – be duly portrayed before he dies!

Stolpe had just 24 hours for the collection work, and proceeded methodically along the little village street with its avenue of coconut palms. Of the island's 150 inhabitants a third were out pearl fishing. The outcome was a collection of household utensils and instruments for making mats and hats. He got hold of a hundred or so objects, the finest being a canoe fashioned from a hollowed-out palm trunk with its oar, jib and bailer made of coconut. Oscar Ekholm was vigorously active, and in the photographs one sees the canoe with its former owner. Another photograph shows one of the island's specialities, mat-making, watched by Hjalmar in a white safari hat. On the back of the picture he noted 'mats for house walls being woven for me'.

Nukuhiva was the French governor's place of residence, and on his veranda one can see visitors from the Vanadis. Thanks to Hjalmar having sent a copy home to his wife and having written names on the back, we know who they are. From left to right: the meteorologist Gottfrid Fineman, Second Lieutenant Arvid Lindman, Stolpe in a white suit behind the hostess, Lieutenants Gustaf Lagercrantz and Wilhelm Dyrssen, and the governor himself by the pillar.

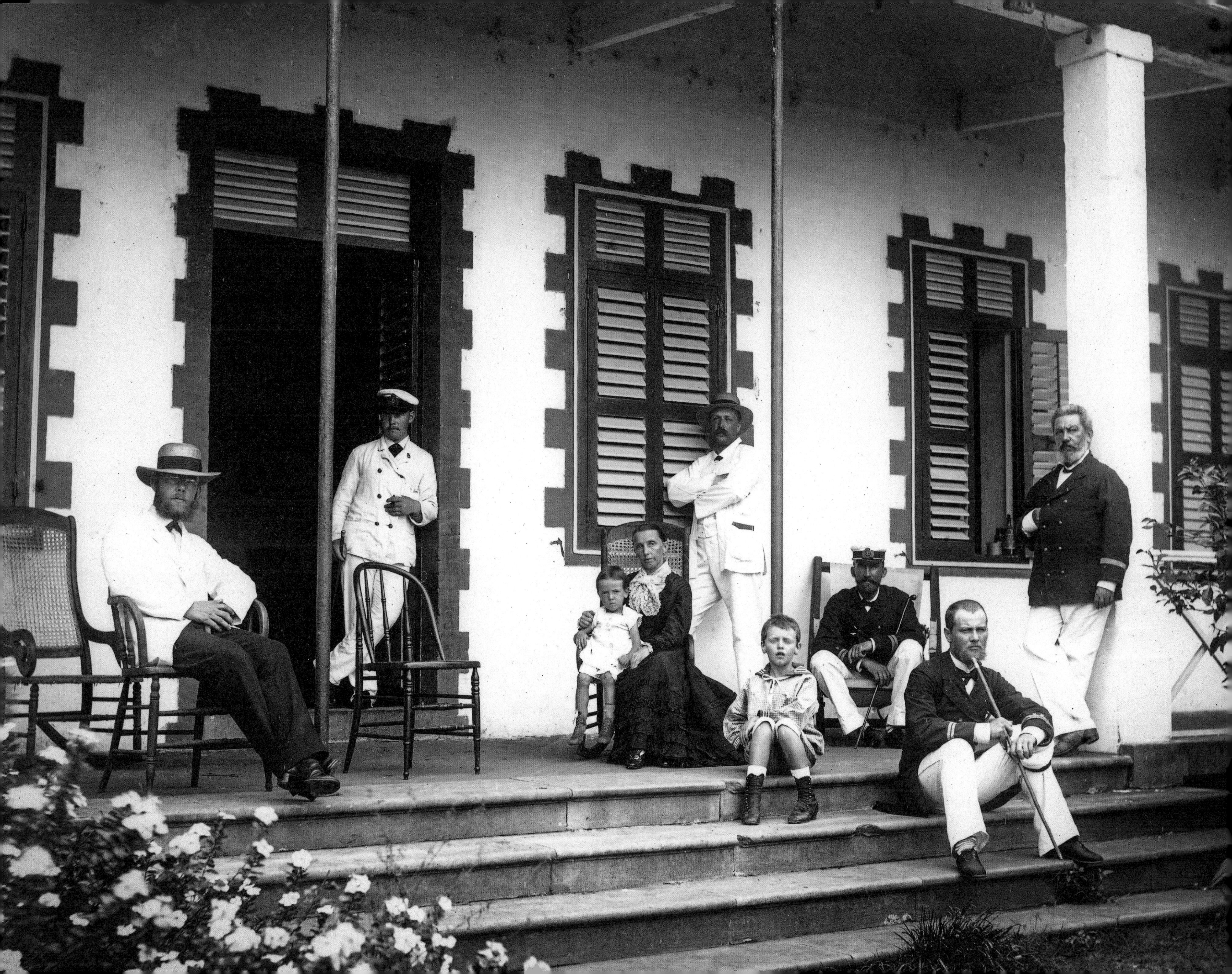

Relationships among fellow passengers on the Vanadis were constantly more or less problematic. The officers had for years been trained to put up with cramped conditions and extreme proximity, but this work environment was totally foreign to Hjalmar Stolpe, accustomed as he was to privacy at the Museum of National Antiquities and the archaeological excavation sites, and setting great store by integrity. He found it hard to adapt to life in the cramped cabin and the little swaying gunroom that was his living room for weeks and months on end. Hjalmar had planned to contribute entertainment during the voyage – to sing, and to talk about ethnographic and archaeological objects of note. But that did not come to pass, as he recounted to his wife while the Vanadis was shuffling its way across the Pacific:

> You mention its having been said that I'd succeeded in interesting the officers. I did for a time, but when I found that this interest resulted in increased zeal for collecting on their part (!) I stopped talking to them about ethnographic subjects. I'd intended to hold lectures for them, but I've dropped that idea. It would merely have meant my going against my own interests, both regarding the collections on behalf of the State and by drawing the gentlemen correspondents' attention to matters with which they could fill their letters. I've never thought I'd be mean about the little I know (you know best how unselfish and stupid I used to be in foreign museums), but here I've totally lost the desire to in any way expand my fellow passengers' outlook.
>
> Moreover, the Priest (a time-server of the first rank) intends to publish a description of the voyage, and I've absolutely no wish to play into his hands. He's deemed to be such a bright spark that he's being allowed to hold lectures for the crew (all the officers and the captain piously listen to him!) on the ports we are to put in at. He writes for *Blekinge Läns Tidning* [a Swedish daily], and as far as I can see produces amazingly asinine writing. But the worst is perhaps Ehrenborg's dull-witted contributions to *Östgötakorrespondenten* [a Swedish daily]! But you should not conclude from this account that there's any kind of animosity between us. Far from it. We are affable with each other in all manner of ways, even friendly, but I'll certainly be glad to return to my old friends back home.

The pastor's 'amazingly asinine writing' constitute many an article in *Blekinge Läns Tidning*. His contributions reflected the attitude of an occidental tourist, with an elevated and moral view of the world about him, the Pacific Ocean's disturbingly erotic dances and the Fuegian natives, who 'seemed to be more animal than human' and whose faces 'expressed the greatest crudity ... These redskins are wild and barbarous, and do not hold back from killing white men as soon as they get to them'. Pastor Kiellman-Göranson was of course a child of his time, and on board the Vanadis he was not alone in his values and his view of primitive peoples as wild beings. For Hjalmar Stolpe this was a totally alien attitude.

TAHITI

Following the brief visit to Fakarava they anchored off Tahiti's capital Papeete on 19 May, and stopped there for two weeks. Pomare V, in his French admiral's uniform, was received with pomp, a 21-gun salute and champagne on board the Vanadis, despite having been a mere titular king with France's blessing for a few years.

The weeks on Tahiti were filled with a string of festivities, dinners, fireworks and flares, vocal recitals and balls organised by the French governor and the French military club. The Tahitian women were attractive, and Seaman Bengtsson absconded for the second time. Oscar Ekholm wrote to his brother Fredrik as follows about one of the festivities:

> The Governor held the first party for us on Tahiti at his house in the country, which was situated 1 1/2 Swedish miles from Papeete (the town off which we anchored). We travelled there in the company of the Governor and the town's nobility, making up twelve carriages in all. At every village we went past we pulled up the horses, as the people were standing embellished with leaves and chaplets of flowers on their head, singing their national anthems in our, or rather the Prince's, honour. At the first farm they decorated us with chaplets and the horses and carriages with garlands of flowers, so the whole thing looked like it chiefly was – a procession.
>
> Upon our arrival at the destination we were presented to a multitude of princes and chiefs, who partook in the dinner, which comprised Soup, Prawns, Crabs, Oysters, Pork, Veal, Poultry, Pudding and Dessert. Wine in copious quantities.

Another festivity was organised in Papara on the south side of the island by the local chief Tati Salmon, who was to provide Stolpe with ethnographic services. Tati was a brother of the queen of Tahiti, and a son of the rich English merchant Salmon and a Tahitian princess. Prince Oscar described the journey to the festivity:

> After a two-hour journey we stopped at a cave, where we took our lunch in an entirely native way, holding the food in our fingers and dipping it in a white and rather tasty sauce. The pit-roasted pig was excellent. Stolpe took a photograph of us lying on the ground on spread-out mats – our heads were adorned with large chaplets ... By about 3 we had reached Papara.
>
> At the district's boundary 5 riders with flags and flowers had come to meet us, and when we drove into the courtyard we were greeted by our host-to-be Tati Salmon and

> nearly all the inhabitants of the district. Their choir is said to be the best on Tahiti, and it really was extraordinary to hear how precisely they sang their songs. The choir was well over 100-strong, and Stolpe asserted that they sang perfectly in tune ... Stolpe in turn sang several Swedish songs, which seemed to make a strong impression. The songs alternated with dance, the inhabitants becoming increasingly lively, and continued nearly until dawn.

The series of festivities was rounded off with a ball on board the Vanadis, as the boatswain Humbla described:

> On Saturday 31st May there was a ball for our Officers on board, to which the town's most distinguished gentlemen and ladies were invited. The Vanadis was regaled from the Mainmast all the way to the stern with flags, signals and flowers. In the evening Electric lighting was switched on, giving the whole thing a lovely appearance. At 8 o'clock, when the visitors were to go ashore, there were fireworks. Rockets were discharged into the air right up until the steam sloop reached land and put in. The Electric lighting was then turned landwards so that from the Vanadis we could clearly see those standing on the bridge of the ship. They then clapped their hands in delight at the bright light, and joyfully loudly applauded and hurrah-ed.

A few days later a small ethnographic expedition started: 'It was just Ekholm and I. In the long run he's the best company – quiet, decent, hard-working and interested in his task. I'm most satisfied with him'. After a while a search for a so-called catacomb began:

> We first went into the forest, and as we were walking along the path I was rather shocked to see a fairly well-preserved stone god, about 1 ell [60 cm] high, standing in front of me! It was immediately purchased, of course. The route continued through brushwood and over rivers, which I crossed borne by my semi-naked Hercules. He was a good-looking, well-built chap. Pity I didn't get a photograph of him. The path finally started to bear almost straight upwards. We had to climb on all fours up an escarpment, dotted with pieces of lava, which rolled around us. It was extremely laborious. One sometimes had only the grass to hold on to.
>
> The guide lost his way and took a long time looking for the cave, which we in the end discovered very close to us. The opening was of such a size that I found it difficult to enter. To start with it was barely one ell high, and I found it hard to make any headway. We'd unfortunately forgotten to take any illumination with us. I just had matches. Had I had some illumination I'd have squeezed through a tight opening and gone into a spacious inner cavity, where I'd have been able to walk upright. But now I didn't dare, as one encountered deep clefts now and then, and it was in any case pointless, as I was unable to see anything.
>
> On this exploration I merely found a single skull and a few bones. The cave is said to extend deep into the mountain. A real pity I didn't get any further! Just as I was creeping out with the skull Ekholm took a photograph of the entrance, which would have been a really entertaining depiction had the native sitting outside not got a fly on his nose and moved.

The picture was amusing. A conscientious ethnographer in a stained shirt crawling up from the underworld, his eyes staring into Ekholm's camera and his right hand gripping a skull. Next to him sits a native, out of focus and batting away a fly. The following day the trip continued:

> I'd had to wade across several rivers, with mosquitos mercilessly biting my bare arms. Chatted with the dear chap in the evening, and the day's collections were labelled. Slept soundly. At dawn the following day I had to set off back. Went over the mountains once again, and enjoyed the solitude and beauty around me. A photograph of the composition at my feet will remind me of one of the most enjoyable times of my whole trip ... Ekholm and I sat up half the night labelling the things. We had more than there was room for in the carriage. It was a good yield, but could have been all the more so if I'd had time to go up into the valleys instead of now having to stick exclusively to the coastal area.

The ethnographic yield was sent home on the Norwegian ship Rollo, from Arendal, and on 2 June the Vanadis left Tahiti, Stolpe noting: 'I wanted to return there, and be able to go from house to house throughout the island, for 3–4 months! There's much to do that will probably remain undone. And now goodnight my friend! Got to crawl into my bunk'.

The Vanadis's next destination was the Hawaiian Islands, and during the voyage there Oscar Ekholm told of his experiences on Tahiti and life on board:

> You can imagine how crestfallen we were when we got to Tahiti before the post, but it arrived a week later. Stolpe and I weren't on board that day; we'd just got back from a trip into the interior of the island when we spotted a sailing vessel approaching the harbour, and in response to our question as to what kind of vessel it was, we were told it was the post. We were so happy that I ran for the Equipment and photographed it. An hour later we sat quietly with a bottle of wine, gulping down news from friends and good old Stockholm.

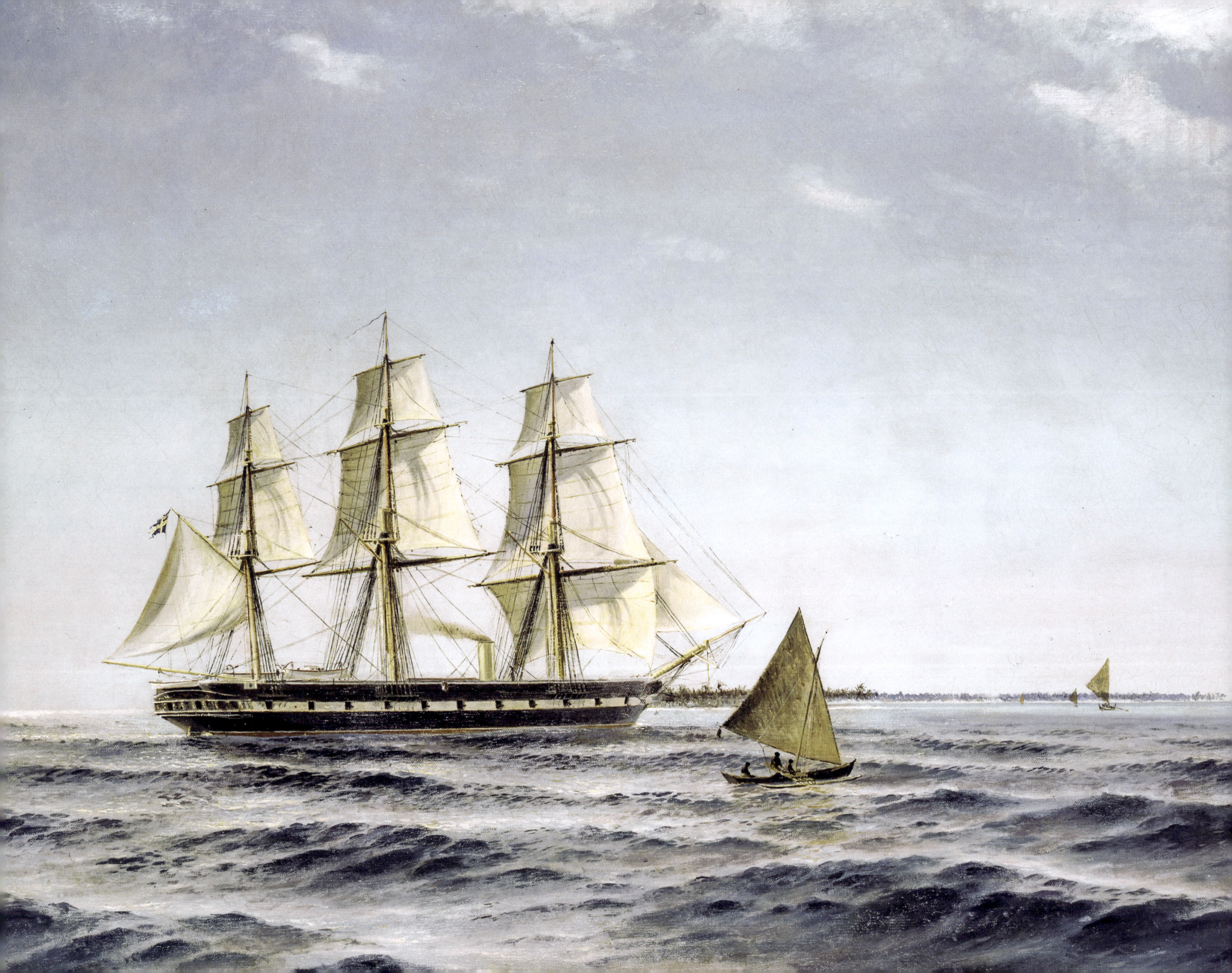

> You ask with whom I socialise. I socialise with everyone, and my situation is not always the best one, as I'm a missing link (not Darwin's). As you know, I eat in the Gunroom, but hold the rank of non-commissioned officer – something that's never happened before. Ranks are rigidly adhered to on board, but on land I'm always included in all parties and dos, of which there have been a good many here on Tahiti.
>
> There are five people on board whom I most value: Capt. Nissen and Klintberg, Lieut. Lagercrantz and Carlsson (our Mother) and Consul Ehrenborg. They're otherwise a haughty lot, each of them wanting to be more important than the next. The commander-in-chief is both stupid and haughty!

THE HAWAIIAN ISLANDS

For the three weeks following their arrival on 20 June 1884, the Vanadis lay at anchor off Honolulu in the Hawaiian Islands. In the capital a magnificent reception had been organised for Prince Oscar, who thereafter paid his respects to King Kalakaua. Outside the castle stood '100 or so men in German uniforms and with a pretty good bearing'. The king and his foreign minister were later greeted on board the frigate with a salute, manning of the yards (when the crew climb up into the rigging) and a cannon salute.

But the Vanadis's logbook for Tuesday 8 July reveals that it was not just foreign royalty to whom tribute was paid: '12 o'clock. The national flags [Sweden and Norway] were run up to the mastheads, and two 21-gun salutes were fired 5 minutes apart to celebrate HRH Crown Princess's happy confinement with a prince'. The ships in Honolulu harbour and the battery on land took part in the cannon salute. The person to whom tribute was being paid in the middle of the Pacific Ocean was Crown Princess Victoria, who was in confinement with Prince Wilhelm, her second child with Crown Prince Gustaf. But not everyone participated in the tributes. Four crew members took the opportunity to flee the ship in Honolulu.

After a few days in a hotel Stolpe moved to the residence of his new-found friend Eduard Arning, a German doctor and bacteriologist who was carrying out research into leprosy on the Hawaiian Islands, an 'extremely decent, kind and happy chap who japes me into laughing all day long. I've rarely encountered a more playful and spontaneous wit'. From an ethnographic standpoint, however, the Hawaiian Islands were a disappointment, being too Americanised in Hjalmar's eyes. But they did embark on a collecting trip to the Kalihi Valley near Honolulu, Stolpe writing to his wife on the morning of their departure:

> My little Sparrow, it's now 1/2 past 5 in the morning, and in 2 hours I'm going off mountaineering with Dr Arning, 2 Germans, an Englishman and an Italian naval officer. It's Sunday, and there's nothing else to undertake here except go out into the woods. The others are riding, and I'm going in a carriol [horse and two-wheeled carriage] as far as possible, and will then walk. I shall convey the photographic equipment.

In his notebook Hjalmar describes in detail – in writing and drawings – the purchase of household utensils and visits to local inhabitants' grass houses. The expenses are also recorded, e.g. 'For Ekholm's horse and carrying of his equipment – $9'. Idols were photographed, likewise the wine-quaffing excursion party resting in the grass, their horses grazing in the background.

Stolpe also carried out the anthropological collecting assignment for the Royal Swedish Academy of Sciences, and together with Arning, Ekholm and their hired guides he went off in search of skeletons and crania:

> We climbed in the mountains all day in search of catacombs, and found two, which we looted. But the bodies were modern ones. One already even smelt. One sees how Christianity has only put down weak roots, as the natives prefer to bury in holes rather than Christian churchyards. Though to be on the safe side the man had the New Testament buried with him – plus a tobacco pipe. My joints are aching from the climbing. It bucketed down for a while, and I was soaked. Have photographed my way through the museum here. Bought a load of things, and have also got things through and from Fornander, but it's not like on Tahiti! I'm really disappointed, and wish I had this time on some other island instead.

Hjalmar Stolpe returned to this visit to the grave holes much later, in a lecture to the Anthropological Society in Stockholm. He explained that it was with a certain anguish that the hired natives showed the way, up in the mountains. A total of four coffins were opened, of which 'one held a body only a few months old. During these explorations the Kanaks started to show signs of great anxiety, thus the finds were hidden, for collection after nightfall'. The aim of the grave looting was to find 'remnants of a pure race', i.e. to determine whether the deceased were 'unmixed' natives. The grave desecration and its scientific objective barely troubled the society's audience. Physical anthropology and the spirit of the time permitted such excesses. But to anyone with some knowledge of Stolpe as a person, the behaviour was surprising.

Stolpe's most important contact on Hawaii was Abraham Fornander, mentioned above, the son of a priest from Öland who went to sea and sailed the world's oceans before settling in Honolulu in 1842. He worked there as a journalist and newspaper owner, went off to California in search of gold for a spell, swore an oath of allegiance to the Hawaiian King Kamehameha III and married a chief's daughter. He eventually

In mid-May 1884 the Vanadis was sailing off Fakarava, one of the Tuamotu Islands in French Polynesia. The following year Jacob Hägg signed this watercolour.

became a judge and a member of the Kingdom of Hawaii's highest court. But Fornander was also an ethnographic collector and self-taught researcher who had devoted decades to studying Polynesia's peoples and languages.

He was a corresponding member of the Anthropological Society in Stockholm, and closely followed what was happening in his beloved homeland. He had read with dismay in the London *Times* how close the Vanadis had come to being crushed against the stone pier near London. And when the frigate appeared in Honolulu he had just received by post the journal *Ymer,* containing an essay on Easter Island written by Hjalmar Stolpe. The latter's name was thus well known to him. Fornander wrote to a childhood friend in Sweden:

> I was naturally in close daily contact with Dr Stolpe. I liked his frank, amiable, jovial manner. But I'm afraid his archaeological successes in Honolulu were not quite as he had maybe expected ... Throughout the time the frigate was in Honolulu the doctor was fully occupied with examining, describing and photographing objects from the public and private collections there and in the vicinity.

Fornander spoke appreciatively of the Vanadis's officers, who took part in the balls King Kalakaua held in honour of Prince Oscar: 'The officers were immediate favourites in Honolulu society, and rightly so. Everyone spoke English fluently'. Fornander took part when they were invited back to a party on board the Vanadis, and was overwhelmed:

> I have attended many such on different national ships, but for completeness of arrangements, taste of decorations, roominess of deck for dancing purposes, incomparable and apparently inexhaustible supply of the choicest refreshments, the really superior music of the Ship's Band, and – what seemed to me and every one present the greatest charm of all – the spontaneous, unaffected, ever vigilant, never obtrusive, courtesy and attention of all, from the Commodore down to the youngest Officer; – for all this, and I say it with pride, the Swedish Frigate Vanadis and her Officers will bear away the palm and be long remembered by Honolulu Society.

Kung Kalakaua was said to have got really drunk at times, and among the party's participants was Hjalmar Stolpe, who drank wine privately with the island king one night.

On Hawaii Stolpe also found an old acquaintance, Christian L'Orange, a Norwegian sea captain and prominent member of the small and energetic Scandinavian colony. A few years before the Vanadis's arrival he had organised an 'emigration expedition' from Norway and Sweden. Around 600 people, mostly Norwegians, were given free passage to Honolulu, against the promise of work on the islands' sugar plantations for three years – an odd episode in the history of Swedish/Norwegian emigration which attracted great attention at the time.

Ekholm and Stolpe travelled by steam sloop to the little community of Waimanalo, 30 miles or so from Honolulu:

> My personal interest was in seeing a so-called battlefield, and furthermore a magnificent mountain landscape: vertical rock faces, several 1,000 feet high. Waimanalo sugar plantation is situated on a narrow strip of land between this insurmountable rock face and the sea. Took a few photographs in the afternoon ... We stayed at Waimanalo overnight. I was regretful, but was rewarded the following day when I investigated the 'battlefield'. It's an ordinary burial site, windblown sandy soil, almost like Ancón, and in it the bodies have been placed in a sedentary position. I scraped out 3 whole skeletons and several skulls with my bare hands. No objects. Further photography of stone formations.

A good century later, the crania Stolpe scraped out in the windblown soil at Waimanalo and the ones he had previously collected in the Nuuanu Valley's grave holes were to make the long journey back to Hawaii. Representatives of the indigenous Hawaiian population wanted them back, and following a government decision they were handed over in 2009 at a ceremony at the Swedish History Museum in Stockholm. The recipients were the organisation Group Caring for the Ancestors of Hawaii. Of the 22 crania returned, 15 could be linked to Stolpe.

Stolpe was able to send 275 objects to Stockholm from Honolulu. The collection included purchases from New Britain, the Solomon Islands, the New Hebrides and the Kingsmill Islands (Gilbert Islands), and it included house models, decorations, fishing tackle, paddles, sleeping mats, drums, various bowls and kava root for preparing intoxicants.

JALUIT

On 10 July the Vanadis left Honolulu, and after two weeks' sailing reached the coral atoll of Jaluit (formerly Bonham), one of the southernmost in the Marshall Archipelago. Further north in the archipelago is the atoll of Bikini, which became famous as the site of American nuclear weapons testing after the Second World War. Some time after the Vanadis's visit the archipelago, comprising over a thousand islands in all, was declared a German protectorate, and it had a German consul – Franz Hernsheim. But the Germans had good relations with the king of the Marshall Islands, Kabua, whose kingdom chiefly comprised the western chain of islands, known as Ralik.

But those on board the Vanadis had differing opinions about the islands' sovereign. In all his writings, Hjalmar Stolpe refers to him as a king, acclaiming him as 'King Kabua' in respectful accolades, but the ship's commanding officers evidently had other views. It was clear that Kabua was the area's highest-ranking leader, surrounded as he was by advisers and courtiers, and furthermore with the most wives – six in all. He was invited to dinner on the Vanadis, which was clearly a success including an exchange of gifts. But he was accorded no formal royal reception. For the previous royal visitors from Tahiti and Hawaii – Pomare and Kalakaua respectively – the salute cannons thundered out 21 shots per king. But this formality was not extended to Kabua. The most surprising thing, however, is that there is no mention at all of his visit in the Vanadis's logbook, which otherwise recorded the visits of everyone right down to captain's level.

Oscar Ekholm reported as follows on the visit to Jaluit:

> There were a few stores, but everything was very expensive, e.g. a bottle of beer cost 1 dollar. The inhabitants were handsome, peaceful people, but very grubby and uncivilised. Their clothes comprised nothing but a short grass skirt for both men and women. We arrived there at a happy time, as they were expecting the islands' ruler King Kabua, who otherwise lives on a different island, and he came a few days later.
>
> He was, believe me, a comic figure, attired in a black tailcoat and trousers, a three-cornered hat, a dirty, coloured starched shirt without a collar and white shoes. When he walked it was supposed to be with dignity, his arms stiffly hanging by his sides and his nose stuck out in front of him, but his dress nevertheless appeared that of a stevedore. He was on board a few times, once bringing with him a couple of his wives, of which he had many, and on another occasion escorted by his first chief, Lojak.
>
> Lojak was bare-headed, wearing a new, white starched shirt with a stiff high collar that dug into his neck, with his shirt outside the white trousers, and on his feet heavy, stately shoes that he switched to on board, though they were clearly too small for him, as he walked with difficulty.
>
> There's someone else I must mention, namely the Kanaks' Minister and navigator. He was a stocky chap with a nose like a scurfy potato, also in a tailcoat, three-cornered hat and dirty trousers, but without any shoes or shirt, and his tailcoat was adorned with two decorations, of which he appeared very proud – though on closer inspection I found they were ones you can buy at Leja [a Stockholm department store] for 25 öre.

The frigate stopped at the island for a week, and Hjalmar wrote enthusiastically to his wife:

> I've collected loads here. It's without a doubt my best place. And I have copious notes. Have received written essays from Germans here. Also a fortification map of one of the northern islands, drawn by a carpenter with a description! I'm staying with Messrs [Max] Thiel and Naht, Hernsheim being in China at the moment. Having a really enjoyable time. Wonderful collections.

The man reported to have travelled to China owned the island's German trading house, Station Hernsheim & Company, where Stolpe had pre-ordered a large collection of ethnographic objects from surrounding archipelagos. One of the firm's employees, Herr Naht, mentions the Vanadis visit in his diary – a rare on-the-spot account by an outsider:

> Saturday, 26th July. In the morning a Kanak seeking to catch fish with a dynamite cartridge completely shot off one of his hands through lack of caution. On such occasions we always have to play doctor, i.e. bandaging etc. ... The Vanadis doctors amputated part of the wounded Kanak's arm.

When the Vanadis appeared in the lagoon, Naht and his colleague Max Thiel had sent out a pilot, who led the frigate to an anchorage. They were immediately invited to dinner, and were presented to the officers and Prince Oscar:

> A Dr Hjalmar Stolpe is accompanying the ship, to collect curiosities for the ethnographic museum in Stockholm. Dr Stolpe – a charming gentleman – was already known to Thiel from Hamburg. The meal was simple but good. During the dinner an excellent band played well-known melodies (the opera The Beggar Student, The Blue Danube etc.) Afterwards we went ashore with all the officers. It was very cramped in the boat, and I sat fairly squeezed up against the prince, who by the way is extremely charming and socialises with the other officers (he is a first lieutenant) in the most cordial of manners.
>
> We subsequently went to our station, and on the veranda drank first Swedish *punsch*, which they had brought from on board, then Mosel wine. The prince and several officers left us at 8, whilst the remainder stayed with us for a long time. Charming company, these Swedes. You don't always find people so warm-hearted and obliging ... In the morning Dr Stolpe photographed various Kanaks, and was also so kind as to photograph Thiel, myself and our [white] employees.

The Germans also arranged a memorable meeting between King Kabua and Hjalmar Stolpe:

When we arrived, the King of the Marshall Islands, Kabua, had 3 days previously sailed home to his island accompanied by over 20 big sailing canoes. Great sadness. But my friend Thiel from Hamburg ... immediately sent for them. The day before yesterday 11 came at one go, and yesterday Kabua himself with the rest. There are now around 25 of them, with over 300 men, women and children camped on the beach in temporary huts made of coconut leaves. Amazing sight! Nice people.

Yesterday Thiel and Naht held a dinner for all of us, after which the king himself appeared in war apparel and danced, while his 6 wives drummed. Other women also drummed – 10 to 11 drums in all. Deafening noise! Wild shrieks and an ecstasy the like of which I've never beheld. Things here are still almost wild. But in everyday life the people go around dressed in European garb, pretending to be Christian. At parties the wildness comes to the fore.

Kabua's attire is magnificent. He is the only person allowed to wear it. Flowers in his hair. Feather embellishments on his arms, a skirt of a special kind that only he wears when at war, and a big mat with edging spread over his knees. Dances sitting down, twists his body, shakes his arms with their luxuriant black feather bushes, deflects his eyes (the wives too, but not the rest of the company), bellows and sings. Elegantly tattooed all over his body. The scene is illuminated by big torches made of coconut leaves. You just can't imagine the effect! And the whole of the seated horde of natives sing a monotonous song that works its way up chromatically, a wild scream and everything is suddenly silent as the grave. The king is constantly fanned with leaves – which is necessary, as sweat runs over his oiled dark-brown body. Another chief (vassal) also danced, and was then clothed for the occasion with Kabua's attributes.

When the whole thing was over I tried to get all the paraphernalia. Difficulty. Then gave Kabua a big knife (pejn nejf = fine knife), and got the skirt and the feathers! And now there was the question of the mat. He said he didn't want to give it to anyone other than the captain. I immediately whispered: You give me. By me King of Sweden look see mat. (It's always 'look see' for 'see'). He hesitated. The captain [Lagerberg] at once asked me to sing a Swedish song (the first time that happened). Yes, I said, but forgive me if I introduce Kabua's name into the song so he thinks I'm singing a song of praise to him. And I then struck up without further ado:

'Thou ancient, thou free, thou mountainous north
Kabua, thou joyful beauty.
I greet you, finest king on our Earth
(King =) Irûd Kabua, your islands so green.

Thou art enthroned on memories of great ancient days
When honoured your name traversed the earth.
Kabua, you are and you remain what you were
Irûd Kabua, hear these words.

Rapturous applause and – the king immediately rolled up his mat and handed it over to me, his face beaming with joy! I thus got everything, and that is deemed to be a great favour.

After Kabua visited the Vanadis, Prince Oscar recounted in his diary:

Kabua came on board, as well as several courtiers and natives – men and women. Kabua was dressed in an old tailcoat with flashy yellow galloons and brash trousers. They were all extremely surprised at what they saw. A courtier named Lojak understood a little English. He had a good grasp of things, and was delighted to be allowed to fire the machine gun, as well as a shot with a cannon, the loud bang from which seemed to frighten him and his companions a little. The machine greatly surprised them, as did all my clothes and footwear. They stayed on board for a long time, and appeared to be having an excellent time. Kabua gave me 8 mats, for which he got a knife, my old uniform hat, a number of bits and pieces and my photograph. He clapped his hands in delight. He was invited to dinner by the captain, at which we saw many demonstrations of his intelligence and quick powers of observation. In the evening the music was on shore, which of course created a great commotion.

Oscar Ekholm was very busy with his camera on Jaluit. The result was 80 or so photographs, and Stolpe's photo lists and notebook explain what they represent. The collection includes portraits of the island's inhabitants and visitors: King Kabua in a tailcoat and hat, and his adviser, the 'prime minister' Laona. Also there is the crown prince Lamorro – the highest-ranking leader on the island of Ebon, described by Hjalmar as 'young but a drunkard'. The proud princess Tui from the Samoan Islands is leaning against a palm. Other portraits show the 'sailing-chart maker' Langedju and the towering and well-to-do owner of the Union Hotel, Tom Tilton, who liked imbibing his own hotel's drinks and gave Hjalmar many objects. The chief Lojak is showing how to throw a spear, and a man is climbing up into a coconut palm. We also see the small German colony of traders with bottles and accordion on 'a hot and thirsty morning on Consul Hernsheim's veranda', where Hjalmar Stolpe also sits in his straw hat. In one of the photographs a young boy is demonstrating Kabua's 'seated dance', and it is apparently his son who is playing the part, while his mother beats a drum. His attire is like that which Hjalmar received as a gift when he had sung 'Du gamla, Du friska' ['Thou ancient, Thou free'].

The photographs seem remarkably vivid and close, despite their being far removed both in time and space – 135 years ago and on the other side of the globe. This is even more true of Ekholm's series of photos documenting Kabua's return to the island. When the Vanadis anchored at Jaluit he had, unfortunately, just left it. But he was called back, and it is this return that a score of Ekholm's photos depict. On the horizon beyond the Vanadis a collection of triangular sails can be seen approaching; a score of canoes have soon lined up on the beach, and several hundred seafarers in the royal retinue are setting up camp. The camera shows the slender canoes with their outriggers, masts and stays, and small huts on board – the extraordinary sailing vessels of the Pacific Ocean. Fire and smoke are visible on the beach as the camera is directed at a young girl drying breadfruit in a fire hole, the people mingling outside their temporary coconut-leaf huts.

Wandering amongst the canoes on the beach and holding a little boy by the hand, Hjalmar declares it an amazing sight. In another photograph he is sitting in a hut together with Marshallese people and writing in the ever-present notebook, as usual wearing his big sun hat.

This is excellent and technically mature pictorial reportage that was unfortunately never published and thus did not take up its rightful place in the early history of documentary photography. Oscar Ekholm was a pioneer – a globe-trotter and professional photographer who sailed around the world and collected a large volume of pictorial ethnographic material. Stolpe was extremely satisfied with him, both personally and as a photographer, and reserved particular praise for his work on Jaluit: 'The group of photographs from this island is one of the most lavish and most interesting in the collection'. Hjalmar thus went out of his way to catalogue those 84 Jaluit photographs and furnish them with explanatory texts.

In one of Ekholm's photographs, Kabua is seen flanked by two tall, austere gentlemen: Stolpe in white tropical clothing and a broad-brimmed straw hat, and the captain in uniform. There had been yet another confrontation between the two of them, this time about a trivial matter. Back home, the Swedish Academy's permanent secretary, Carl David af Wirsén, had published an article in the newspaper *Posttidningen* on 'the circumnavigation now being carried out by the frigate Vanadis on an ethnographic assignment, which has been entrusted to Doctor Stolpe, a skilled Swedish ethnographer'. There was much laughter about this in Norrköping, according to Emmy, who wrote: 'So it's you whom the prince is accompanying, and you who are heading up the expedition!'

But there was no laughter in Stockholm. Reducing the royal fleet's military and tradediplomacy journey around the world to the status of a scientific expedition was just not on. The indignant minister of naval defence, von Otter, wrote to Commander Lagerberg on the other side of the globe, expressing the suspicion that it was Stolpe himself who had spread this 'story', prompting Hjalmar to write to his wife:

> It came to an exchange between the Captain and myself yesterday, when I requested a delay here. I told him a lot of home truths. He cited von Otter's stupidity in order to get the better of me. I energetically protested, and got the chance to further clarify my position. I boldly went for him, and he went his own way, with an emphatic explanation that he would not give any consideration to my request. 2 hours later he came here to dinner, took me aside and explained that for my sake he would delay until the Saturday, if (the meat) the provisions were good. I replied that I was absolutely delighted to hear that 'the spirit was willing' and that I hoped the 'meat' would not be poor. With a little banal agility one can sometimes win him over. He was extremely amiable, and that evening we talked more than during the entire expedition hitherto. So all's well! I gained the upper hand.

Hjalmar thus succeeded in slightly extending the stay on Jaluit, and there was a rich yield in terms of photographs and objects. From the firm Hernsheim & Co he bought a pre-ordered collection including 100 ethnographic objects from the surrounding Marshall Islands and more remote archipelagos such as Kingsmill and New Britain, and the Caroline, Hermit, Solomon and Admiralty Islands. On Jaluit he collected around 150 objects through purchases and bartering. Many of them were linked to the area's distinctive sailing canoes, bailers, oars, hooks, prow decorations, fireplaces for the canoes and a sort of sailing chart made of lengths of wood and shells. These 'sailing charts', Stolpe wrote, are 'the islanders' own independent invention going back a long way, and this, together with their vessels' perfect design, makes them incontrovertibly the Pacific Ocean's most distinguished navigators'. For sailing between Jaluit and Ailinglapalap, King Kabua's own island, there was a special chart.

The other objects included mats, raffia skirts, various implements and decorations. And the crowning glory was Kabua's complete accoutrements for the performance of his seated dance. From the Kingsmill Islands came a rarity: 'cuirass and full armour for warriors; the helmets are made of the skin of a horny fish; the neck protectors for the cuirasses serve to provide protection against mistakes by the women, who fight with slings in the second line.'

The sailing canoes not only aroused Stolpe's interest but also impressed the Vanadis's crew, above all Arvid Lindman:

The natives here are an excellent seafaring people, and their canoes are extremely well constructed and maintained. They are sewn together with various pieces of wood, after which the holes are made in such a way as to make the canoes completely watertight. The attachments on which the outriggers rest serve as supports for surfaces that also extend a long way on the other side of the canoe, and on which there is space for people and possessions, and the actual canoe also often has a proper deck. On these surfaces one frequently sees small mat huts erected, in which there is room for the crew and protection from the sun ... A peculiarity of their canoes is that they are curved on the side facing the outrigger but straight on the other side, to avoid drift and counteract the outrigger's resistance, and thus improve control. They have sea charts for the whole group, comprising battens attached to each other, forming a lattice. The islands are marked with small attached shells.

The ethnographic collection from Hernsheim & Co was expensive: 50 English pounds (about 55,000 kronor in today's money). A dozen boxes of objects were sent to Hamburg on board the barque Alona, owned by the firm. Before the expedition began, Stolpe and the supporting Vanadis committee were already well aware that the budget for acquisitions was insufficient. And the freight costs that arose during the course of the journey had confirmed this prediction. A bailout had thus been launched, resulting in Oscar II personally donating 6,000 kronor (380,000 in today's money) in order to save the day and keep this element of the expedition going. In a letter to the Royal Swedish Academy of Sciences' permanent secretary Georg Lindhagen, Stolpe reported the results of his work during the first seven months, including the visit to Jaluit. He stated that 'the collections have grown to a size that neither I nor anyone else could have imagined when we started. At no less than four locations I have thus had to leave significant collections, to be sent home at the first opportunity arising'. These four locations were Callao in Peru, Papeete on Tahiti, Honolulu on the Hawaiian island of Oahu, and Jaluit. The cost of the transportation worried Hjalmar:

I did not think about the transportation at the time. Even less did I imagine that pretty much nothing would be carried by the Vanadis. I'm now coming to the most expensive areas. Hope I won't be left in the lurch as regards money ... You don't now get any showpieces on the South Sea Islands. There are only practical things left, and they have pretty much never before been sent to Europe. Even if my collections are not showy, they give so much better an idea of the people's lives. For all objects for which it has been possible I have precise designations, and have thereby collected contributions for a dictionary of technical terms that are totally missing from existing dictionaries.

I've had to completely abandon the anthropological investigations, however. Time doesn't allow! Measuring naked people, including men, is extremely difficult and involves spending a huge amount of time persuading them. I've many a time been in despair about the time wasted on this in vain. Prudery is incredibly deep-rooted in Oceania, partly as a result of established religious ideas about not exposing oneself, and partly thanks to the contribution of missionaries. On top of this, I have just one *poorly made* and useless craniometer, which just hurts people, so I hope my decision is justified ...

Another obstacle for me is the way in which I'm travelling. Other anthropologists come on their own, calmly mix with the people (e.g. Finsch) and gain their trust. I arrive with an entire frigate crew, which means a great stir is created. Almost no-one even has the time to sell anything. They're stormed from all directions by people wanting to buy 'stuff'. Everyone collects, from the captain down to the lowliest ship's boy – some lamely, and others much more energetically. This greatly hinders my efforts. Not to mention working in peace and quiet, and pursuing a path I've embarked on. You just have to be present everywhere in order to try and pre-empt, so as not to be forestalled. And how this general competition drives the prices up!!

The researcher Stolpe mentions in this letter to the Royal Swedish Academy of Sciences was Otto Finsch, a German ethnographer and zoologist who was in the region at the same time as the Vanadis, visiting New Guinea and the Marshall Islands for an extended period and amassing significant collections there. Stolpe saw them in Vienna on his way back from the Vanadis.

In some of Ekholm's photographs taken on the last day of their time on Jaluit, one can see a two-masted schooner with a gaff rig at anchor a short distance from the Vanadis. Stolpe says the following about the owner in his notebook:

On the last day most of the canoes had sailed off, but Kabua's schooner remained. He was in the process of selling it to Black Thom [the hotel owner] for $200 in order to buy a better one, and as we passed by the outlet he was to be seen together with Laona, leading the cheers from a number of natives who had lined up along the shore in an orderly fashion. The Loiak family was there too. At the Hernsheim pavilion Naht and Thiel were to be seen hailing us with a flag.

The Vanadis departed Jaluit on 2 August 1884, and set a course for Japan.

'4 natives at the burial site, Taiohae, Nukuhiva', Stolpe wrote on the back of the picture, later describing in a travel report to the daily *Stockholms Dagblad* 'the giant Kahupihau (178.8 cm) from Dominica', second from the left. Together with Ovai (175.3 cm) they were 'two fellows who had been interned on Nukuhiva for having been involved in beating to death and, according to some, eating an American a couple of years ago. They were towering and ostentatiously tattooed warrior figures, and didn't look nice at all. I realised Kahupihau had a tendency towards bad moods when one day, having forgotten his name, I had to ask after it. He first refused to answer, but then brought his mouth right up to my ear and shouted "Kahupihau" in a thundering bass voice, which must have had a hilarious effect in combat!'

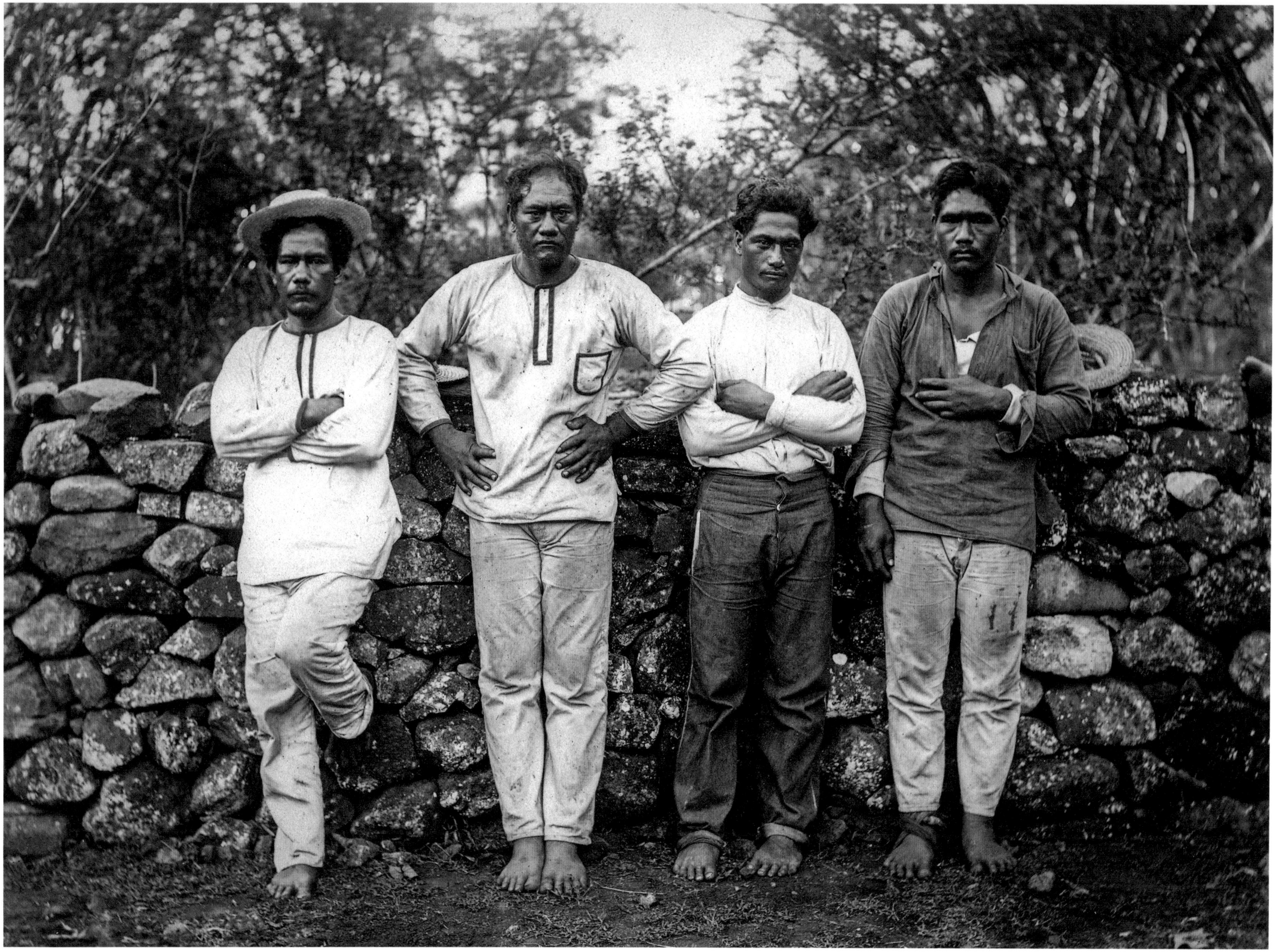

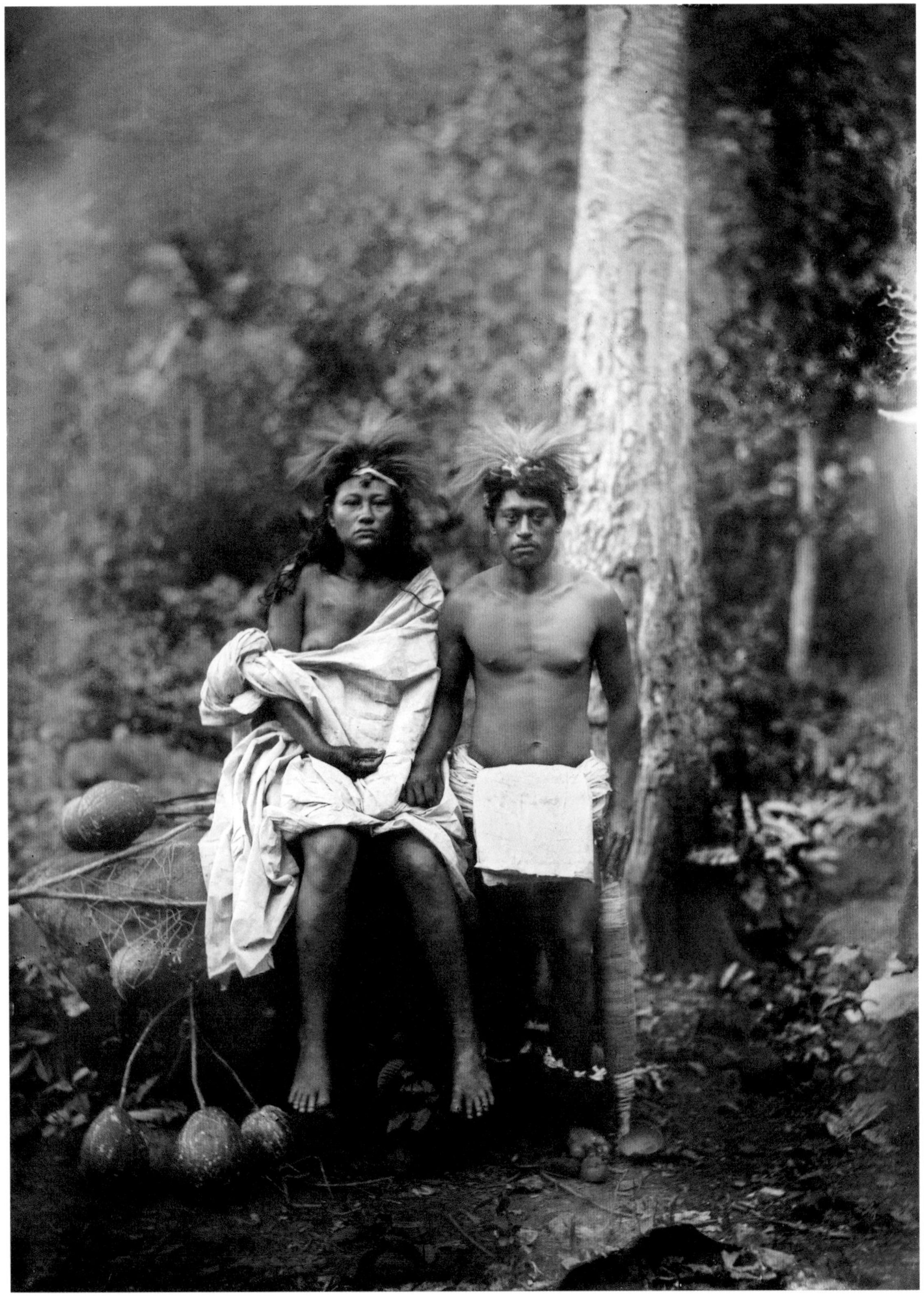

Oscar Ekholm's remit included photographing 'ethnic types' – something the anthropologists of his day were greatly interested in. The couple in the picture are, however, unknown – as is the place where it was taken.

Opposite page: The company John Hart & Co on Nukuhiva helped Stolpe find ethnographic objects and tattooed islanders, who for Hjalmar had become a special interest as part of his study of ornaments.

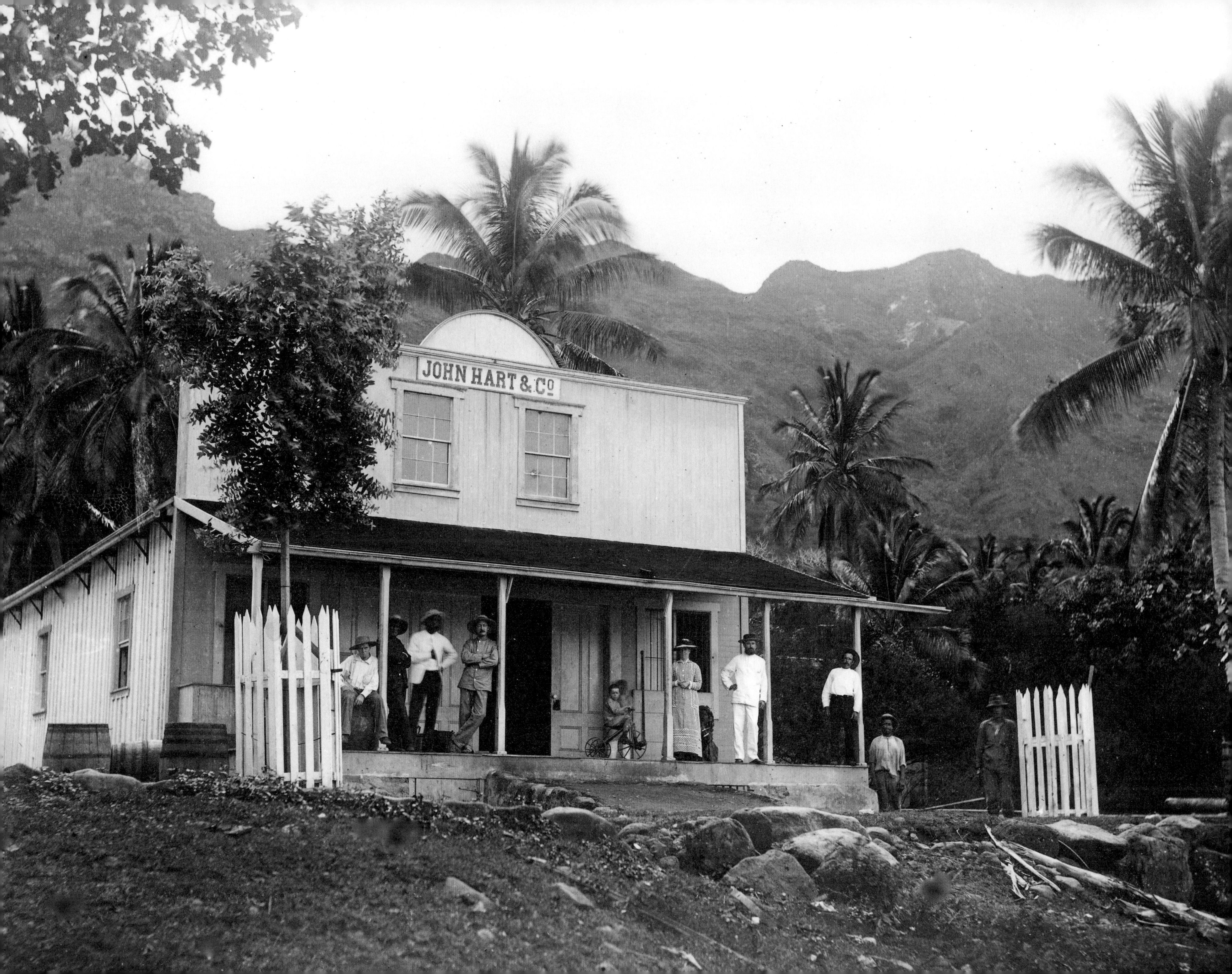
JOHN HART & Co

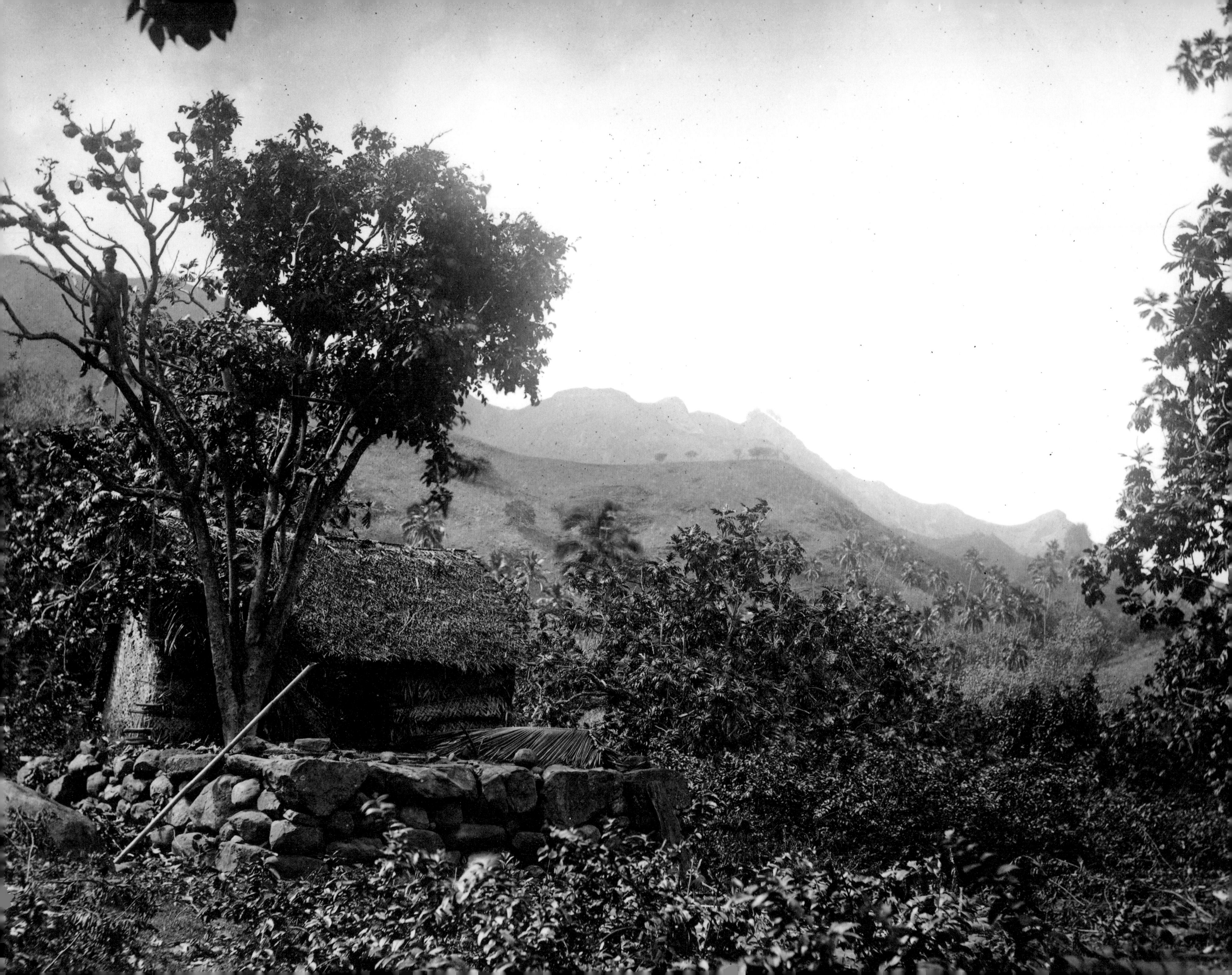

Opposite page: High up in the tree on the left, Ekholm and Stolpe suddenly discovered a man standing on a branch. Above him hung small balls of copra – the coconut's endosperm. Stolpe noted: 'It was only on Nukuhiva that I saw this way of processing coconuts – removing them from their shells and cutting them into pieces, then tying the pieces together to form bunches and hanging them on rods arranged horizontally in a tree growing next to the house with lopped branches. Everywhere else I saw the copra simply spread out to dry on the ground. In this dried state it's shipped to Europe and America in enormous quantities, where it's refined to make oil, soap etc., and many of my female readers (if I have any) maybe use insignificant quantities of coconut-oil preparations, whose first step en route to the boudoirs of the civilised world is this hanging process on one of the Marquesas Islands. ... The photographic equipment in constant use that day was of course directly focused on the man in the tree, and we thus gained a true impression of life on Nukuhiva. It was not the last time such scenes were photographically captured without any organisation or grouping. The good fortune of having the assistance of a skilled photographer on a voyage such as this cannot be emphasised enough.'

Right: Precisely where Oscar Ekholm took this portrait photograph of the 43-year-old Hjalmar Stolpe during the Vanadis voyage is unclear, but on his visits to the islands in the Pacific Ocean Hjalmar often wore this big sun hat.

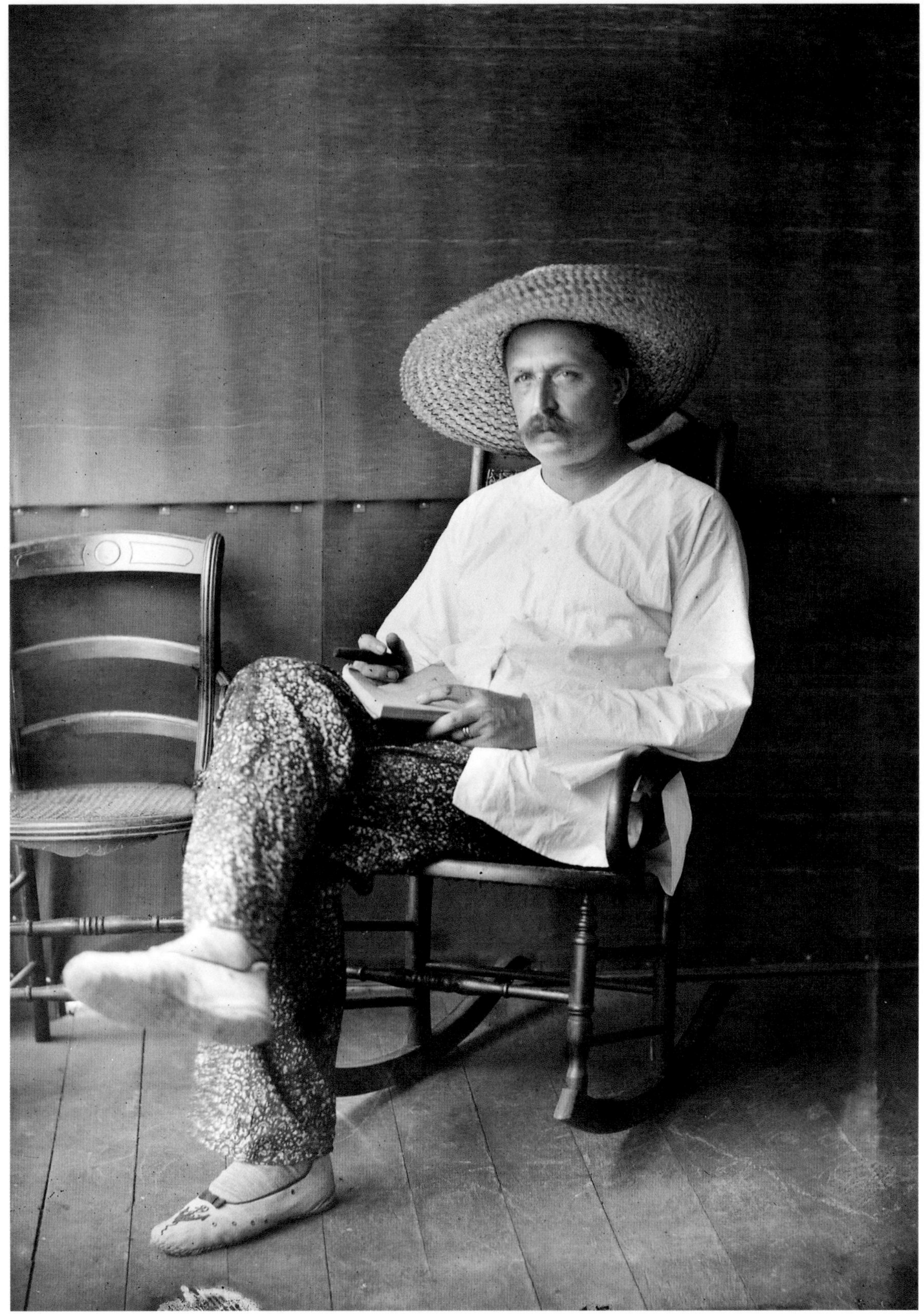

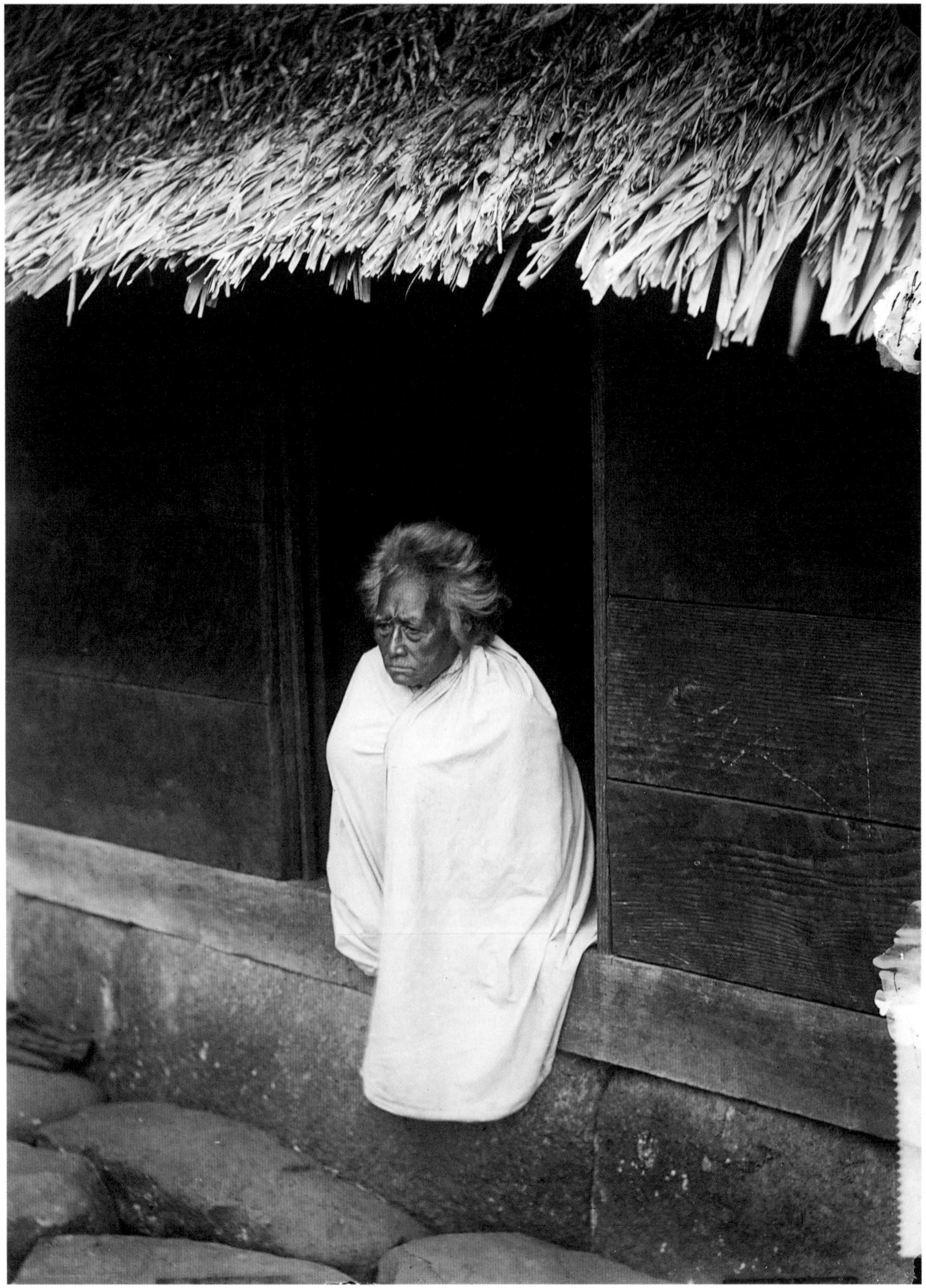

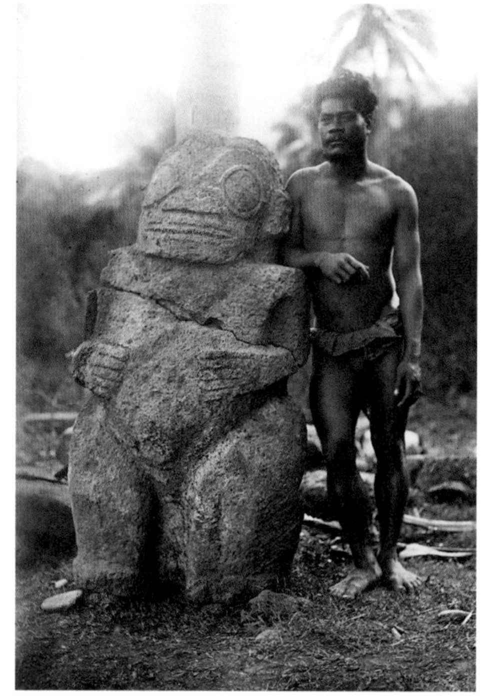

Left: 'Old man in doorway, Hakuai, Nukuhiva.'

Above: In Taiohae on Nukuhiva, Ekholm photographed Tiki, who was '158 cm tall'. Stolpe makes no comment about the figure he is leaning against – possibly a holy stone statue.

Opposite page: On the back of the photograph Stolpe has written: 'Poi ovens in the mountains above Taiohae, Nukuhiva.' In these ovens the islanders prepared their basic food, made from the crushed fruit of the breadfruit tree mixed with coconut milk. 'Preparation involves first lighting a fire in the pit, so the walls get hot. It's then scratched clean and the walls are coated with carefully plaited mats made of green coconut leaves, after which the pit is filled with breadfruit, and it is all well covered with coconut leaves. The breadfruit soon ferments and goes sour. After a fairly long time it is removed, and is pummelled to form a dough using a well-made and strangely shaped pestle. In this state it can be kept for a long time, and when it's to be eaten you take an appropriate amount with your fingers and place it in water in a wooden bowl.' The interpreter and guide Harone offered his favourite dish to Stolpe, who 'stuck his fingers into the dish and tried to look undaunted. The first mouthful tasted repulsive, but the second and third ones were better. You can get used to this taste, which is unlike any traditional taste in Europe, and if you keep at it (definitely for a very long time) it's said you'll even get to like the taste'.

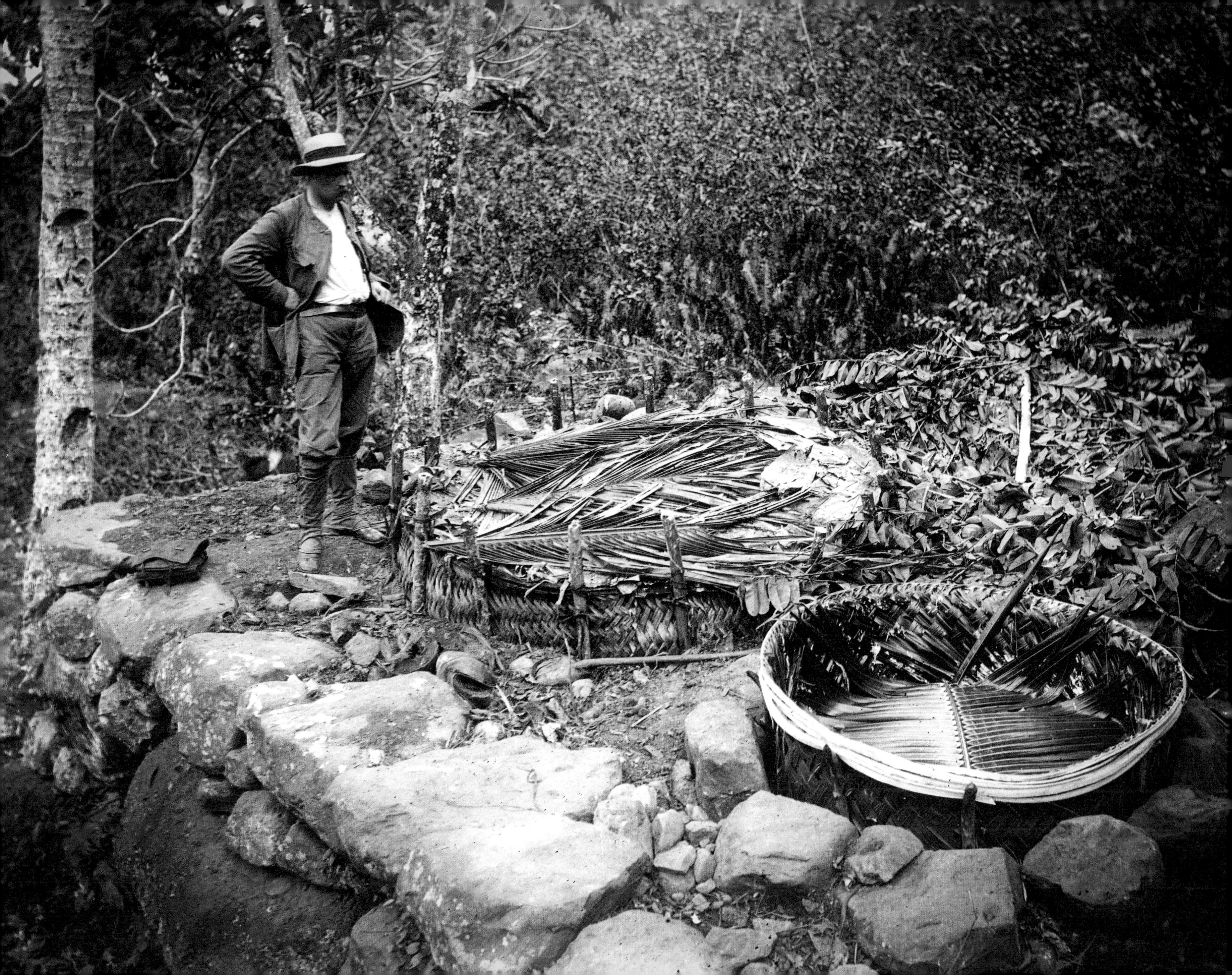

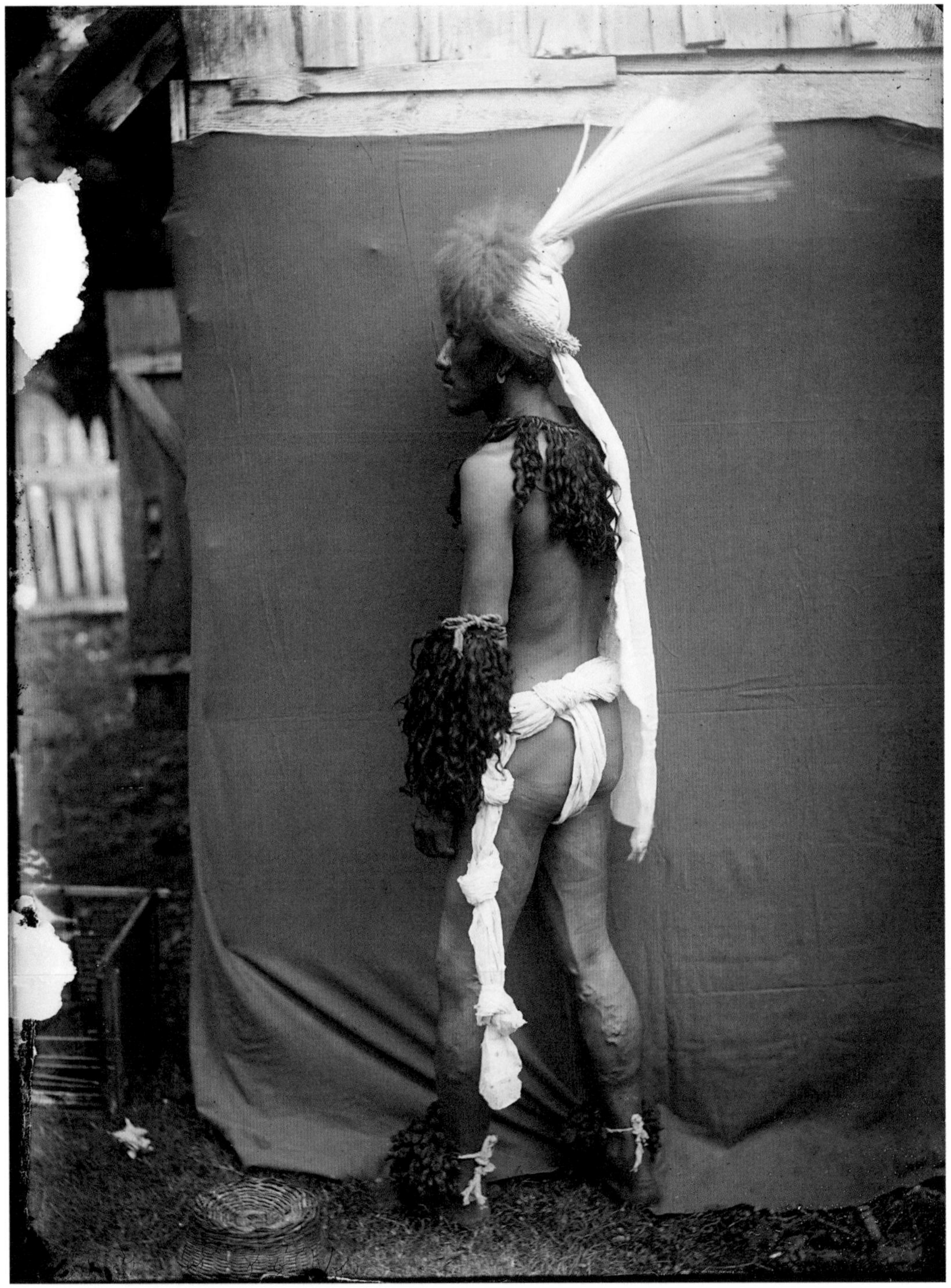

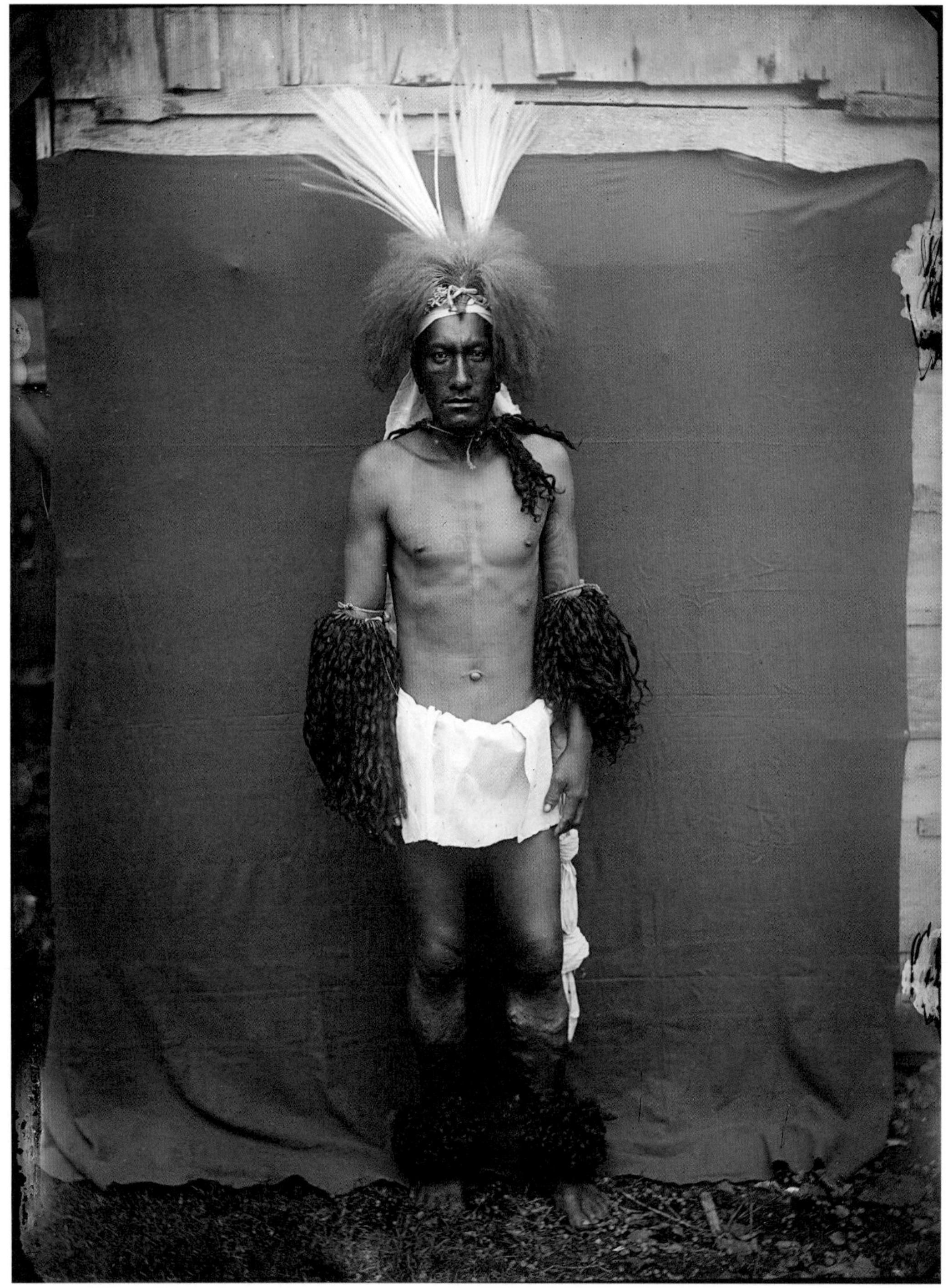

Opposite page: 'Harone in old native warrior dress', it says in one of Stolpe's lists of pictures – adding that Harone is 178.8 cm tall. He was Hjalmar's interpreter and guide on Nukuhiva during a long expedition into the island's dense forests. But, writes Hjalmar, 'despite his attractive tattoos and magnificent attire, my friend Harone was a poor representative of the warriors of former times. Liquor had ravaged him. In his younger years he must have cut a fine figure, with his abundant matted hair, big sparkling eyes and crooked nose. But – an enfeebled drunkard. He did, however, have some good qualities, as it transpired on the following days. ... In one instance Harone was excellent – he gave me the names of a whole host of plants and insects, describing things with a clarity and differentiation that showed an above-average intelligence.'

Right: Stolpe's anthropological remit from the Royal Swedish Academy of Sciences was to photograph and measure the various ethnic groups – 'races' – he encountered on his voyage. On Nukuhiva he found Akata (168.6 cm), 'a by no means attractive girl. But the only one on offer.' Physical anthropology also prescribed creation of plaster casts. 'I first tried with a hand, but it was impossible to get her to keep her fingers still until the mould had set. I then tried twice with her breast' – with such unsuccessful results that he now completely abandoned the activity. This delighted his wife Emmy, who acerbically wondered 'Why not measure male individuals too?' She evidently thought her husband had sometimes spent a little long on his female study objects.

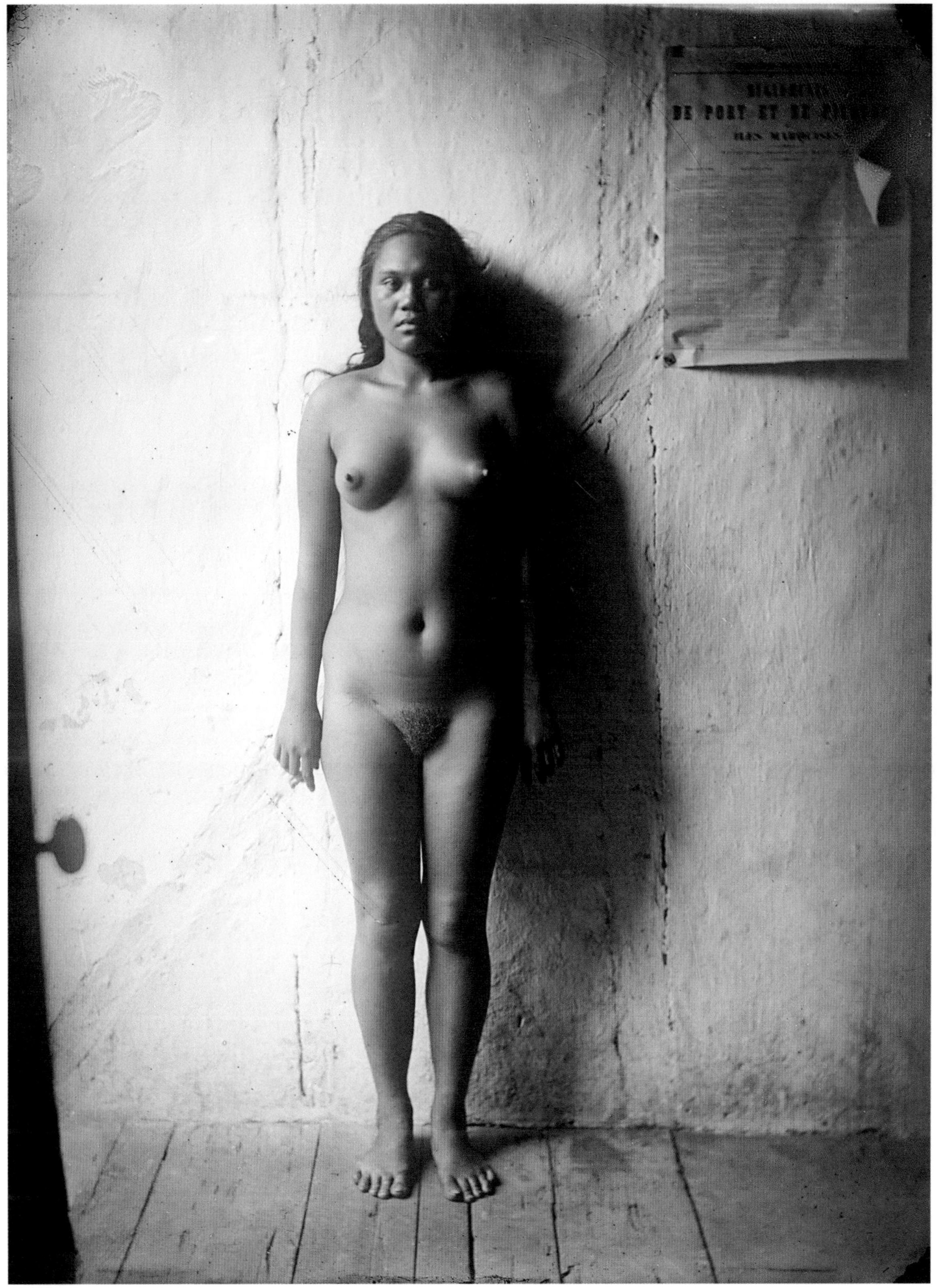

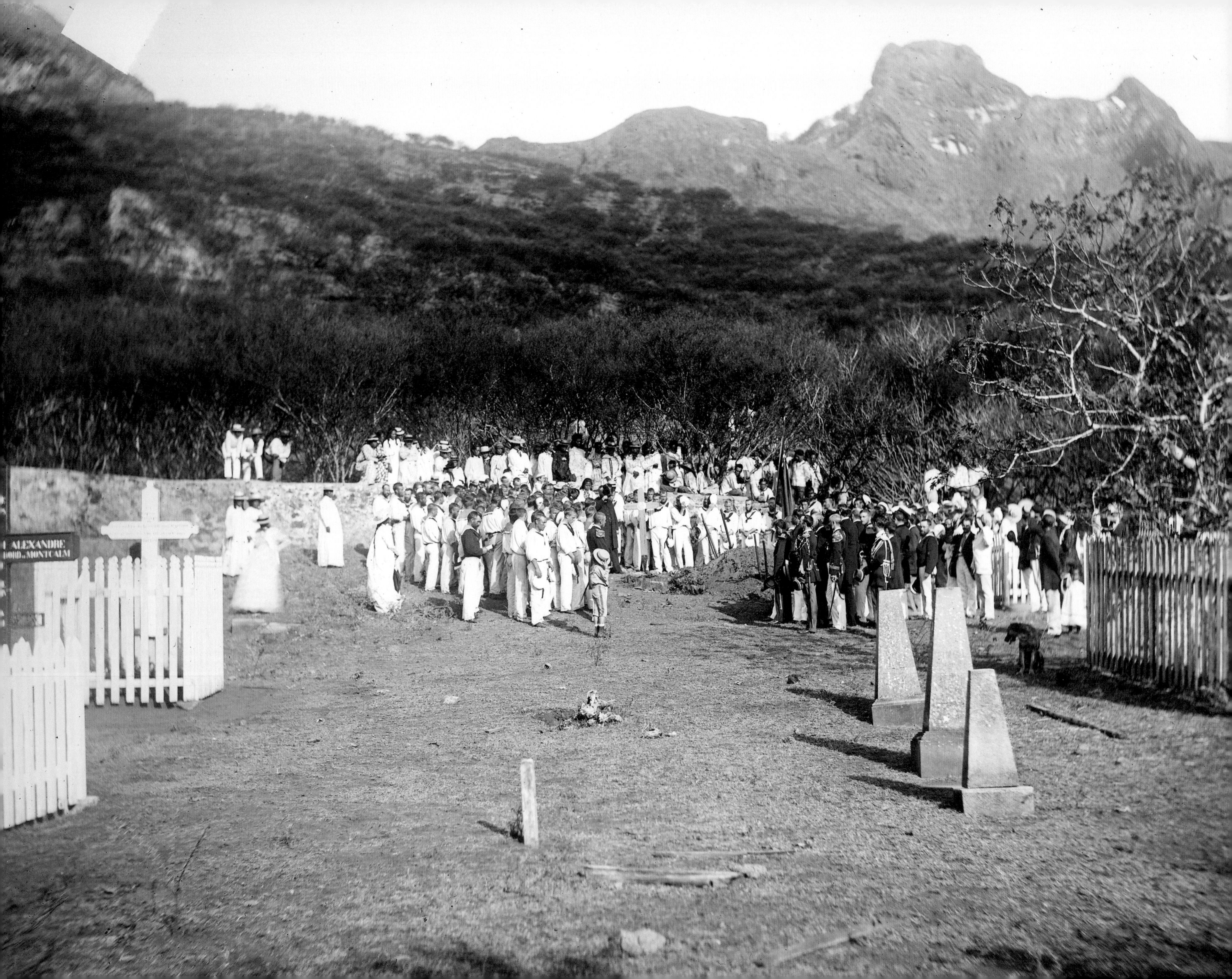
ALEXANDRE
MONTCALM

Opposite page: Otto Larsson's funeral took place at the French churchyard on Nukuhiva, with military honours and in the presence of the Vanadis's crew in parade dress and the island's royalty – primarily Prince Stanislaus with family.

This page: The deceased seaman Otto Larsson by a coffin on the deck of the Vanadis. The previous day he had been involved as a driver in a hunt on Nukuhiva, but fell from a cliff and suffered 'injuries to his brain and cervical spinal cord'. Hjalmar Stolpe did not take part, but was highly critical in letters home to his wife: 'one of those wretched battues that poor hunters are so keen on'. Hjalmar was himself an experienced hunter, primarily devoting his time to hunting hares in Kolmården's forests around the family's summer house in the county of Östergötland.

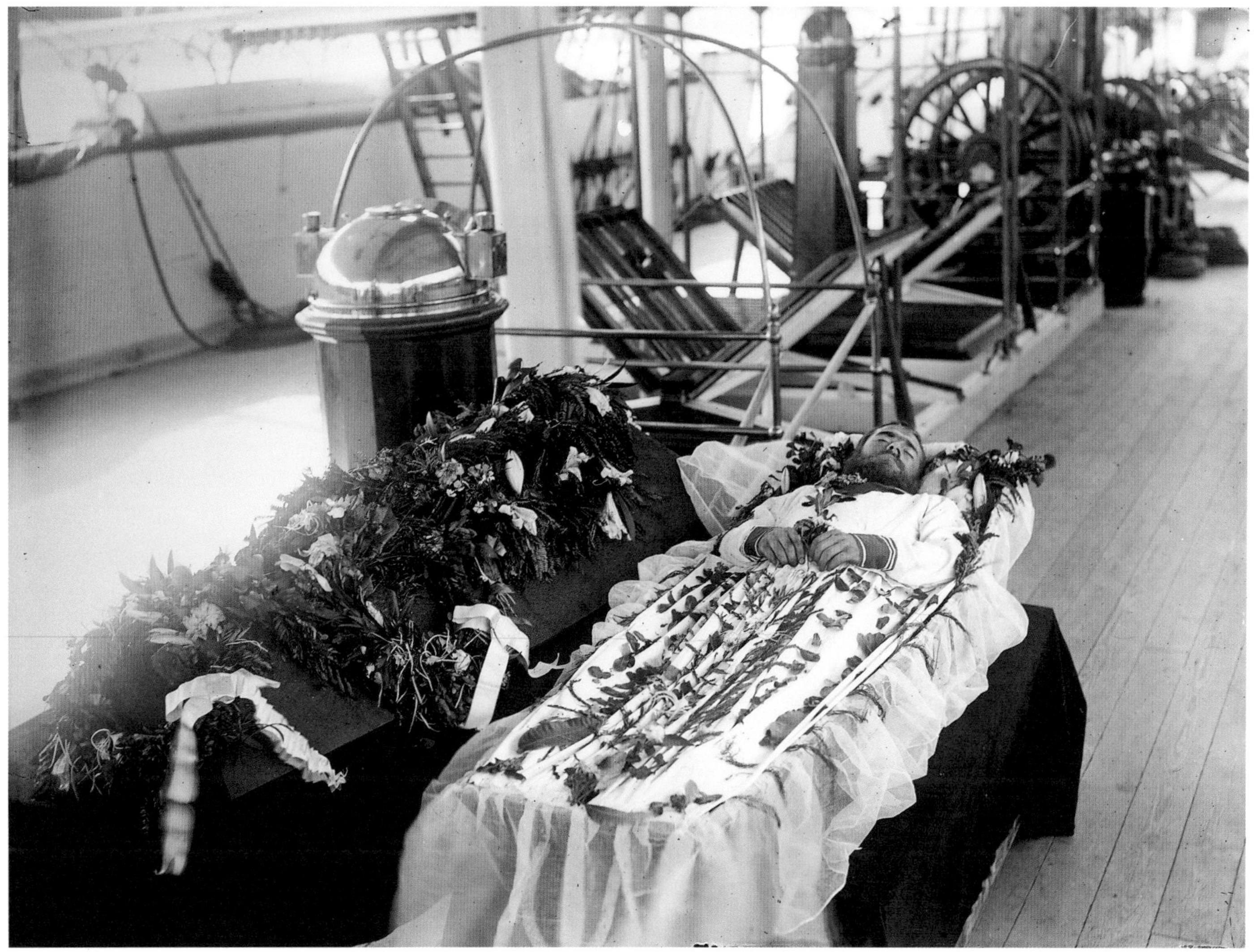

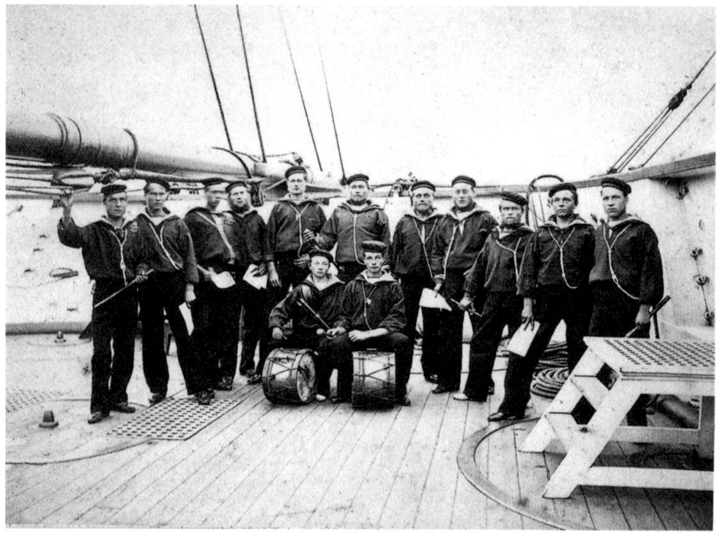

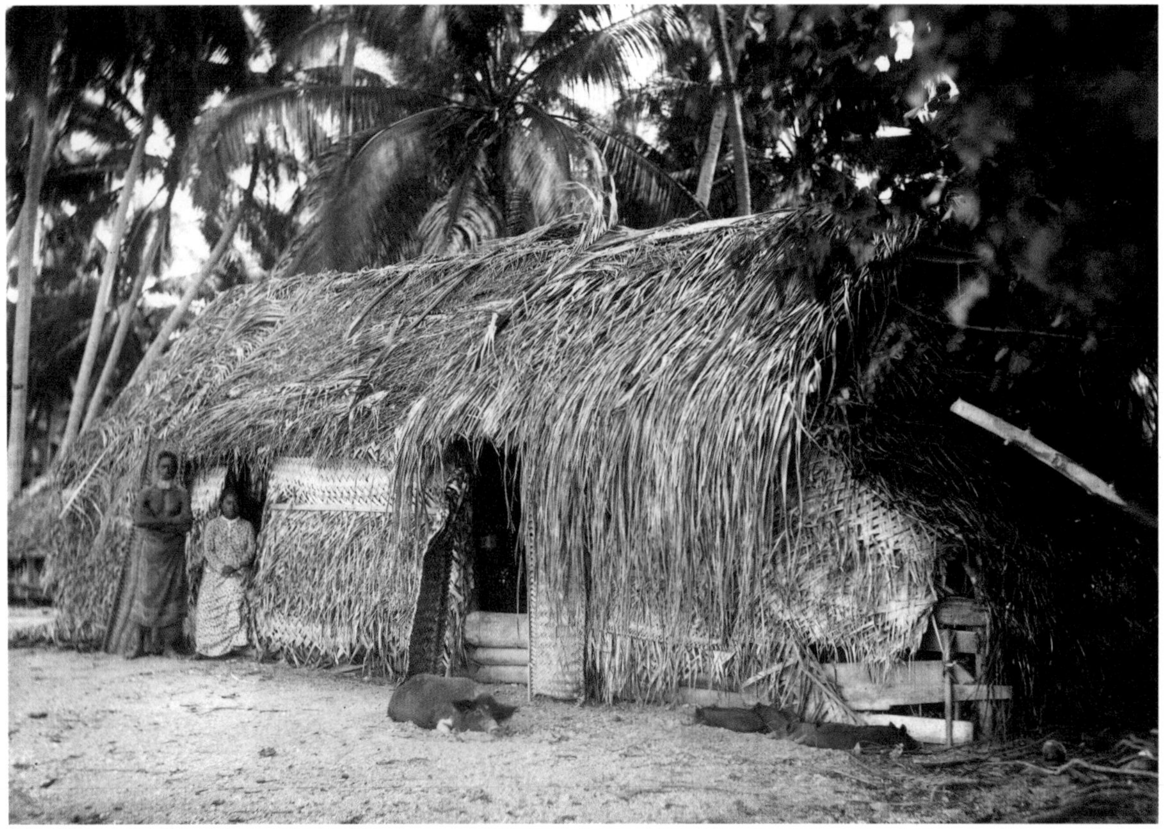

Left: On the back of the copy of the photograph Stolpe has written 'Tairi's house. Tairi (age 60, 167.8 cm) and his daughter Taio (age 25) at the door. Fakarava' – but she is possibly his wife. The family's black pigs are lying in front of the hut. Stolpe was constantly on the lookout for tattoos, and he found some attractive ones on Tairi's body: 'I attempted to copy the complicated pattern, but time was so short that I was far from finishing the work. May the old boy Tairi, the fortunate owner of a large two-cornered hat, tailcoat and trousers with galloons (yellow strips) – the *admiral, as we called him – be duly portrayed before he dies!'*

Above: According to the caption to this old photograph of the Vanadis, the motif is 'the corps of seamen's band during the 1883–85 circumnavigation', though the normal ensemble during this journey was far smaller, five wind players and a drummer often being mentioned. Their appearances were extremely popular, and they had a wide repertoire including folksongs, operetta melodies and marches. On the South Sea beaches they also struck up for dances for the local population.

Opposite page: On Fakarava, observations were performed to facilitate adjustment of the ship's chronometer, which was important for navigation. On the far left is the meteorologist Gottfrid Fineman, in white uniforms are the second lieutenants Henning von Krusenstierna (sitting) and Arvid Lindman, and in a dark uniform jacket is Captain Rudolf Nissen, a navigation officer and the person responsible for the (modest) scientific activities on board the Vanadis. In one of his picture lists Stolpe places 'the magnetic observers' on Fakarava, but Nukuhiva has also been cited as a location.

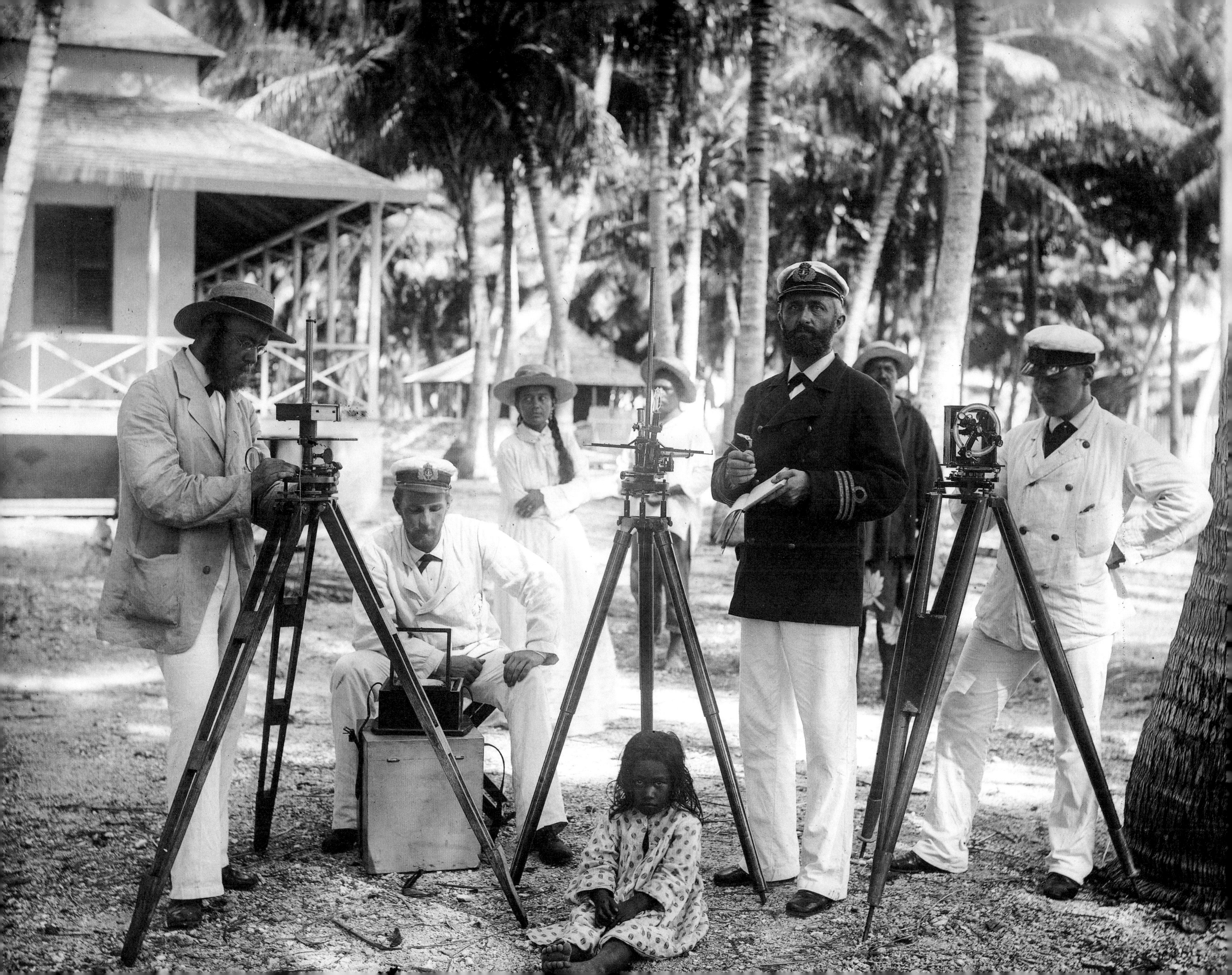

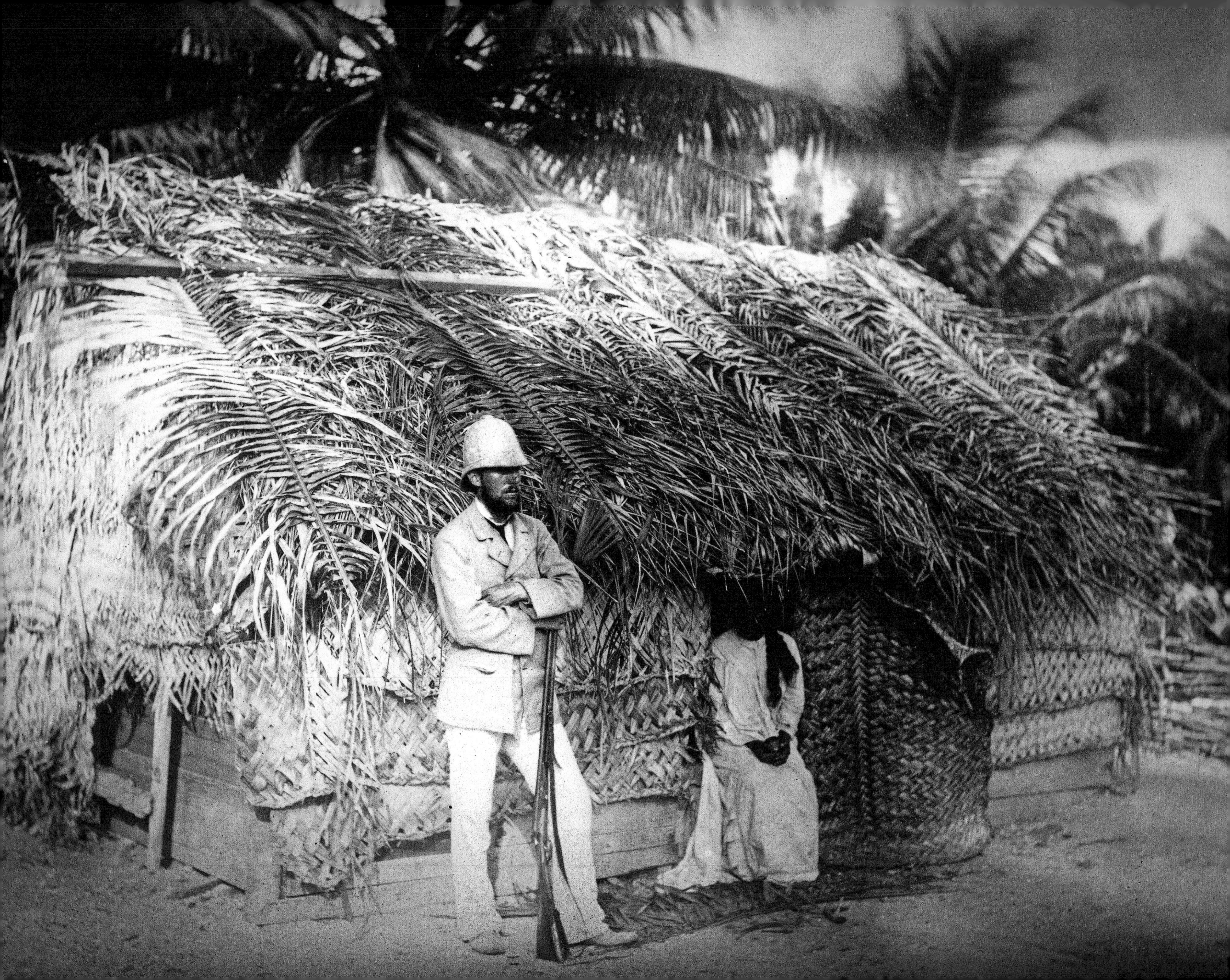

Opposite page: This photograph could well be seen as an illustration of colonialism. White man with rifle and safari helmet, black native by simple hut. But it is the 24-year-old Prince Oscar who is posing in conjunction with a hunt on one of the Pacific islands (Nukuhiva or Fakarava – the information varies). He travelled round the world on the Vanadis at the behest of his father Oscar II, in order to become a naval officer – and so as to forget an inappropriate love affair. He became a captain. But he soon married his beloved, Ebba Munck af Fulkila.

Right: 'One of the most valuable things acquired on the island was a canoe, such as was used within the lagoon. It was made of a hollowed-out palm trunk and was adorned with a strange animal's head at the pointed stern. It took great persuasion to get the owner to release this gem for 100 francs,' noted Stolpe. The owner who was persuaded to sell his 'gem' was probably Tairi, who crops up in a few of Oscar's other pictures. Hjalmar mentions the find in a travel report from Fakarava. The back of the picture bears the following note: 'Original in Vanadis collection'. Transportation of the canoe was difficult – first to Tahiti and then to Copenhagen on a Norwegian vessel, eventually arriving at the Museum of Ethnography in Stockholm.

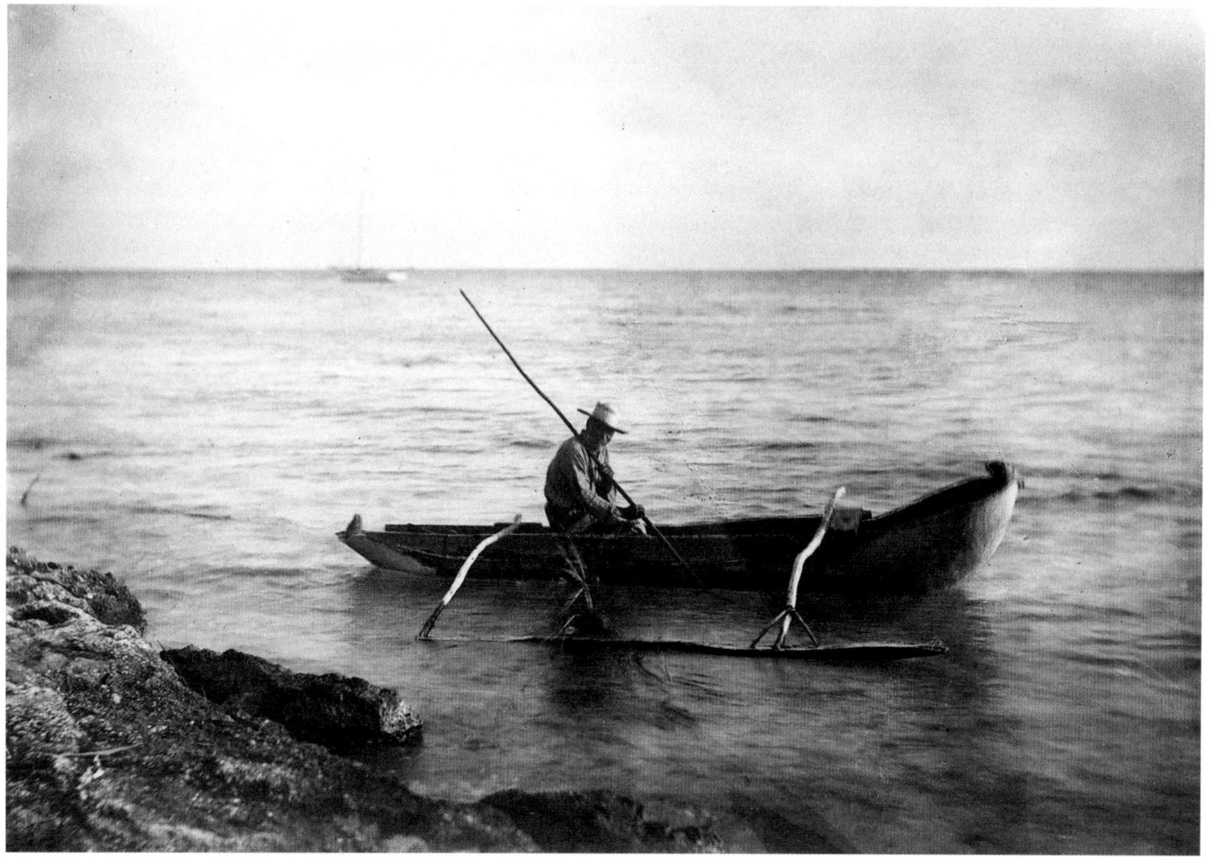

Below: 'Tahiti. Catacomb in the mountains above Paea. The native got a fly on his nose'. The picture shows a conscientious cave researcher in a stained shirt crawling up from the underworld, his eyes staring into Ekholm's camera and his right hand gripping a skull. Next to him sits the guide Herkules, out of focus and batting away a fly. Over 130 years later, in October 2016, ten skulls were presented to a local organisation in Papeete on Tahiti that is engaged in recovering human remains removed from French Polynesia. The ceremony took place in Uppsala, seven of the skulls coming from the Karolinska Institute in Stockholm and three from Uppsala University. The latter had been collected on the island of Nukuhiva in conjunction with the Vanadis's visit there, but it is unclear which skulls can be linked to Stolpe and his collecting activities. The only person mentioned in conjunction with the ceremony was the assistant ship's doctor, Carl Otto Holmer.

Opposite page: On Tahiti Hjalmar found this 'bathing cave', and 'had the cave photographed whilst I was swimming in it. See photograph, which gives some idea of how beautiful it was.' The letter and photograph were sent home to his wife. Her reply reached him two months later in Japan: 'Your bathing venture in the cave on Tahiti was bold. Knowing now that you emerged unscathed from your escapade, I'm glad to have this photograph, which particularly amuses me. But I want an attractive portrait of my Friend – send me one with the next post! And a lot of other photographs too!'

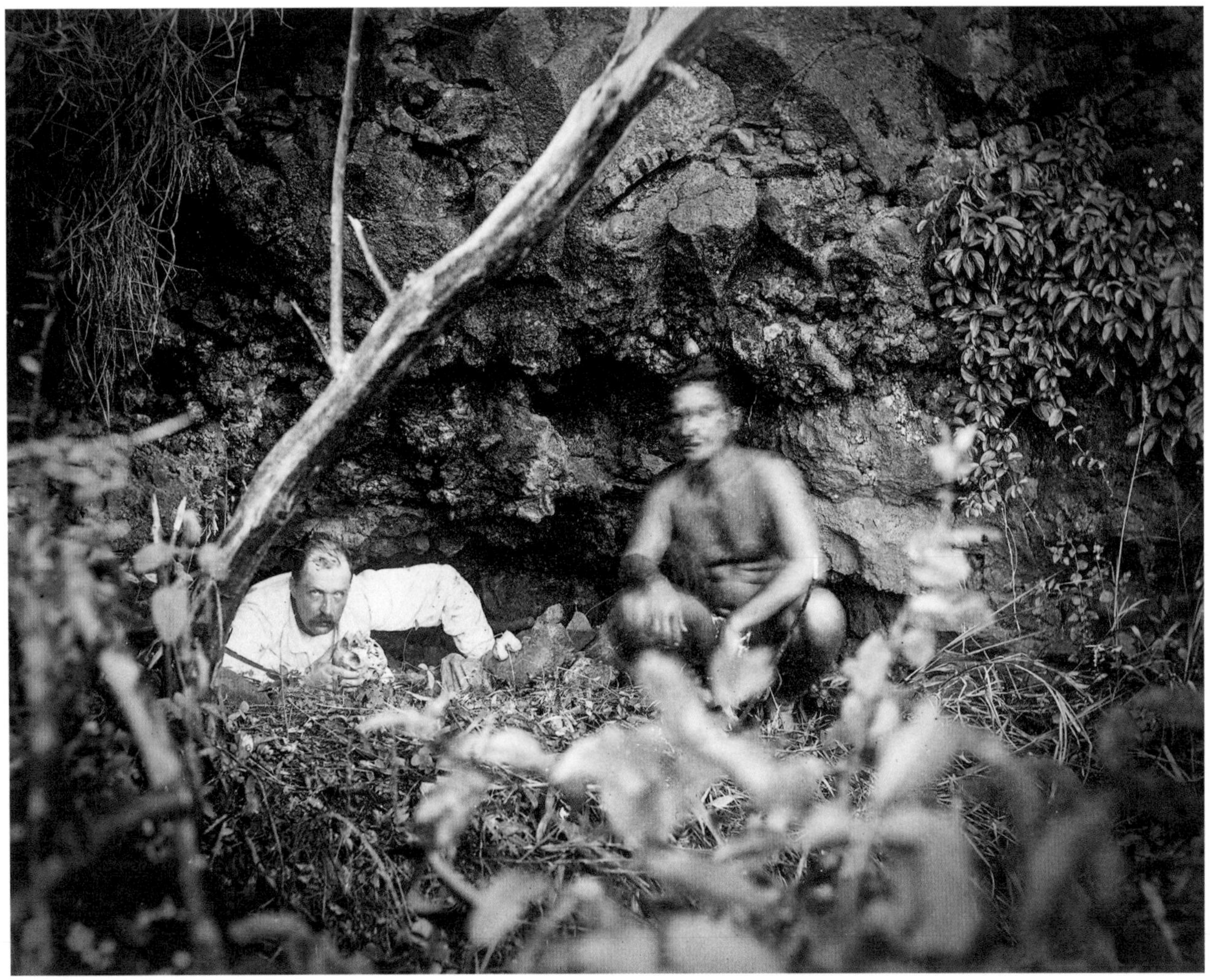

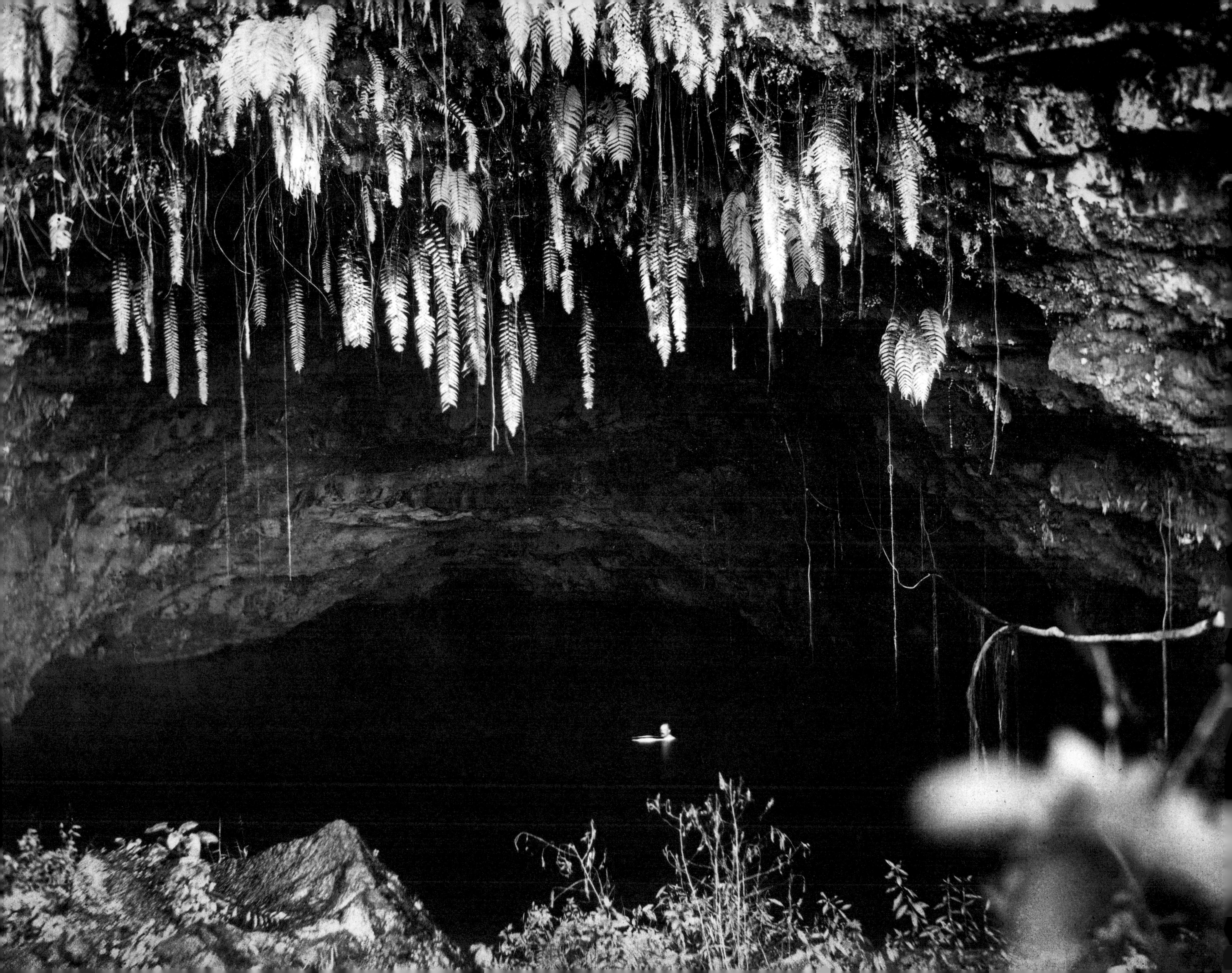

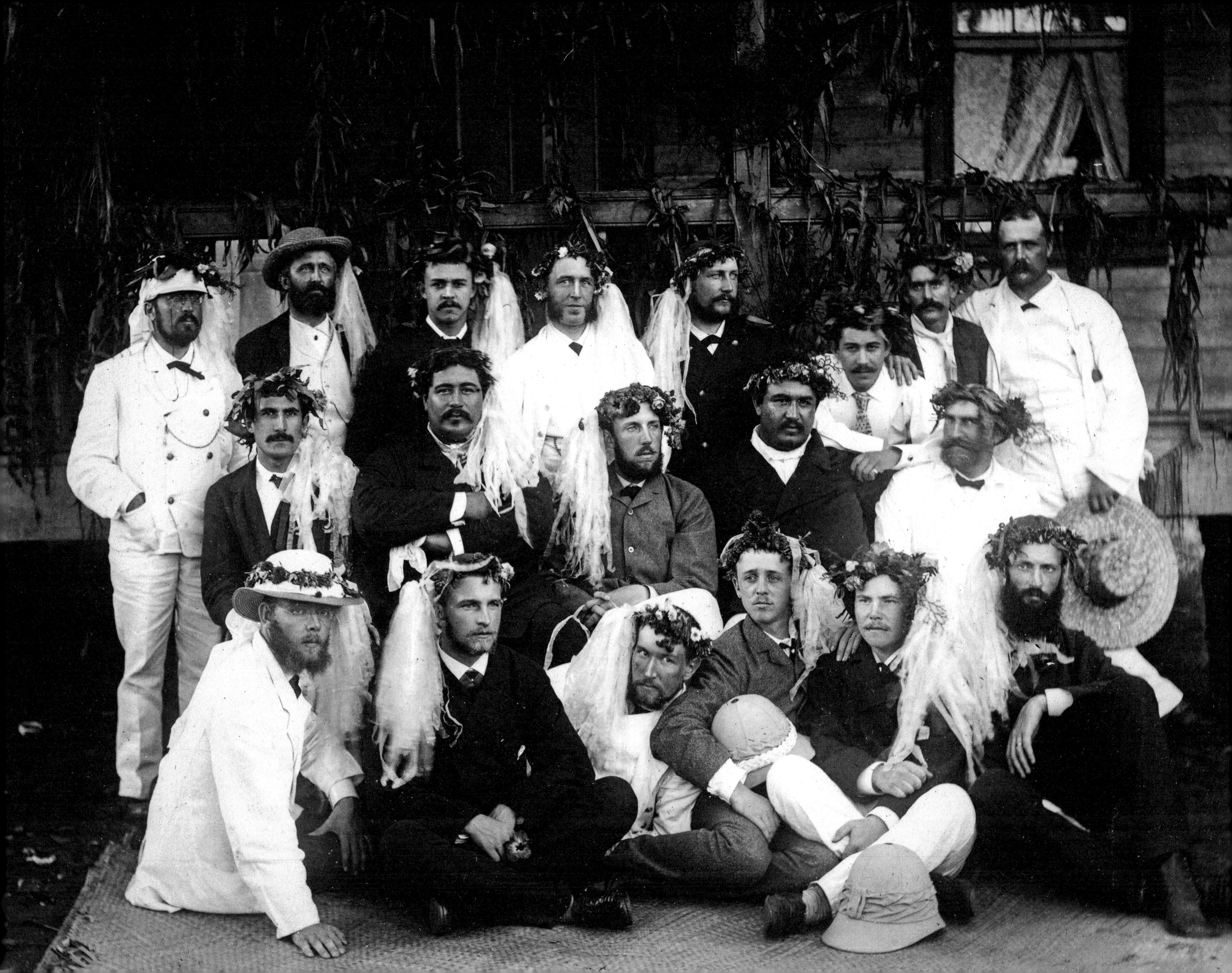

Opposite page: In Papara on Tahiti, the chieftains Tati and Alexander Salmon put on a huge party for the Vanadis's officers. A hundred-strong choir sang in four parts: 'Stolpe asserted that they sang perfectly in tune', Prince Oscar noted in his diary, adding that 'Stolpe responded with several Swedish songs, which seemed to make a strong impression'. In the middle of the picture, between the Salmon brothers, sits the prince, with a garland of flowers on his head. At the top right is the singer himself, as usual without a garland of flowers, but with a big hat in his hand. Much later Stolpe explained that he had 'for more than one night sat up and listened to the endless choral singing in the big elliptical meeting houses in southern Tahiti, and had tried to commit the different parts to memory. The choral singing of Polynesia is an area of study one hopes will not remain uninvestigated for too long. In Papeete they were already singing choruses from Offenbach's La belle Hélène in 1884 (!), and the old songs are dying out'.

Right: On Tahiti Stolpe met a man called Tepano from the remote Easter Island, whose tattoos so fascinated him that he wrote a scientific essay about him. But it was not the facial tattoos that were the most important ones. Of greater interest were the tattooed figures on Tepano's right arm: a recumbent statue is being dragged by ten people, and right next to them are two men, one of whom is holding a long object in his hand. A small figure is dancing above the statue. In response to Stolpe's question about the tattoo's meaning, to his surprise Tepano said it depicted what was for the Easter Islanders a sensational event in November 1868, when two of their stone statues were taken on board the English naval vessel Topaze, ending up at the British Museum in London. Ten seamen are dragging the four-ton stone statue down to the beach. The man with the staff in his hand is the First Officer, and above the statue a chieftain is dancing. This was, according to Stolpe, a striking example of the peculiar nature of ideography in the form of a living chronicle. Through Tepano the story of the dramatic event had migrated 2,500 miles across the Pacific Ocean from the isolated Easter Island to Tahiti. Stolpe's essay was published in Berlin in 1899, leading to new research over a hundred years later. The tattooed man had been identified as Vaka Ariki, and was given the name Tepano when he converted to Christianity. The abundant tattoos show that he was a high-ranking person back on Easter Island. It is assumed that he was left-handed, and that he did the tattoos on his right arm himself.

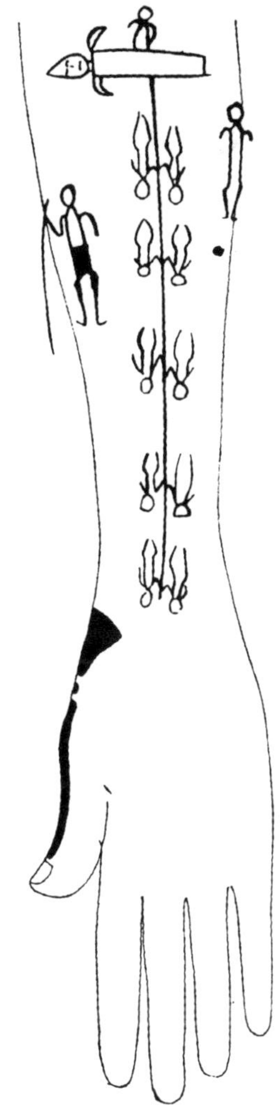

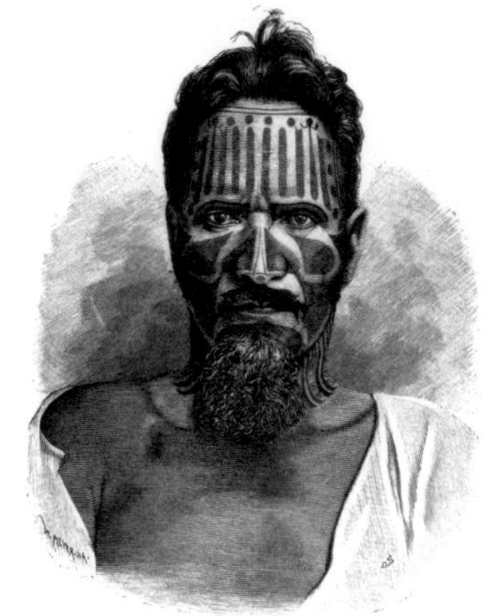

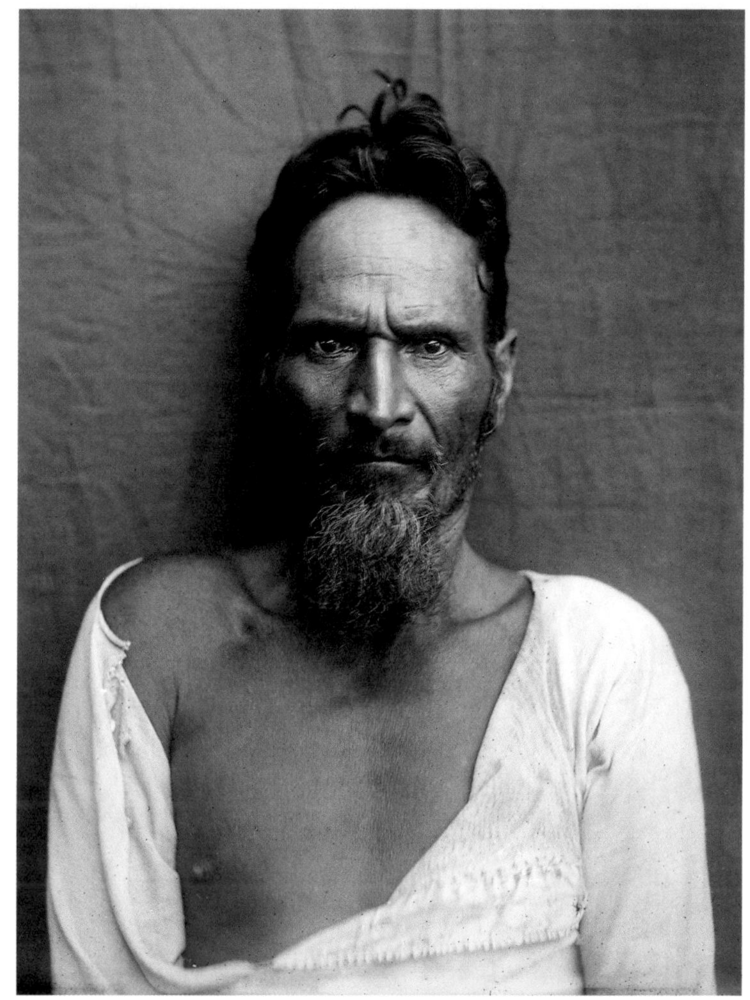

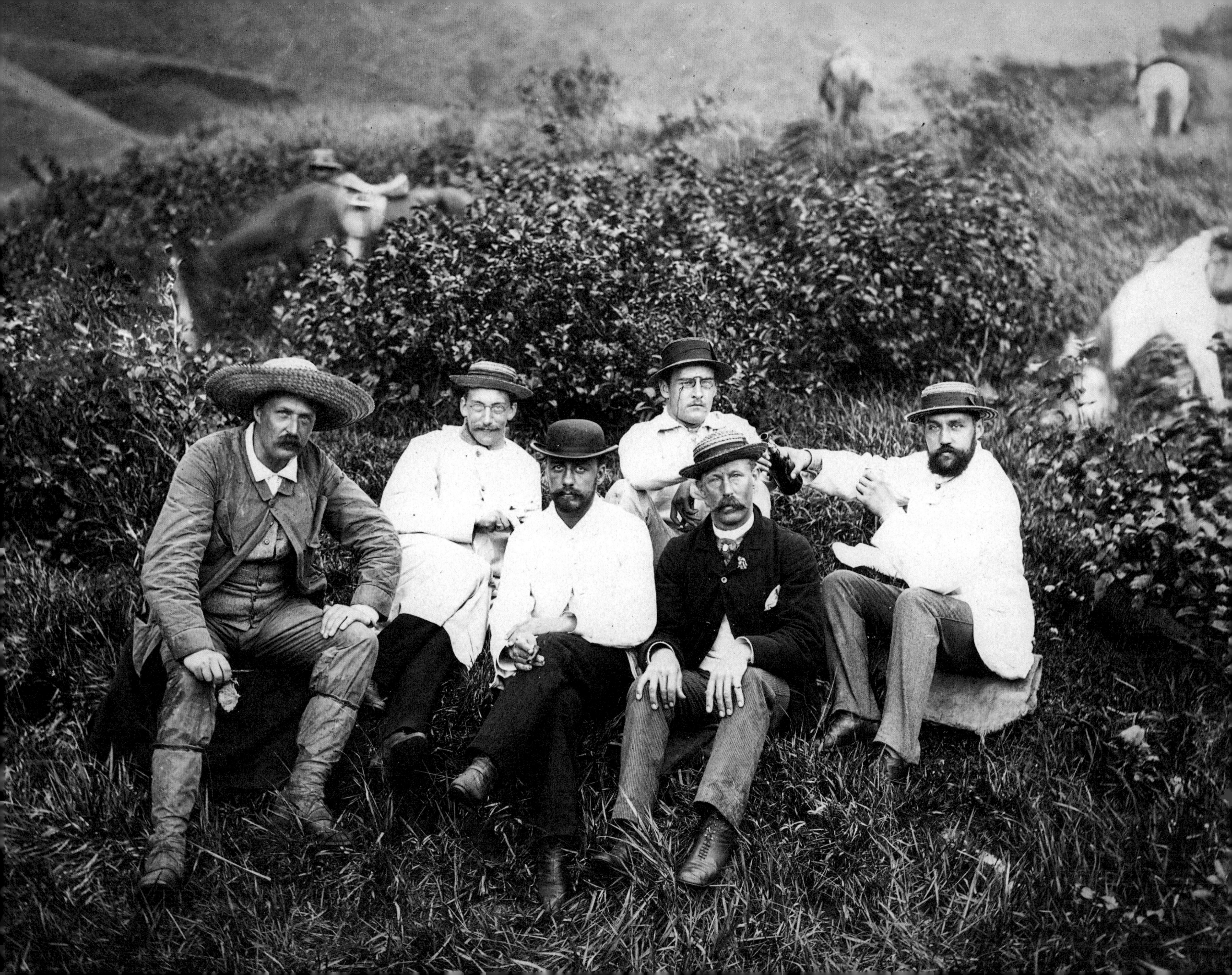

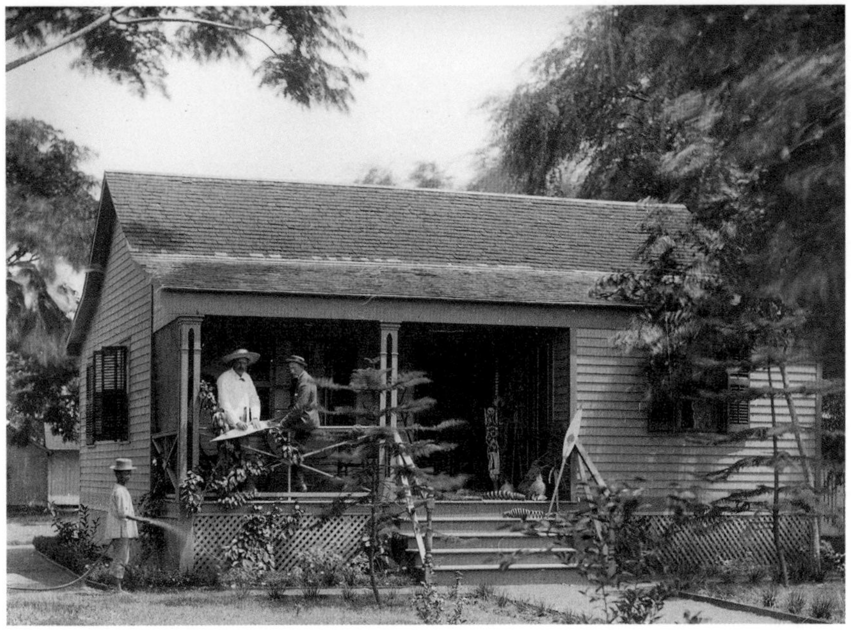

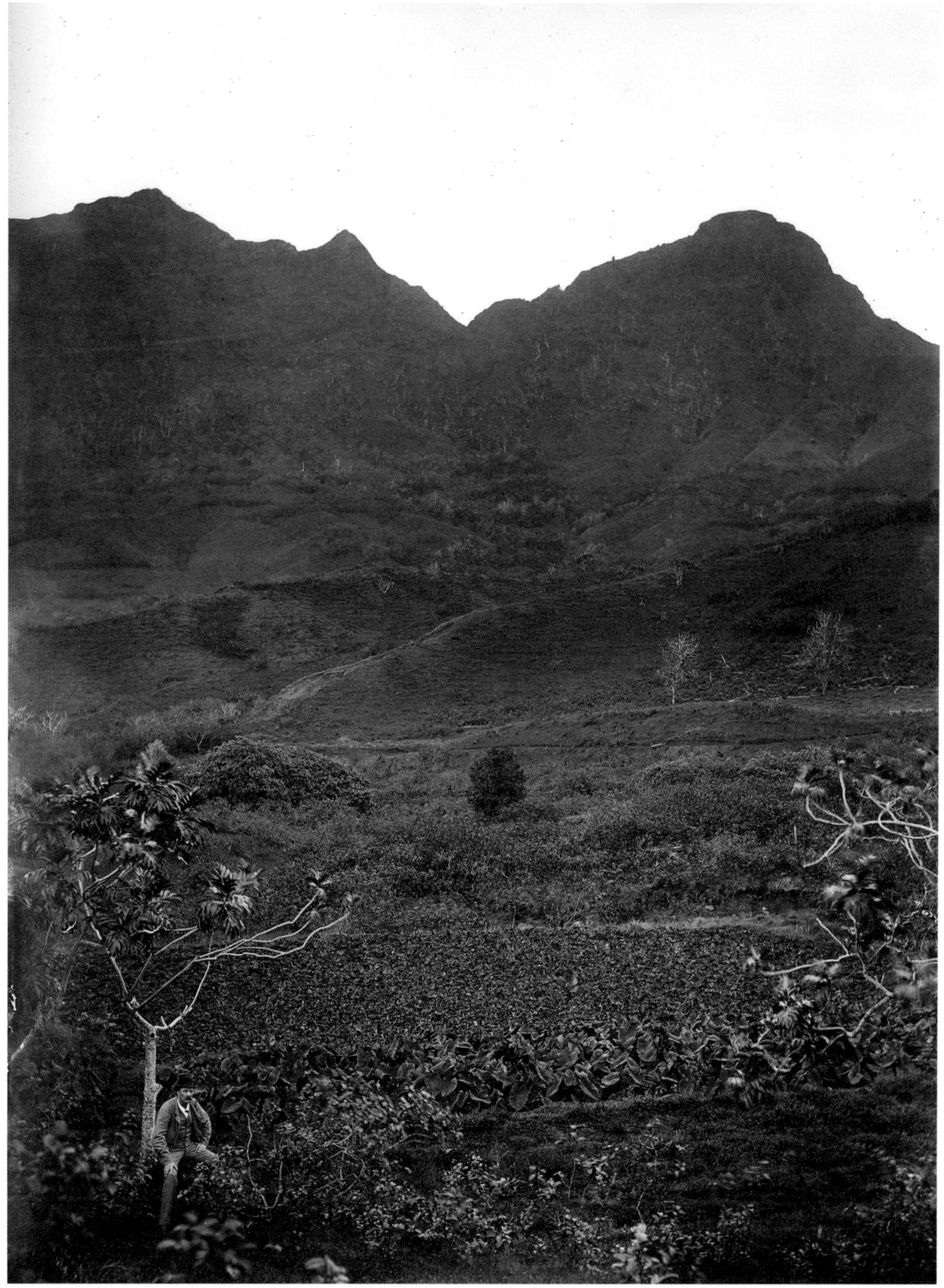

Opposite page: On 6 July 1884, Hjalmar wrote to Emmy from Honolulu: 'My little Sparrow, it's now 1/2 past five in the morning, and in two hours I'm going off mountaineering with Dr Arning, 2 Germans, an Englishman and an Italian naval officer. It's Sunday, and there's nothing else to undertake here except go out into the woods. The others are riding, and I'm going in a carriol [horse and two-wheeled carriage] as far as possible, and will then walk. I shall convey the photographic equipment.' Ekholm took the picture during an excursion in the Kalihi valley, where Stolpe visited the dwellings and bought household utensils. The letter writer is on the far left, with his friend the German doctor Eduard Arning at his side.

Above: The Vanadis remained in Honolulu for three weeks, and Hjalmar Stolpe stayed with Arning, the bacteriologist who was researching leprosy on the Hawaiian Islands – 'an extremely decent, kind and happy chap who japes me into laughing all day long'.

Right: With mighty mountain peaks in the background, Oscar Ekholm has positioned the well-dressed Hjalmar Stolpe at the bottom left, seated next to the small tree. The picture's composition is based on the contrast effect often used by Ekholm in his photographic work – a surprising small figure inserted to make the landscape or structure in the background especially imposing. Where the photograph was taken is uncertain – perhaps somewhere on Oahu, one of the Hawaiian Islands.

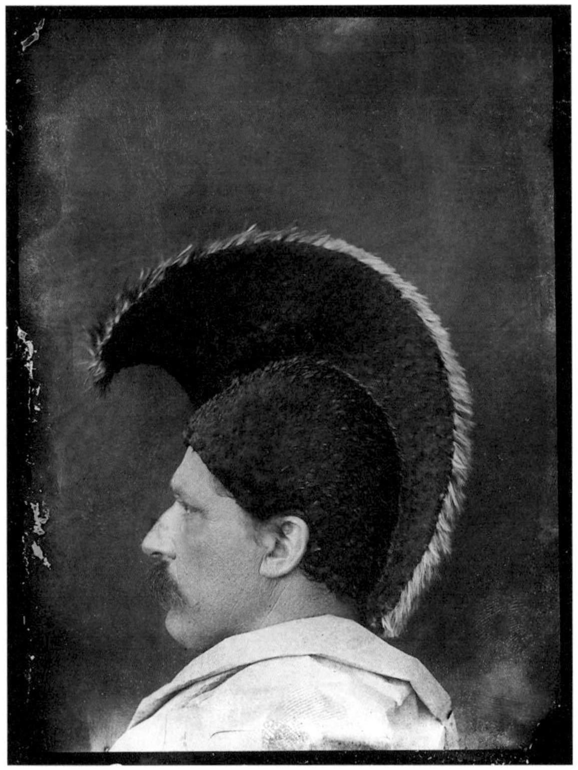

Above: Hjalmar Stolpe displaying royal headgear from the Hawaiian Islands. On the back of the picture he noted: 'The feather helmet (mahiole) belonged to King Kaumoalii of Kauai, deceased 1823. Honolulu Museum. Helmet red and yellow.'

Right: Stolpe bought this soldier's armour on Jaluit through the trading company Hernsheim & Company. It is an ethnographic rarity whose provenance was stated as being the Kingsmill Islands, now the Gilbert Islands. In conjunction with the Vanadis exhibition in Stockholm 1886, he wrote: 'cuirass and full armour for warriors; the helmets are made of the skin of a horny fish; the neck protectors for the cuirasses serve to provide protection against mistakes by the women, who fight with slings in the second line'. The curiously shaped sword is set with shark's teeth as sharp as an awl.

Opposite page: The tall man in the middle is commander-in-chief Otto Lagerberg, captain of the Vanadis. King Kabua is standing on his right, and on his left is the king's 'Prime Minister' Launa. Hjalmar Stolpe is to be seen in a sun hat, and the other people belong to the royal family. Lagerberg and Stolpe found it hard to get on, and just before Oscar Ekholm took this picture they had had a big argument. Back in Sweden Posttidningen *had incorrectly stated that Stolpe was the leader of the Vanadis expedition, thereby infuriating the minister for naval defence, Carl Gustaf von Otter, and Lagerberg.*

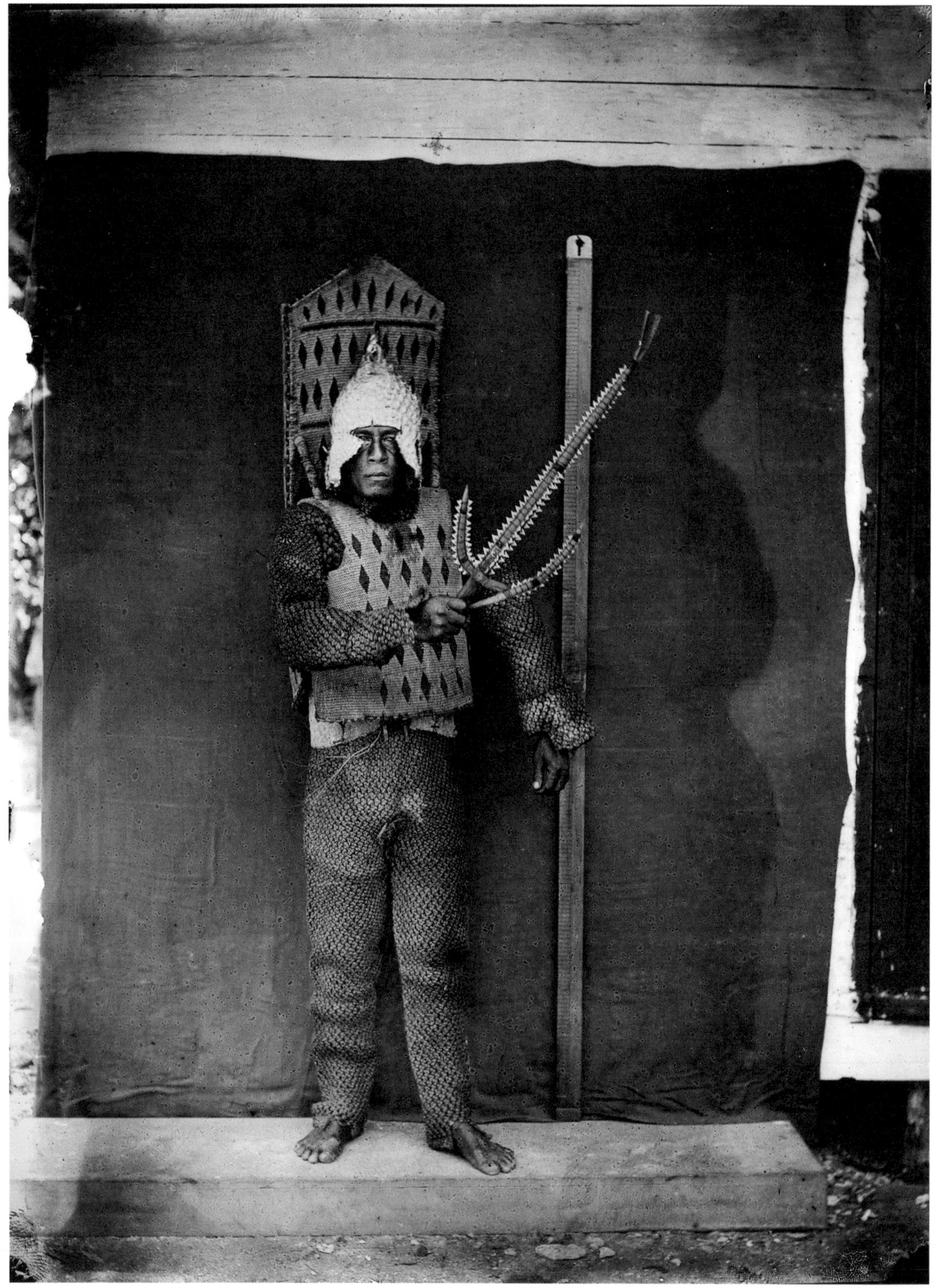

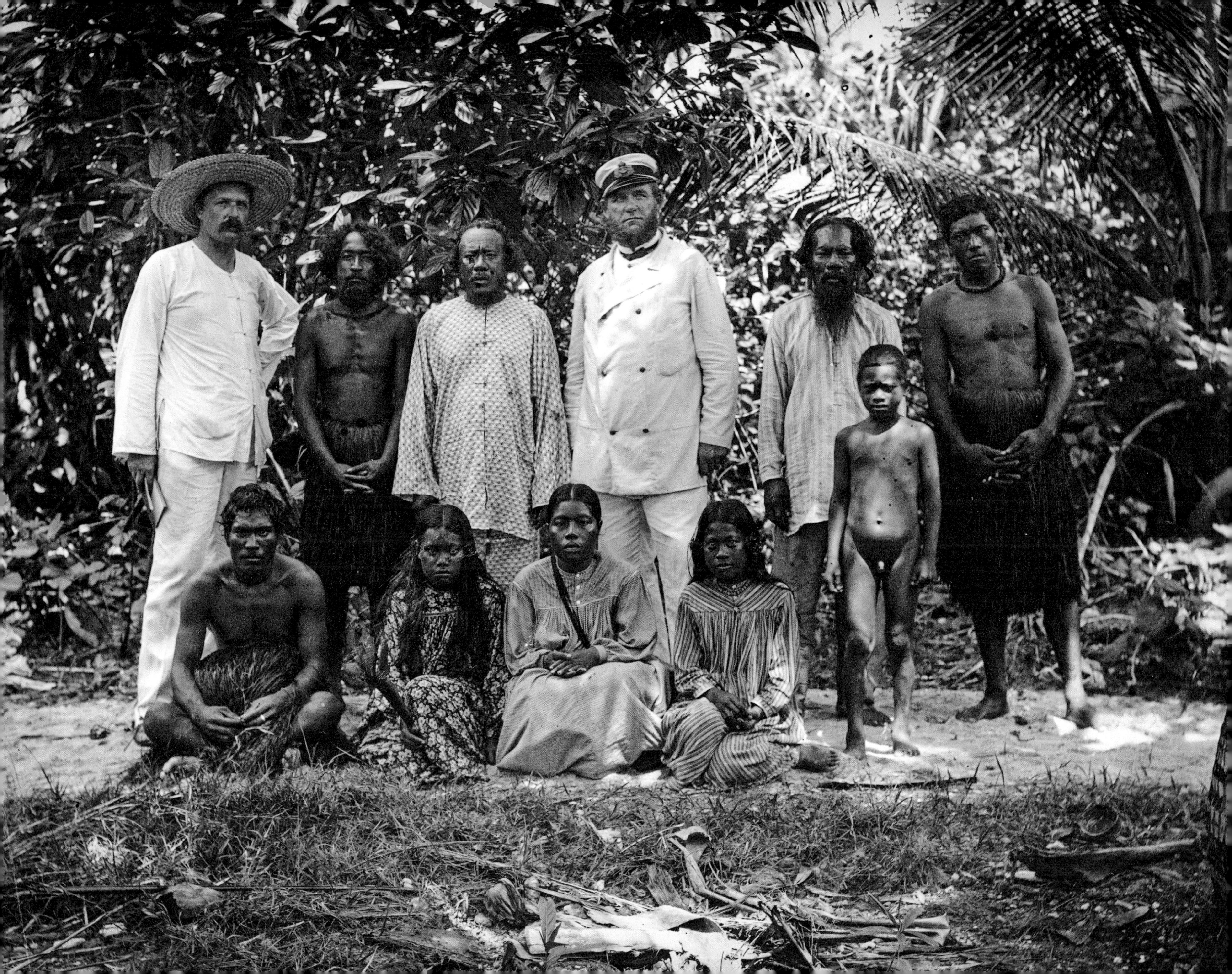

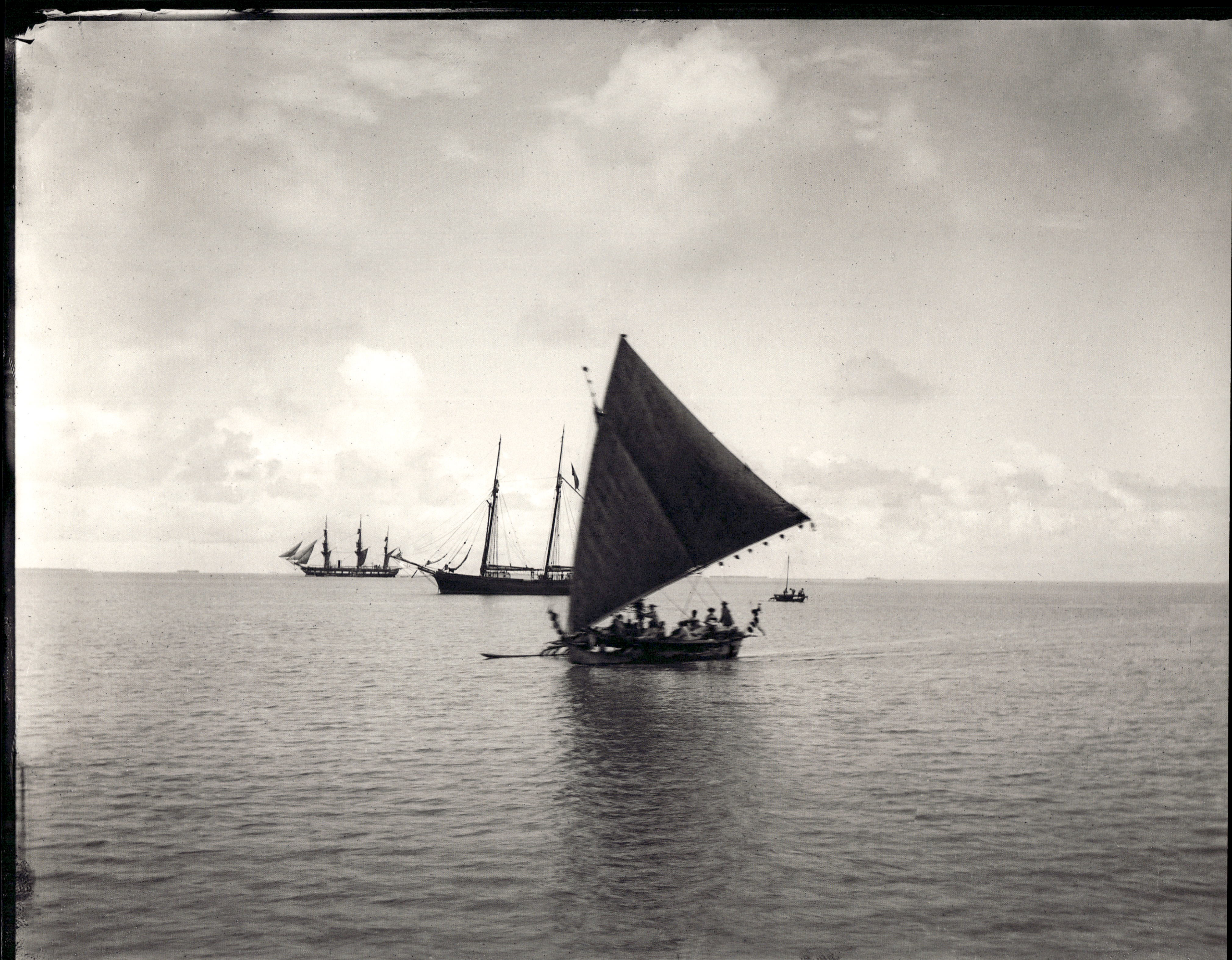

Opposite page: Behind the sailing canoe is King Kabua's two-masted schooner, and on the horizon the frigate Vanadis with set headsail.

Right: King Kabua is mentioned by Stolpe as the ruler of the Marshall Islands. But in reality his realm was restricted to the Ralik group – the western part of the archipelago. Relations with the eastern Ratak group were bellicose. Kabua had six wives (hence the six rings on his fingers), and resided on the nearby island of Ailinglapalap. He very much appreciated the presents he received at the dinner on board the Vanadis, not to mention getting the chance to try out both a machine gun and a cannon. But was he distinguished enough? When the kings of Tahiti and Hawaii visited the Vanadis, royal salutes were fired, according to the logbook. But the logbook makes no mention of King Kabua's visit on board.

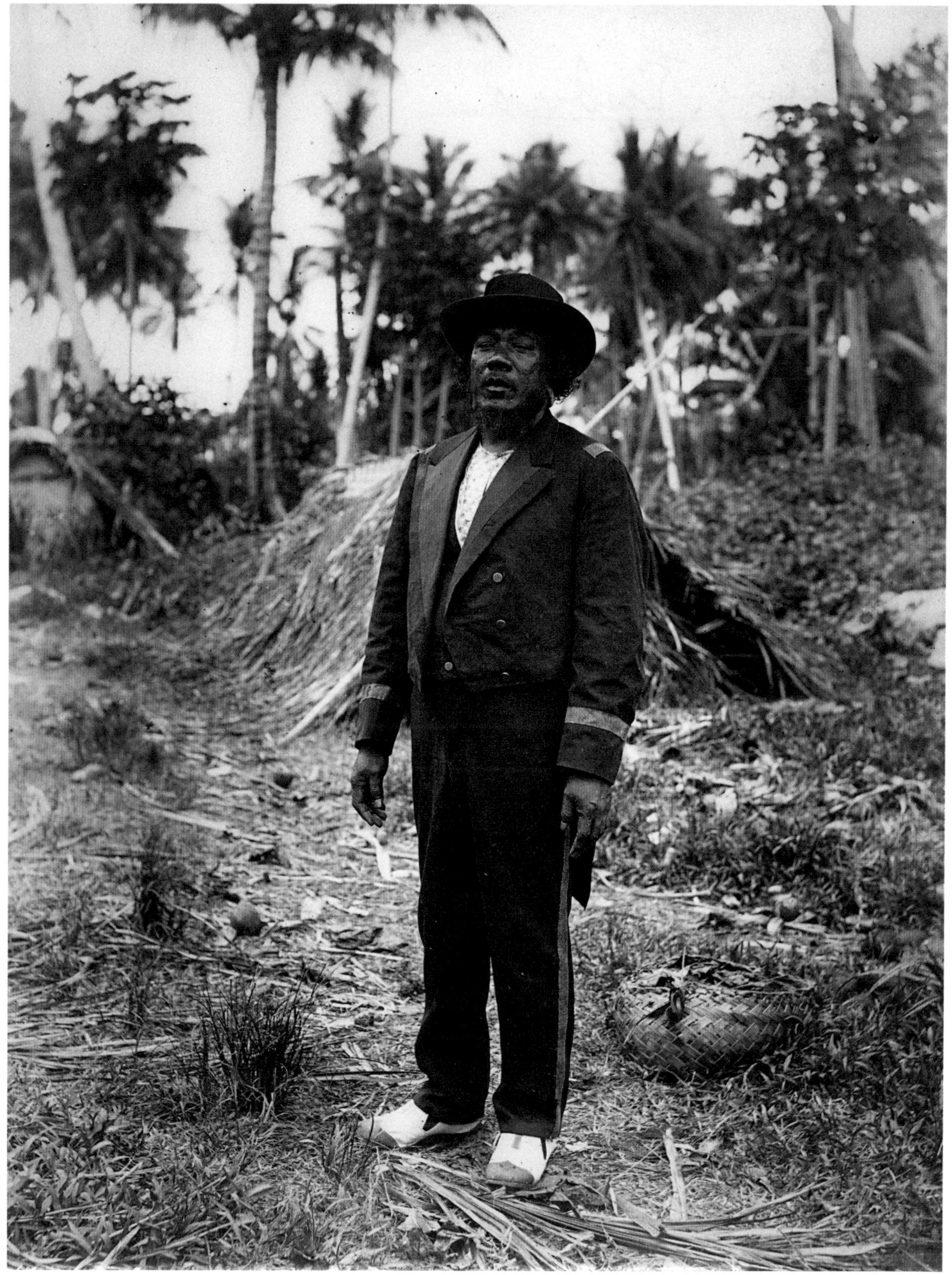

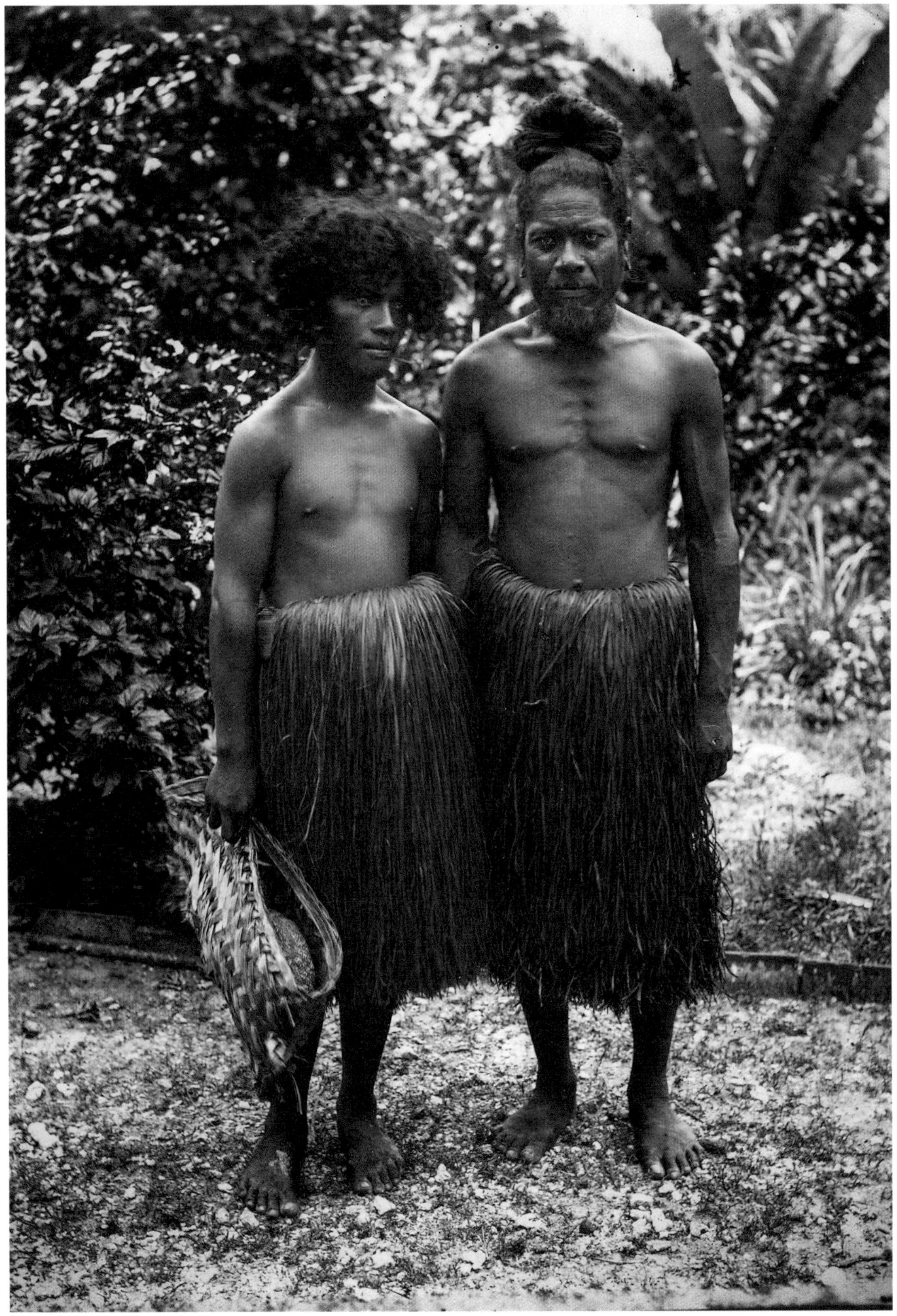

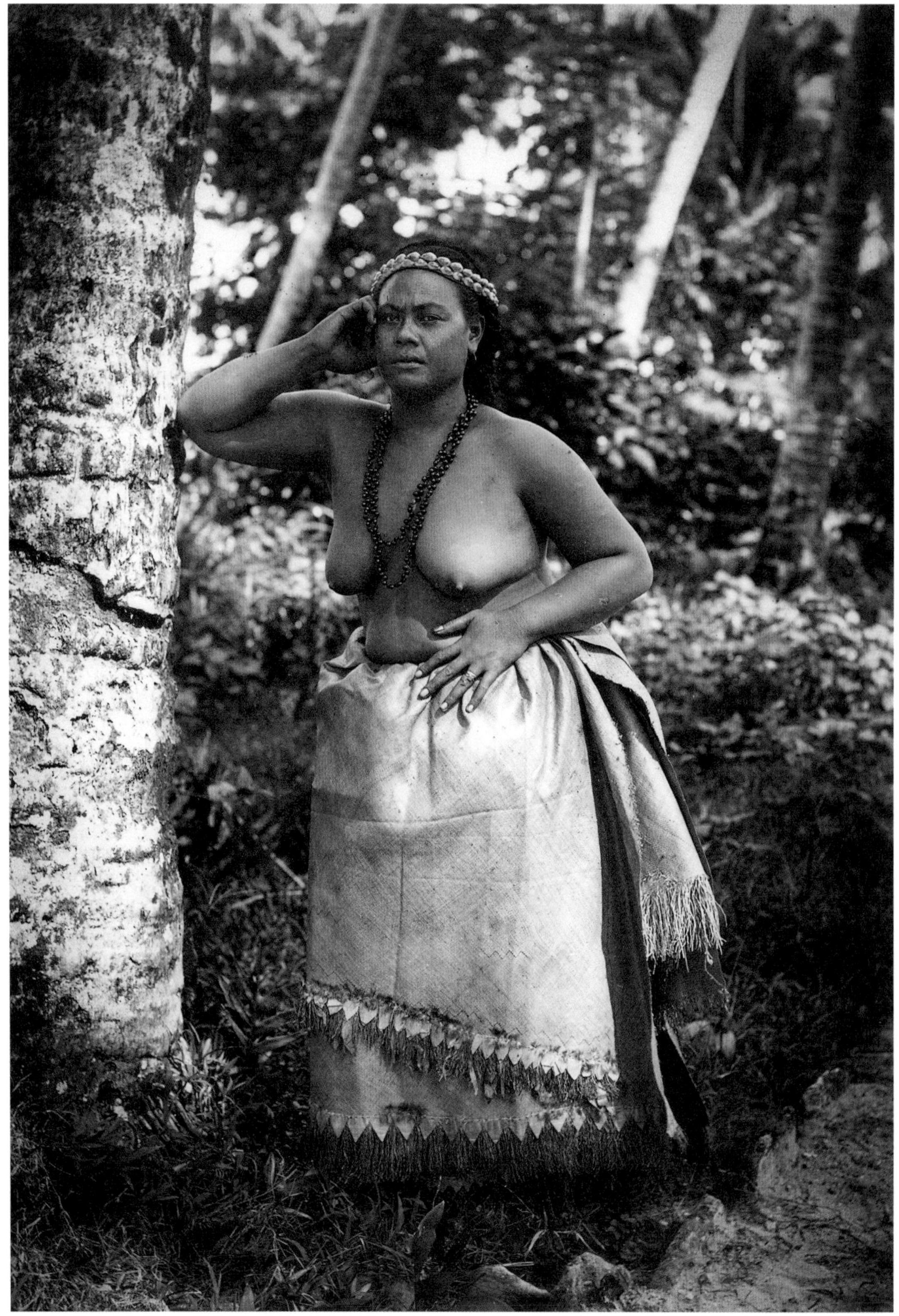

Opposite page, left: On Jaluit, Lagodjen from Kwajalein was photographed with a woven basket in his hand, as well as Radjuruilik from Bikini.

Opposite page, right: Princess Tui (166 cm) from Samoa is dressed in her Samoan attire, Stolpe informs us on the back of the picture. She is also referred to as Kabua's queen, though not by Stolpe.

Right: The chieftain Lojak (Loiak), who sailed in his canoe from his dwelling on Pingelip in the Jaluit atoll. Stolpe refers to another of the chief's houses on Jaluit: 'In this house he shuts himself in with his four wives (4) at night. If he goes out in the day the wives have to stay indoors (not closed), and no Kanaks are allowed to talk to them'. Arvid Lindman wrote: '... after the king, the most distinguished person was Lojak, who was also quite a rogue.' Stolpe's contemporary German ethnography colleague Otto Finsch maintained that Lojak was 'the actual paramount chieftain', and that the spear thrower in the picture had warred with Kabua a few years before the Vanadis arrived. Finsch referred to the latter as 'King Lebon Kabua'.

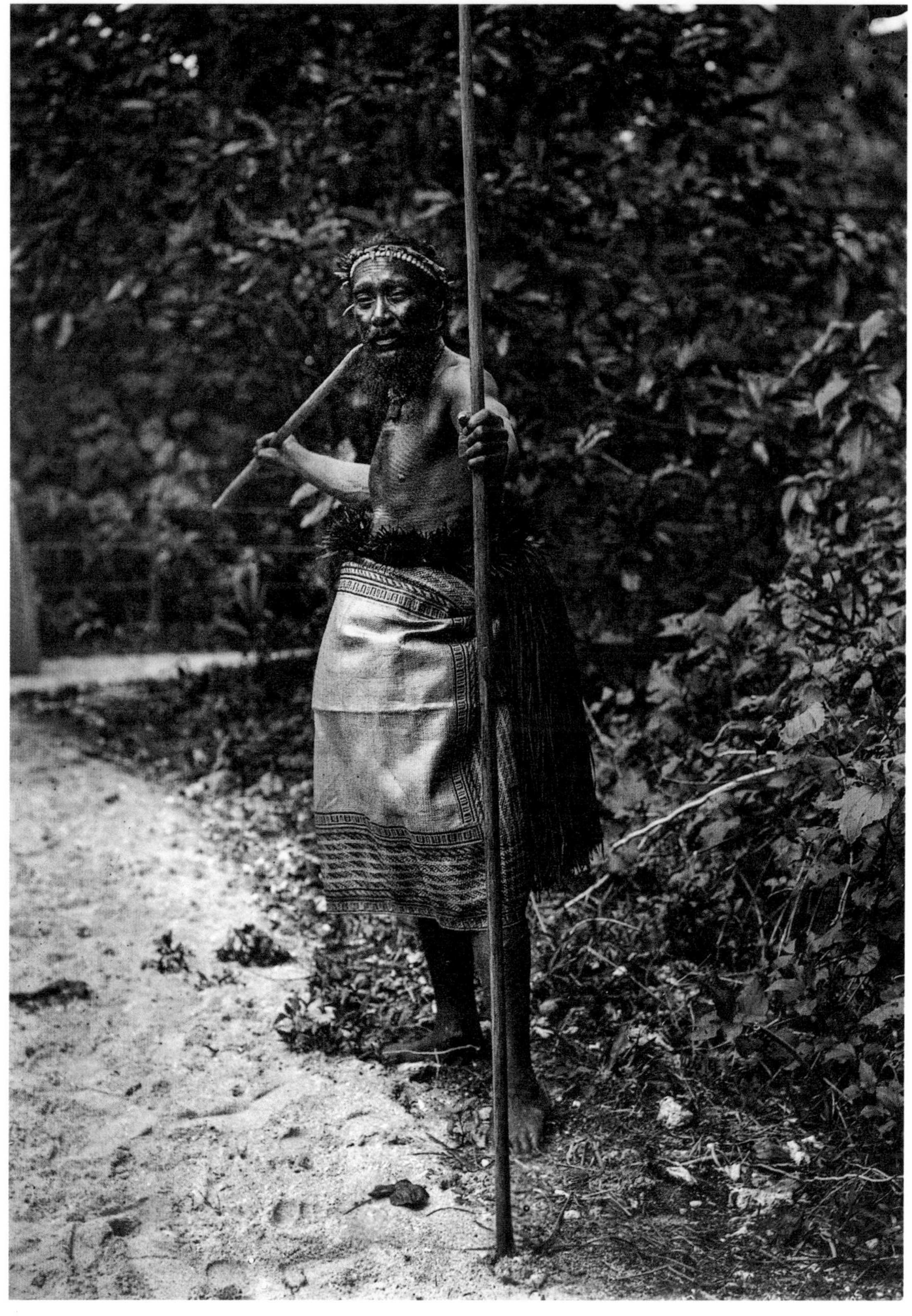

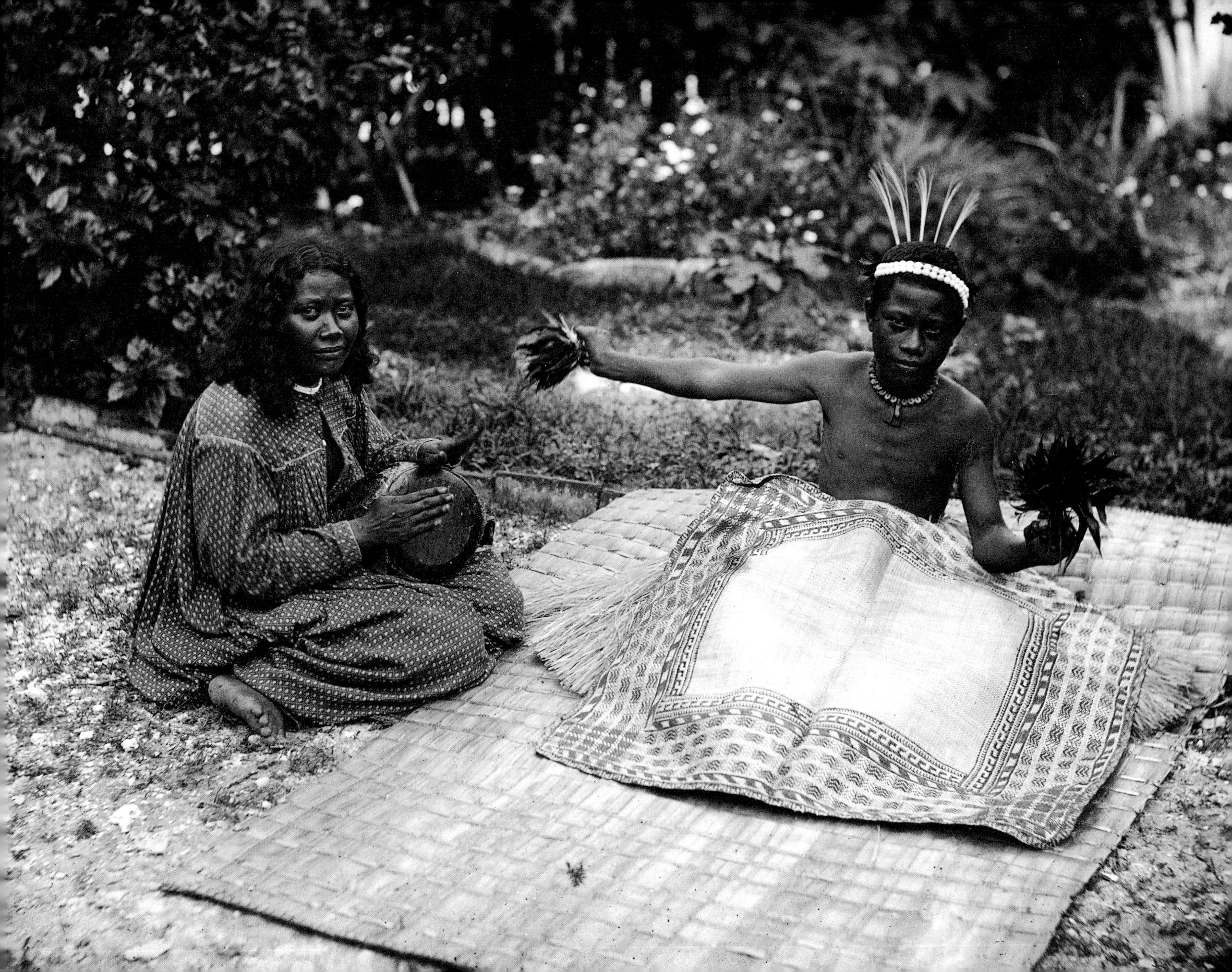

Opposite page: We see here Kabua's son number two, Prince Laelang, sitting with a mat over his knees. But, Stolpe emphasises, he is 'not drinking' – unlike his thirsty big brother Lamorro. The woman drumming is one of the king's six wives, and shark's skin is stretched over the drum. Ekholm's picture illustrates how the king himself behaved during a festive night on Jaluit: 'Dances sitting down, twists his body, shakes his arms with their luxuriant black feather bushes, deflects his eyes (the wives too, but not the rest of the company), bellows and sings.' Emmy received this information from her husband, who immediately after the night's dancing paid homage to the king by singing 'Thou ancient, thou free'. Kabua's own mat is now at the Museum of Ethnography in Stockholm.

This page: In one of the island's huts Stolpe is sitting on the far left, writing in his notebook No. 4, which contains 30 or so pages on the people of Jaluit and its sailing canoes, implements and customs. The hut belongs to Crown Prince Lamorro – 'young but a drunkard' – a chieftain on the island of Ebon, one of the most highly populated islands of the Marshall Archipelago. The woman next to Stolpe is probably one of the crown prince's three wives. He said in a letter home: 'I've collected loads here. It's without a doubt my best place. And I have copious notes.'

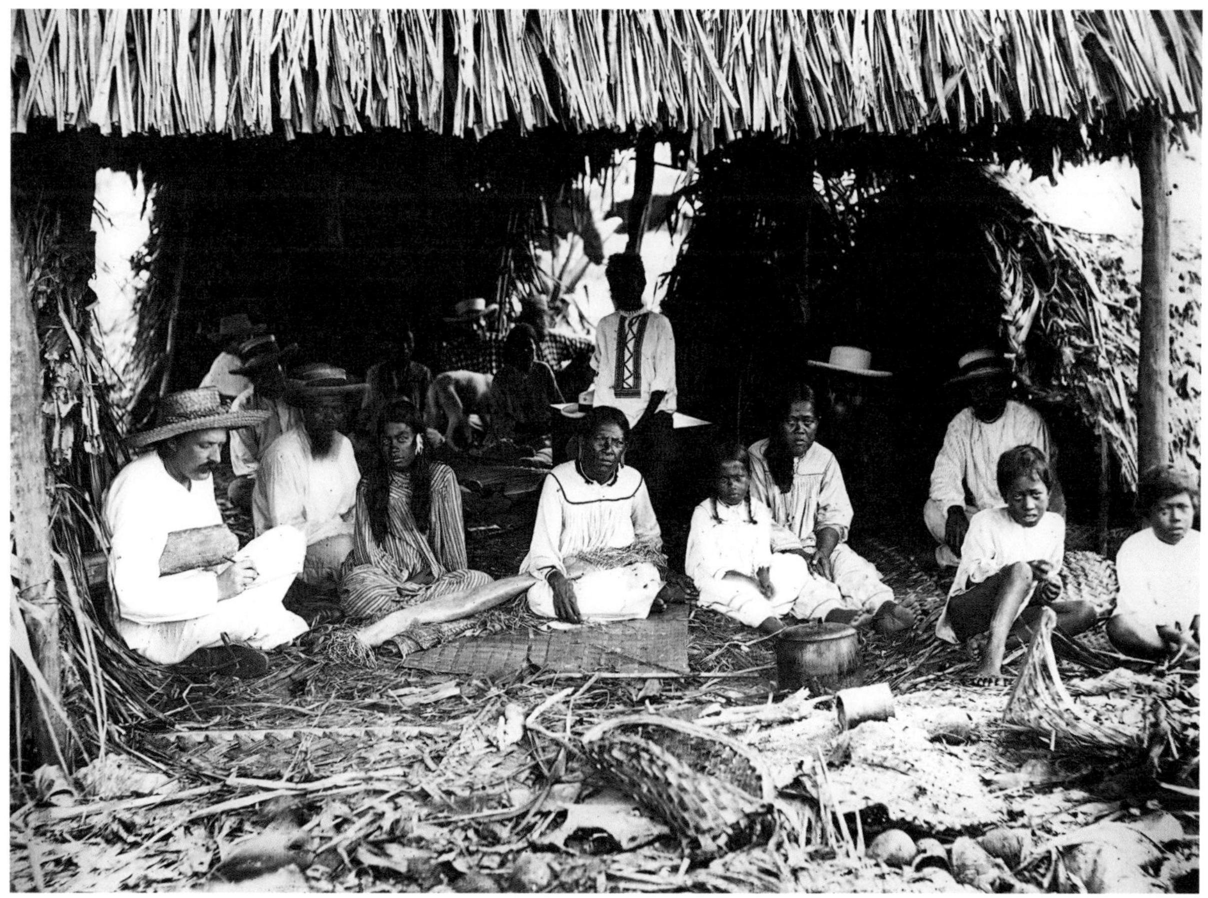

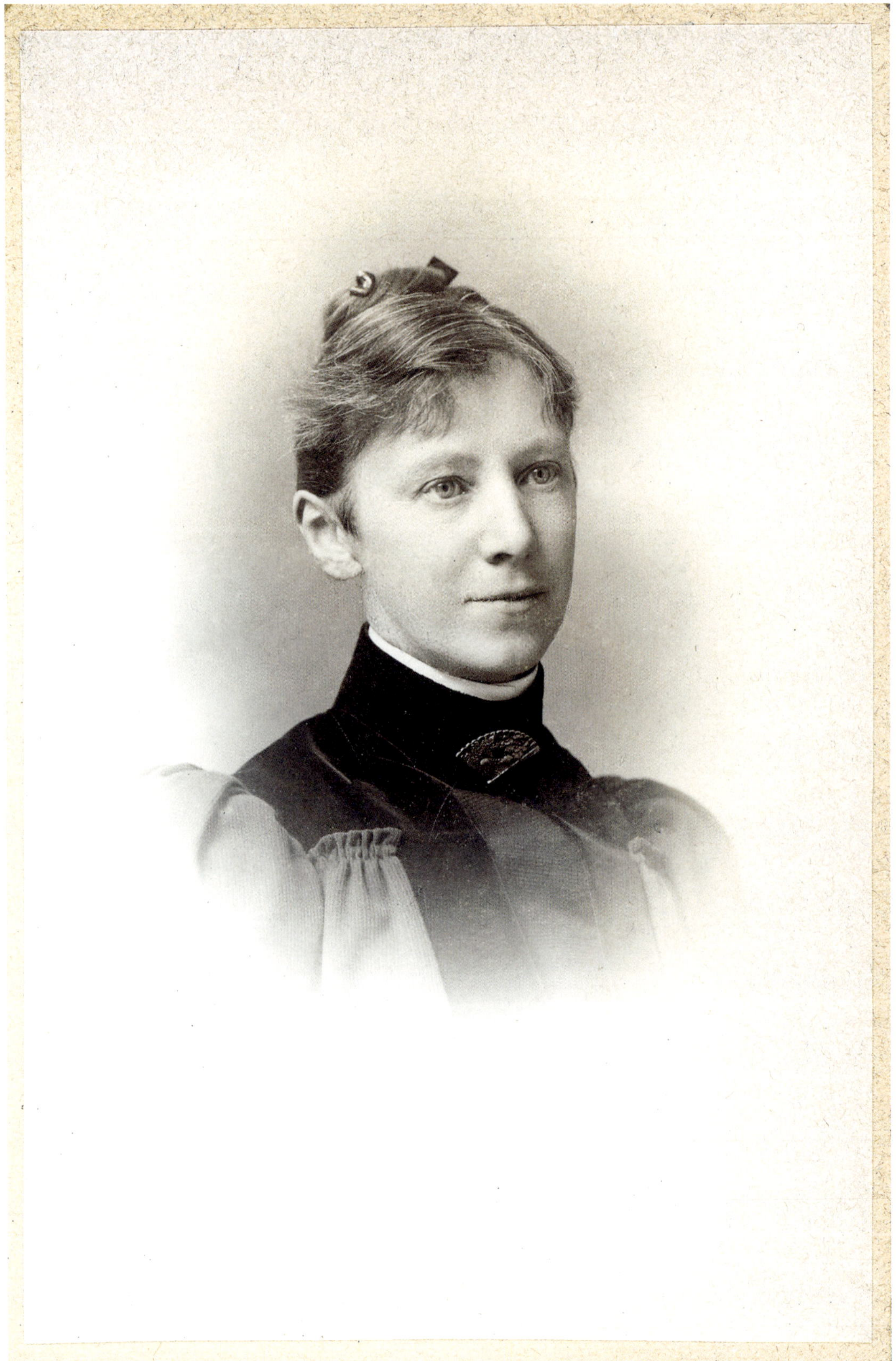

Emmy Stolpe (1846–1908) and Hjalmar wrote each other many loving letters during the Vanadis voyage. She was dubbed 'Sparrow' (Sparfven), and during her husband's voyage she and their daughter Estrid lived with her father-in-law Carl Johan Stolpe, the juridical burgomaster in Norrköping. She considered Hjalmar's voyage to be important to his scientific career, and in her letters she endeavoured to display stoic calm. At times, however, she suffered both physically and psychologically from the long separation, and from being a 'lodger' in Norrköping whilst for financial reasons the flat in Stockholm was rented out. But she followed the Vanadis voyage carefully: 'On one wall in our room there's a nautical chart, and on another a map of the world.' She also informed herself about the places the Vanadis and Hjalmar visited by reading the English author Annie Brassey's story of her circumnavigation in the three-masted yacht Sunbeam. This bestseller was published in Sweden in 1881 under the title En sjöfärd omkring jorden [A Voyage in the Sunbeam]. *The book was also to be found on board the Vanadis in Hjalmar's ship's library.*

Opposite page: Hjalmar Stolpe is standing at the bow of the sailing canoe in a white safari suit. He was not alone in admiring the canoes' design. Arvid Lindman, a second lieutenant on the Vanadis, writes enthusiastically: 'The natives here are excellent seafarers, and their canoes are extremely well built and maintained. They are sewn together using various bits of wood, after which the holes are caulked, to make the canoe completely watertight.' On a structure built across the canoe there are 'small houses made of mats, providing space for the crew and protection from the sun'.

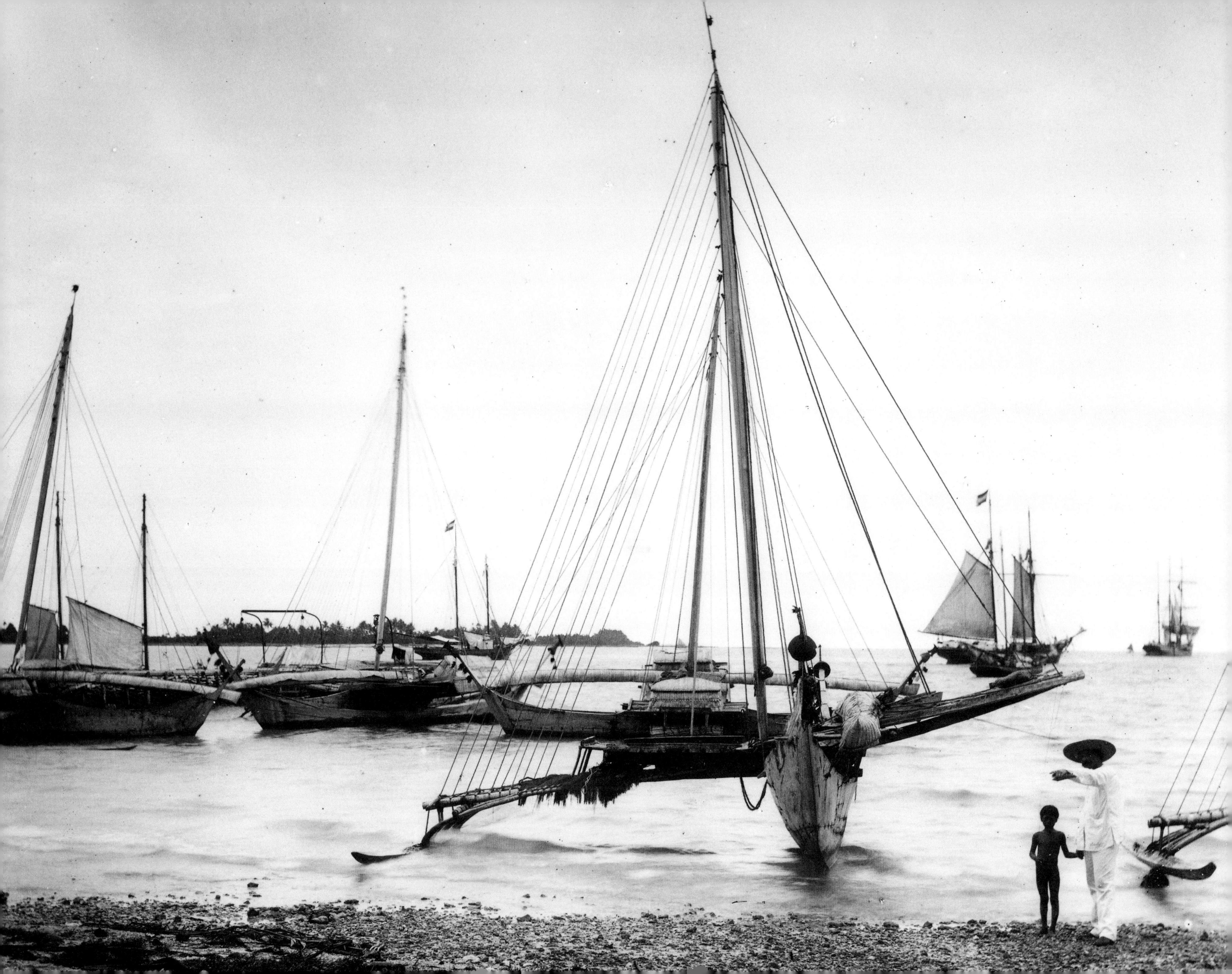

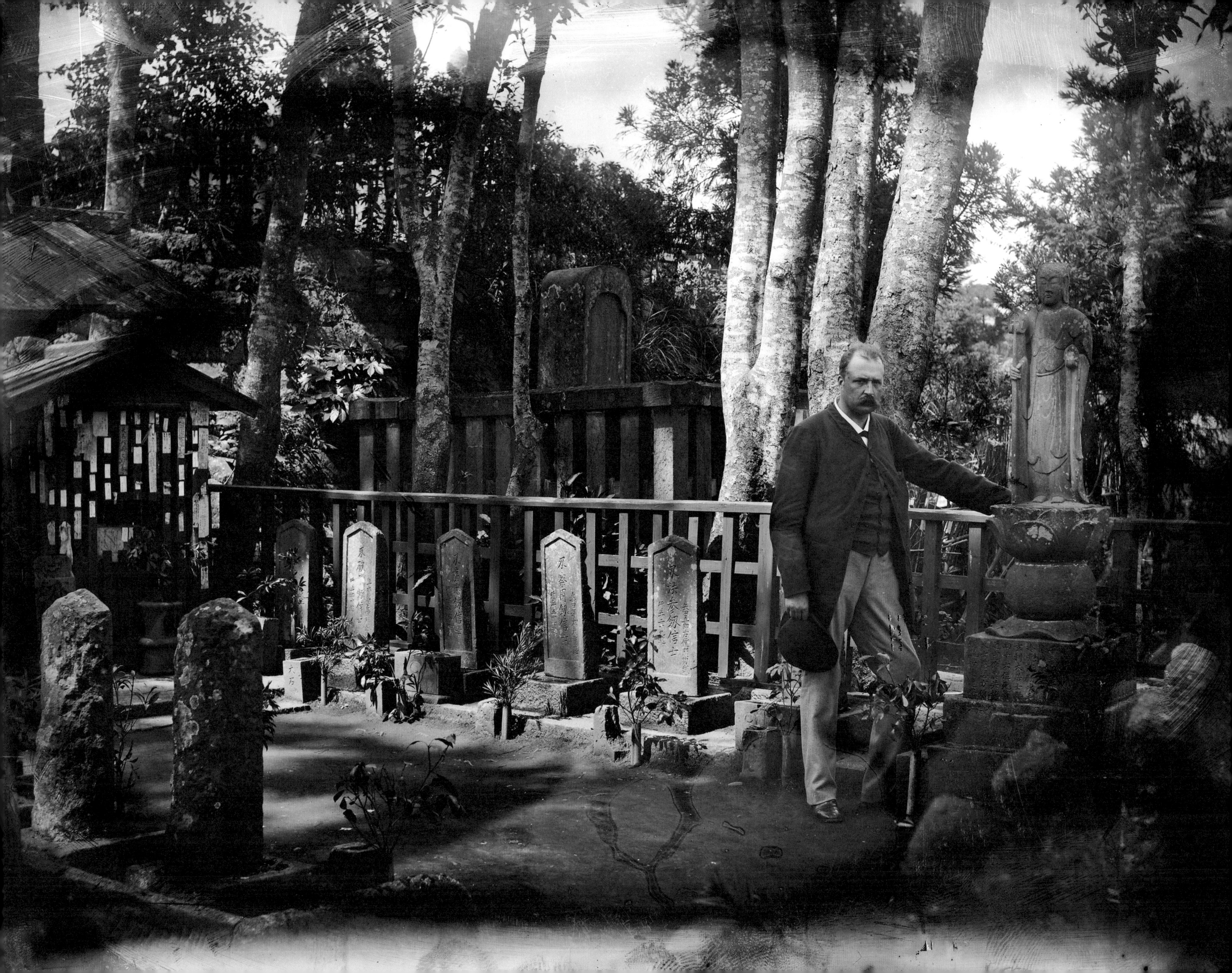

JAPAN

After a little over three weeks at sea the frigate reached Yokohama on 26 August, latterly in a whole gale and with her sails torn to pieces. An inspection was now carried out, the bottom of the vessel was cleaned and the hull was painted.

This meant that, for the first time on the voyage, Stolpe and Ekholm had a reasonable opportunity to make systematic collections and devote ample time to photography. The stop-off in Japan lasted over six weeks, the first two of which were spent in the port city of Yokohama. Despite an exhaustive search they could find nothing to buy at reasonable prices, and they appear to have spent most of their time socialising, taking dinner at the German Club and generally establishing contacts, including with the Dutch diplomatic representation, which dealt with Swedish/Norwegian trade interests in Japan. Not until the turn of the century, in 1900, was a Swedish/Norwegian general consulate to be established in Kobe.

Prince Oscar's presence meant that the Vanadis's officers and civilians were prioritised by the Japanese emperor, known as the Mikado, who placed one of his summer palaces at the prince's disposal. The prince recorded in his diary that there was a huge display of obeisance:

> Inspection for me at 9 o'clock in the morning on the big parade ground. There were said to be 6 to 7 thousand troops. I don't think there were more than 4 to 5. I rode on the left-hand side of the Mikado in full naval-officer's uniform. There followed a huge horde of princes and generals. When we had passed the line at a walking pace (the battalions were standing in columns) the troops filed past us at the front of the company. It of course all went fairly badly, except for the artillery. By 10am I was back home, and then received the Order of the Chrysanthemum from the high master of ceremonies.

Stolpe, dressed in tails, formed part of the Swedish delegation: 'At 9 o'clock I travel up to Tokyo to take part in the audience with the Mikado. It's the first time I've been involved, and I will of course use the occasion for a chapter of my book. We turn out to a man, and the damned priest is of course there too. My tails, which had been left in Honolulu, were here on our arrival for 1 1/2 dollars'. Thanks to the maligned ship's priest Kiellman-Göranson there is a description of the great day, which started with troops on parade and a band playing Swedish folksongs:

> At the back of the big room stood the Mikado, dressed in a richly embroidered uniform of a European cut, over which hung the light-blue sash of the Order of the Seraphim. The Mikado is now 32 years of age, and of a slim build. His hair and his thin beard are black, and as a type he is completely Japanese. From the face one initially discerns severity, indeed hardness, but one soon perceives a melancholy and quixotic expression, whilst the protruding jaw indicates displeasure, maybe world-weariness.
>
> At the Mikado's left-hand side and slightly in front of him stood Empress Haruko. She was attired in a completely Japanese manner in magnificent silk brocade, and her hair, which was arranged over high hair pads, was adorned with jewellery made of gold and precious stones. As per Japanese custom, the whole of her face was painted – one could almost say enamelled. One thus cannot talk about any expression in it, though were her face not distorted in this way she would probably have an intelligent appearance, as she is said to have a good head on her shoulders ...
>
> Also present at the audience were an imperial prince and various high-ranking dignitaries, as well as the Swedish minister van der Pot and the legation secretary van der Polder. The audience only took a short while, and nothing was uttered during it. We merely had to show our reverence with a bow.

Arvid Lindman found the brief audience – lasting all of half a minute – disappointing, and as regards the emperor he noted: 'He was very ugly'. Neither was Prince Oscar impressed: 'Both of their majesties were pretty awkward and shy, and it was thus rather uncomfortable for me'.

Oscar Ekholm visited teahouses and travelled on 'carrets' (rickshaws). 'One Japanese man for each carriage, and as long as he has a passenger he never walks, but perpetually runs, only allowing himself 10 minutes off every two hours,' Oscar told his brother Fredrik, continuing:

> We're still in Tokyo, but I'm now longing to be back in Yokohama again, because I must say it's not great to have been here for 3 weeks now and not received a single line from home, and for as long as we're here in Tokyo I'm always unsure about whether there's a letter lying waiting for me in Yokohama. But I very well know there can't be a letter for me before the 25th, as I wrote so late from Honolulu.
>
> The few days I've spent here have been eventful. On one of the days a powerful Cyclone broke out over the city, destroying trees, plants, planks and fences, and

'The graves of the 47 Ronin near Tokyo', Hjalmar wrote on the back of the picture. And to his wife: 'In 1701 Asano Tukumino Kami, a young nobleman, was so harassed by an old master of ceremonies at the Shogun's court that he once drew a knife on him. Condemned to harakiri for breach of the peace at the palace. His death was avenged by his 47 servants through the murder of the master of ceremonies 2 years later, after which all 47 of them committed harakiri.'

turning several buildings upside down. I first saw one building being lifted right up then thrown at an angle, and then collapsing, though not in a heap, as the storm carried roofing tiles, planks, window frames etc. around. And finally what the storm had not destroyed was crushed and damaged by flying projectiles – I in particular saw a wooden trestle moving at such speed towards a gaslight that it was torn up, and both continued along the street. In the evening the weather was absolutely magnificent!

On one of the days there was a big celebration in honour of the god Shiva, and all day long several enormous idols were pulled up and down the street on big floats, accompanied by the wildest music.

Oscar concluded the letter thus, sending greetings to his mother, his brother John and his uncle (the photographer Carl Fredrik Lindberg).

In Tokyo, Hjalmar succeeded in tracking down his teacher from the Guimet Museum in Lyons, Y Ymaïzoumi, who in France three years previously had taught him about Japanese icons, and who was now working at the Japanese Ministry of Education. He was hired as an interpreter and assistant, and together they raided shops and temples, studied a street peopled by prostitutes and experienced an 'unparalleled' festive procession. The typhoon mentioned by Oscar spread death and misery throughout central Japan. In Yokohama harbour, vessels collided with one another and broke up. The Vanadis coped thanks to its three anchors and the engines being set to Full Speed Ahead. Entire districts of Yokohama were razed to the ground or damaged, and the neighbouring city of Tokyo suffered an even worse fate.

Stolpe's collecting as usual targeted simple everyday objects that were used either by the 'common people' or by the priests representing the two predominant religions, Shinto and Buddhism. Everyday Japanese life was thus reflected, and the selection principle has been applauded by today's ethnographic experts, though it was of course also a consequence of the expedition's scant funds. Lavish objects – if they had not already been snapped up – were too expensive. Some opulent items, however, were irresistible, e.g. an enormous processional statue, bought for a high price and conveyed to Stockholm with great commotion. The background to this exclusive purchase was an experience in the Tokyo district of Kanda. In a letter to Emmy, Hjalmar described the festivity he had witnessed as being 'without peer'. He continued:

And you will get to see it, as 14 photographs were taken. An endless procession with colossal mythological figures pulled along on floats. Tremendous splendour. I don't yet fully understand the significance of the festivity, unfortunately. But will describe it in detail with the aid of the photographs, and for no less than 1,000 francs(!) I've bought an old, very well-modelled [statue]. That is to say, just the head and arms are carved in wood, and the body is made of bamboo and is attired in cloth clothes. He's standing on a high stand [drawing], at the bottom an excellently sculpted balcony for drummers, unlacquered brown wood. After this a red curtain with Chinese lettering on it, then a black lacquered fence, and at the top stands Kintoki, with a wild appearance and magnificent attire, partly made of silk, throwing peas at the demons to banish them. The whole thing is an excellent museum piece, though at 5 1/2 metres it's colossally high. There will, however, be room for it in the Prince's palace, and it will draw people to the exhibition.

It was thus the forthcoming Vanadis exhibition at the hereditary prince's palace in Stockholm that had now begun to assume concrete shape in Hjalmar's head. Prince Oscar had moreover already written home to his father about permission to use the premises. It was there that the mighty Kintoki was to serve as a people magnet – a hero from Japanese mythology who had been elevated to a god. This festival took place in Tokyo every year in mid-September, with Kintoki and other statues of gods being presented in a huge procession. And Hjalmar wrote in the magazine *Ny Illustrerad Tidning*:

Floats bearing gods were drawn by bulls, and they were all surrounded and preceded by men in festive garb. One sometimes saw small children at the front of the horde, walking in front of a float, all pulling on ropes attached to it. If the children were too small to walk amongst the crowd the men carried them on their shoulders. A few floats were preceded just by young girls dressed in splendid warrior's armour. On each float sat a troupe of musicians, whose shrill contributions merged, like sounds from a host of barrel organs at markets back home.

But the crowd itself behaved in an exemplary manner. No jostling, despite its being such a huge gathering of people, no shouting, no drunks – just happy people and friendly faces everywhere. The only visible upholders of order were the tram employees, who ensured that the trams crossing the procession now and then could get through. The entire festivities lasted from about 6 am until way into the afternoon. Throughout this time our photographer had his time cut out doing portrait photos of one group after another ...

It was by sheer chance that an antiques dealer with whom I was doing business happened to have in his possession an old statue that he had seized many years previously on account of an unpaid debt. It was an opportunity that could not be resisted, despite the high price, and our Kintoki is without a doubt the only statue of its kind in Europe.

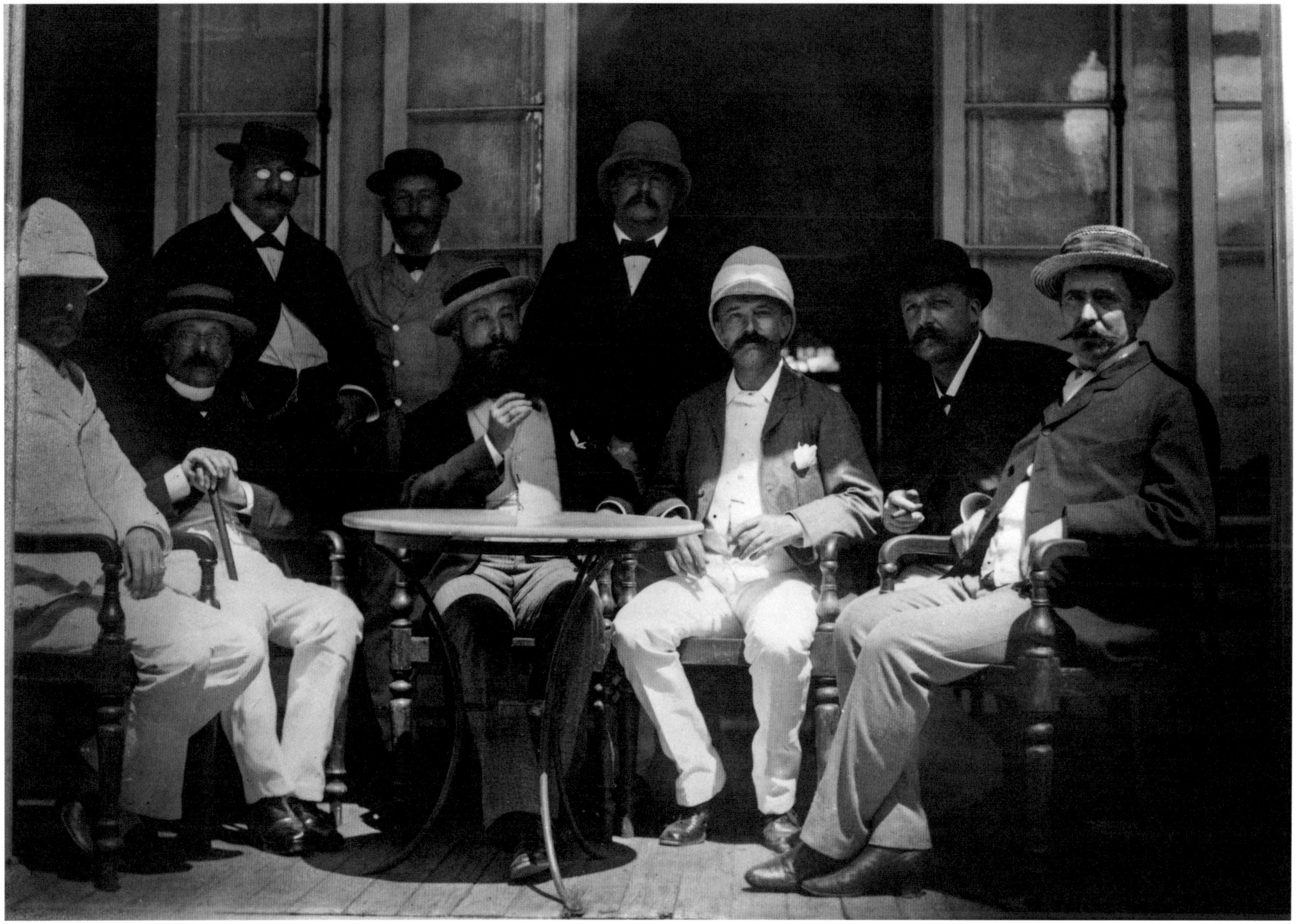

In Yokohama, Ekholm and Stolpe stayed for several weeks, and the fixed point was the German Club, where Hjalmar is to be seen in a bowler hat and dark jacket. Alongside him on the far right sits Kommerzienrat (a title conferred on rich businessmen) Langfeldt, who was a great help to Stolpe in Japan: 'A decent, cheerful fellow. His wife was fat and sentimental. I once sang to the woman.' His new-found friend did business all over the world through the chandler Langfeldt & Mayers. The contact with Stolpe continued for several years.

It was a bumper photographic yield. Around 200 photos from Japan are to be found in Stolpe's private archives, a good quarter of which were taken by Ekholm, whilst the remainder comprised purchases. Among the documentary pictures that aroused his interest is a series showing the so-called Ainu people, who led a languishing life on the island of Hokkaido in northern Japan. They were indigenous people who lived from hunting and fishing, and who deemed bears to be holy animals with supernatural powers. In Tokyo's photography shops Hjalmar also found pictures documenting various professional categories: postman, hairdresser, wrestler, and something as unusual as 'shingle cleaver'. Throughout his professional life Stolpe bought photographs, and his collection ultimately numbered a thousand or so images, often showing cultural environments, monumental buildings and portraits of 'ethnic types'.

During those ten days in Tokyo the trio of Stolpe/Ekholm/Ymaïzoumi explored the colossal city, and the Shinto god Kintoki gained company in the collections:

> I also bought another expensive item, namely a load of complete costumes for Shinto priests: 4 priests, 1 priestess and 1 musician. It cost about 950 francs! Everything is silk and brocade. Here in Tokyo I've ordered the heads, arms and feet for them all. It will be magnificent! But that's not all. I'll have a room for the 'tea ceremony' with a professor of the tea ceremony, 2 gentlemen and a lady. Have got excellent heads. The costumes are simple, but chosen with the utmost care.

The next stage of the Japan itinerary involved a steamer trip from Yokohama along the coast to the city of Yokkaichi. From there, on 22 September, Stolpe, Ekholm and Ymaïzoumi set out on a small ethnographic expedition, sitting in rickshaws with their suitcases and photographic equipment. The destination was Kyoto, about 80 kilometres away, and for three days the ensemble was to trundle along small rural roads, with many stops for buying and photography, and with overnight stays in Sakanoshita, Minakuchi and Kusatsu.

An encounter on one of these small roads illustrates Stolpe's ethnographic curiosity regarding the Japanese people's great interest in symbols and symbolic acts. He had noted that a picture of a snail in various contexts had been marked 'the owner's lassitude in old age'. He found emphatic evidence for this:

> On the way I met a severely hunched old woman, laboriously dragging herself along, supporting herself on a staff. She was of course carrying the inevitable parasol, and close to the edge of the opened umbrella, where the owner's name is otherwise always to be found, the old woman had had a crawling snail painted as an identifying mark. The old lady's statement to the interpreter confirmed that the significance of the symbol was the one described here. The above is one of a thousand examples that could be cited regarding the benefits of studying every Japanese object in detail. There is always a small reference concealed in even the simplest of adornments.

There is otherwise little information about this 'overland journey', as Stolpe called it, but thanks to Oscar Ekholm's photographs we can broadly follow it. In the coastal city of Yokkaichi they visited a small porcelain factory, which was photographed. And one morning in Sakanoshita, Ekholm's camera immortalised Stolpe and Ymaïzoumi outside an inn, clad in kimonos and sandals. Ahead of their departure, Hjalmar was to be seen in white clothes and a bowler hat standing beside the diminutive interpreter, who was only half his height.

Via the city of Kusatsu the rickshaw procession trundled on to Otsu, whence they took the train to Kyoto. Here Ekholm photographed the Japanese tea ceremony – the custom doubtless closest to Hjalmar's collector's heart. He acquired a total of 150 objects from this ceremony, which was planned to be a main feature of the forthcoming exhibition in Stockholm. The tea ceremony was one of the few pursuits Stolpe wrote about in greater detail after he had returned home. He stated that this Buddhist practice epitomised the essential nature of the Japanese people of old:

> No other manifestation reflects the tendency towards the secretive ceremonies, overwhelming detail and stubborn conservatism that were the hallmarks of the old Japanese way of life better than the tea ceremony. It is a science in itself that used to be studied in detail – and indeed still is studied at public schools. As recently as autumn 1884 I sat for a whole morning in a tea-ceremony pavilion belonging to the public girls' school Nyokoba in Kyoto, sharing in the tuition given to a dozen young ladies, a few of whom were apparently even married.
>
> My involvement was not without its difficulties, but thanks to the confident assurance by my excellent interpreter and friend YMAÏZOUMI that I would not make fun of what I saw but would be inspired by an exclusively scientific interest, I succeeded in gaining admittance, and what is more, I got to drink all the tea that is prepared at these lessons using a method totally unlike the normal profane one. The seriousness and precision of the whole thing was such as would be appropriate to the most important teaching subject in the world.
>
> And this ceremonial act has indeed been important to Japan! It has been one of the most important features of social life, has made its mark on the country's art industry and has even played a role in political developments.

What the traditionally five participants are given to consume during the ceremony is extremely simple: green tea ground to a powder and whisked in hot water plus something sweet. But what accompanies the tea-drinking is all the more complicated – endless rituals peppered with aesthetic experiences that were not originally to be interrupted by conversations including gossip and politics but that in practice involved precisely such conversations, and thereby gained political significance.

The ten days spent in Japan's old capital Kyoto were stressful, as Hjalmar related to Emmy:

> Every evening so exhausted that both Ymaïzoumi and I drop off whilst working on the notes about the objects collected that day. Up at 6 in the morning. The fact that I've got to see neither theatre nor dance grieves me. It's really beautiful here!! Leafy parks with temples, but never the time to go for a walk. Have hardly seen the temples! Just traversed narrow streets and delved in dark shops! Just imagine me from morning to evening gadding about in a jinriksha carriage, and after me Ymaïzoumi, usually half asleep. The poor thing couldn't put up with the gesticulating the way I did. But we did also live somewhat differently.

Nagasaki was the last Japanese port that the Vanadis visited. From the mid-17th century Japan had been a closed country, the only place open to contact and trade with the West having been Nagasaki, or rather the island of Deshima (Dejima). Those in power had wanted to isolate Japan from the contagion of Christianity, but allowed Holland to carry out trade on this small, fenced-off island. Life there has been described by the Linnaeus disciple Carl Peter Thunberg, who resided there in the 1770s and had to pretend to be Dutch so as to be able to remain.

When the Vanadis anchored off the island a hundred years later, the winds of change were blowing across Japan. The dominion of the *shoguns* had come to an end, the island's isolation had been broken and the empire had been restored. Since 1867 the throne had been occupied by Mutsuhito, of the 121st generation of emperors from the same family – divine beings and descendants of Amaterasu, goddess of the sun. Hjalmar Stolpe thus encountered a society in great flux, and this situation would benefit his collecting activities. A few years earlier another Swede – Adolf Erik Nordenskiöld – had incidentally found himself in the middle of this modernisation process. In the summer of 1879, the Vega had emerged from the ice masses of the Northeast Passage, and our new global celebrity, who at the time boasted the same superstar ranking as Linnaeus, was feted during his stay in Japan. Like Stolpe, Nordenskiöld was a major collector, and he exploited the favourable situation to purchase ethnographic items.

For Stolpe the outcome of the visit to Japan was a haul of 1,300 objects (the figure 1,500 is even mentioned). The impression given was that he and his interpreter worked around the clock, and he later confessed to his boss Hans Hildebrand back in Stockholm: 'In Japan I overspent. Impossible to do otherwise! It was just *too* tempting'.

In Japan, nine months into their journey round the world, Hjalmar was overcome by a longing for Emmy and home – a longing so strong that some of his letters to her were written in cipher, to keep his excited dreams from his parents in Norrköping, the letters always being read out loud to the family group. He recalled his first rendezvous with Emmy on a hillock in Kolmården's ancient forest:

> Beloved Sparrow! It's 8 in the morning here, so for you it won't yet be past midnight or thereabouts, but may this kiss be bestowed on your brow (!), without it waking you up! Thank you my friend for twenty-one years ago today (!) at last allowing yourself to be talked into – not going and playing hide-and-seek with your Namman any more! You naughty girl! In my thoughts I've been up on the 'hill' for a while, have repeated in my mind what I felt then, and have been caressed by a lovely breath of air from our youth. You never become old if you have memories such as I have, and for all this I've you to thank. You, my dear friend, and no-one else!!
>
> I hope you've also received the photographs. I couldn't send any images other than ones in which I'm visible. The journey here was a bore ... I didn't go ashore until the evening of the day before yesterday, and then first had to visit Kommerzienrat Langfeldt, a chandler with whom we're doing business. He lives in elegant circumstances, and has a fat wife with five bejewelled rings on each finger. I then sat for a while at the German Club, after which I went to a teahouse in Kanogan, where everything was very prim and proper, whence home to the Windsor House hotel.

Liberated from his swaying cabin while in Japan, Hjalmar entered into correspondence with the newspaper *Stockholms Dagblad*, rapidly penning three travel letters about the visits to the Pacific islands Nukuhiva and Fakarava. His writing then ground to a halt.

Indeed, for the remainder of the journey round the world he did not write a single line intended for publication, meaning that the public missed out on reports from Tahiti, the Hawaiian Islands, Jaluit, Japan, India and other locations. Emmy pleaded and begged for more articles, as the readers had liked them and they had brought in income – and she received the fee. A total of nine travel letters were sent to *Dagbladet*. He had anxiously wondered what people thought of his writings, and his wife said: 'Well, they're always referred to as 'Dr Stolpe's interesting letters', and [you] can be completely relaxed about them. The only thing one might perhaps single out is that you possibly reveal your dissatisfaction with the short stays on land too much and too often'.

But the praise did not help: 'I tried to write the correspondence, but it did not work out. I've developed a block in that regard. I've too much to think about. I'm overcome by intractable fatigue if I even take hold of a pen'. When Emmy persisted he responded with a damning review of his own travel letters: 'My letters hitherto are exceptionally poorly written. And *you* haven't noticed that!? This impression is largely because of a number of misleading typological errors, but the style most certainly *is* poor. Haste and disinclination are the reasons for this. The unpleasant atmosphere on board is the same as ever'.

He was thus afflicted by the old Stolpe curse: paralysing self-criticism that reared its head as soon as he put pen to paper. This paralysis also affected his judgement, as the articles in *Stockholms Dagblad* are well written and, unlike many newspaper texts of the time, still make for good reading today. The situation was just as the author Viktor Rydberg wrote in a letter of recommendation to the publisher Albert Bonnier about Stolpe: 'He writes fluently, engagingly and clearly'.

On 11 October 1884, the Vanadis left Japan, en route for Hong Kong. The original plan was to visit China and Shanghai, but because of the war between China and France (1883–85) they instead headed for the then English crown colony, where they arrived after a week's sailing. Thanks to gifts from the Swedish captain, Benedict Carlos, in Newchwang (today's Yingkou) in northern China, Stolpe was able to send home Korean and Chinese costumes as well as 'a lot of small painted clay figures representing ethnic types, genres of drama etc., icons and altar decorations'. In one of his letters from Hong Kong there is an indication, as previously in Peru, that Hjalmar was thinking of leaving the Vanadis early: 'I cannot endure this company all the way round Europe. Oh, how I enjoy being away from them! I enjoy it so much that I don't actually do anything'.

While Hjalmar atypically enjoyed a few days of leisure, and was photographed in a sedan chair by Oscar Ekholm, the latter's namesake Prince Oscar was all the more active. To improve trading links between China and Sweden/Norway he was to pay homage to the Chinese viceroy, who resided in Canton in the nearby province of Kwangtung. On account of the war, Europeans were resented and the assignment was risky. The scope of the expedition must therefore have necessitated thorough diplomatic preparations. The prince recounted in his diary:

> Wednesday 22 October. By 2pm we were all ready, dressed in (full) uniform, to be taken in sedan chairs under military escort to the Viceroy of Canton (considered to be one of China's most intelligent and talented statesmen), whom I was to visit. We were accompanied by our consul, plus an Englishman, Mr Mansfield, as an interpreter, and everyone had their own specially coloured sedan chair, depending on their individual rank. Our loaded revolvers lay beside us. The procession was led by around 25 soldiers in a motley variety of uniforms, and soldiers likewise brought up the rear. The journey took nearly 1 hour, during which time we were tirelessly conveyed through dirty narrow streets bursting with people. A number of soldiers, probably about 1,000, were positioned on the streets we passed through, to protect us as necessary ... The only sound to be heard was the words 'Sweden/Norway', which were written on the clothes worn by our bearers. We have calculated that we encountered at least 100,000 people in the 2 hours we were out.
>
> When we arrived at the palace, or rather house, where the viceroy lived, (400) troops were lined up, and we passed between them. Some of them were armed with old breech-loading rifles, and some with more up-to-date needle-fire rifles.

Prince Oscar described the viceroy as 'a small, thin man with what for a Chinese person was probably a fairly friendly face'. He continued:

> He was wearing a long blue blouse with an embroidered square on the breast and back. On his head he had a comical straw hat with red tufts and a red ball on the top, as well as a pearl at the front ... I said to him that I had come to pay my respects, and lamented the fact that we had not been able to come to a Chinese port, so I would have been able to come into contact with Chinese authorities, but hoped they would not, however, forget us, but would be inclined to establish closer trading links.
>
> After we had conversed for a while we were invited to sit down at a table, where we were served fruits in preserves. The viceroy asked many amusing questions, including regarding my age, and was a little surprised to hear I was only 25. He asked what we ate back home, how many ports we had, and so on. The whole time, both throughout the conversation and during the meal, we kept our hats on, in accordance with their rules of etiquette.

At the end of October 1884, the Vanadis raised anchor and headed for the Philippines.

On 30 August 1884 the men of the Vanadis paid homage to Mutsuhito, Japan's 121st emperor, whose prime mother was the sun goddess Amaterasu. Stolpe almost missed the meeting, the tailcoat he had left in Honolulu arriving at the last minute. Arvid Lindman found the audience short and disappointing.

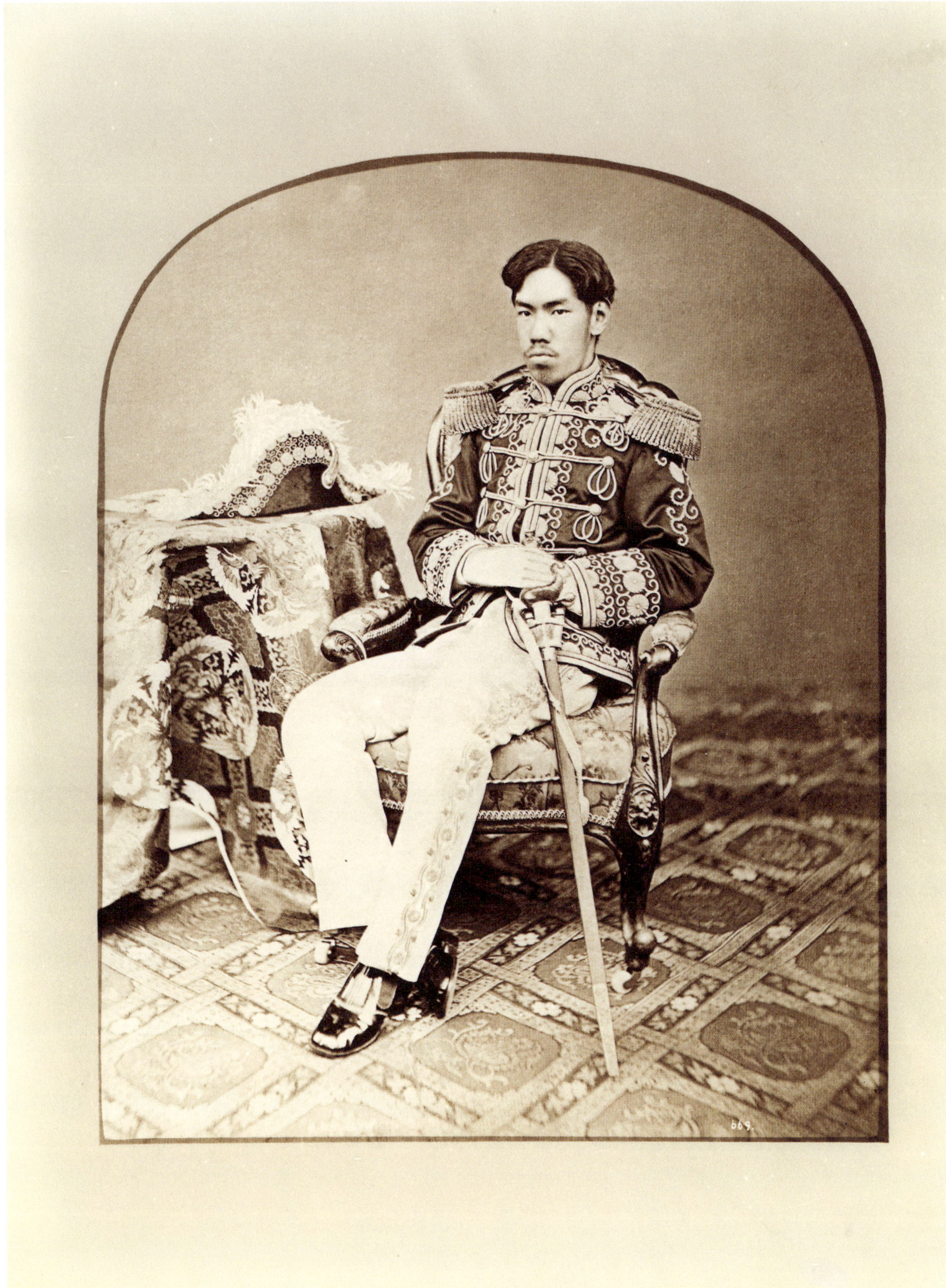

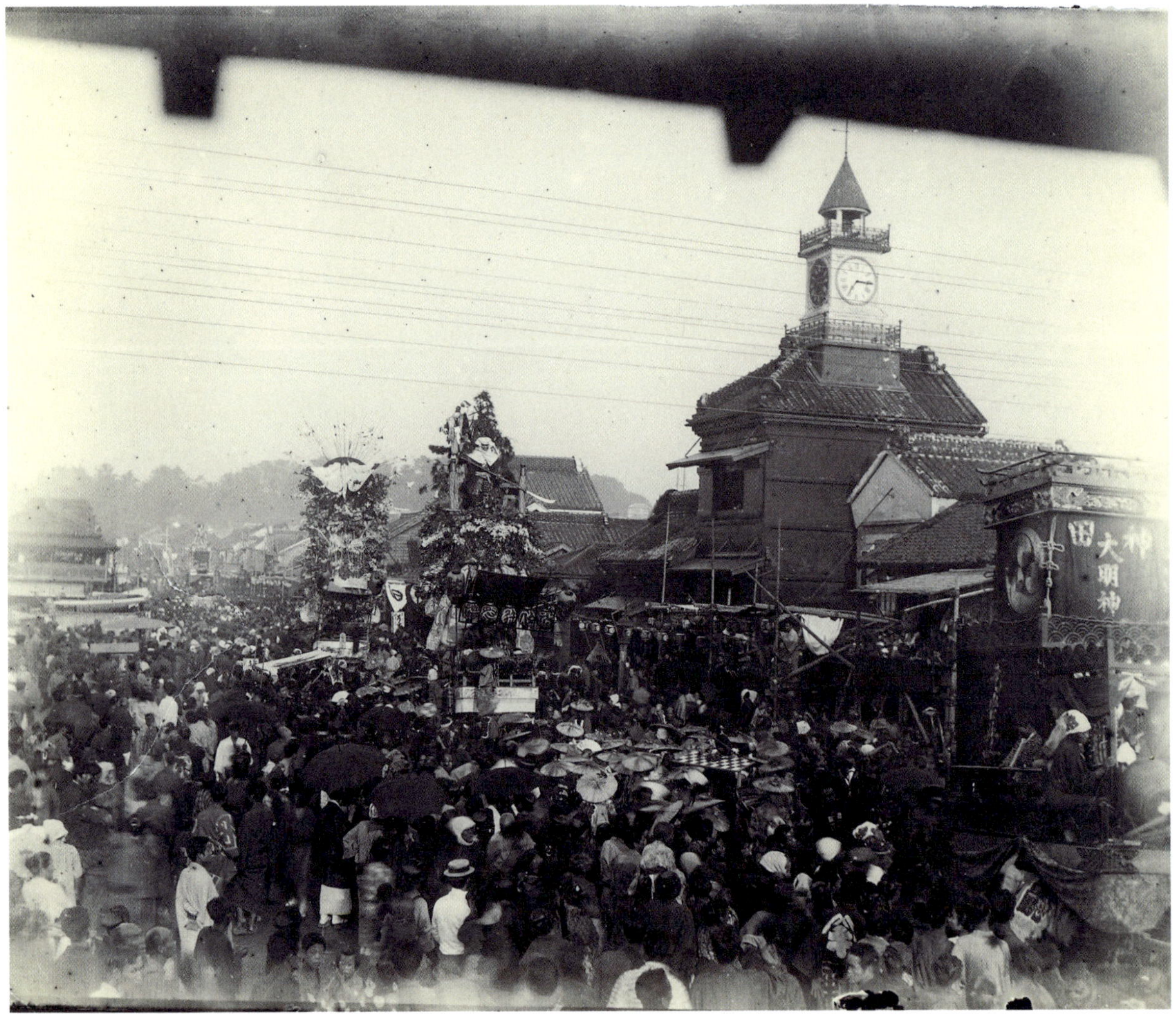
神田
大明神

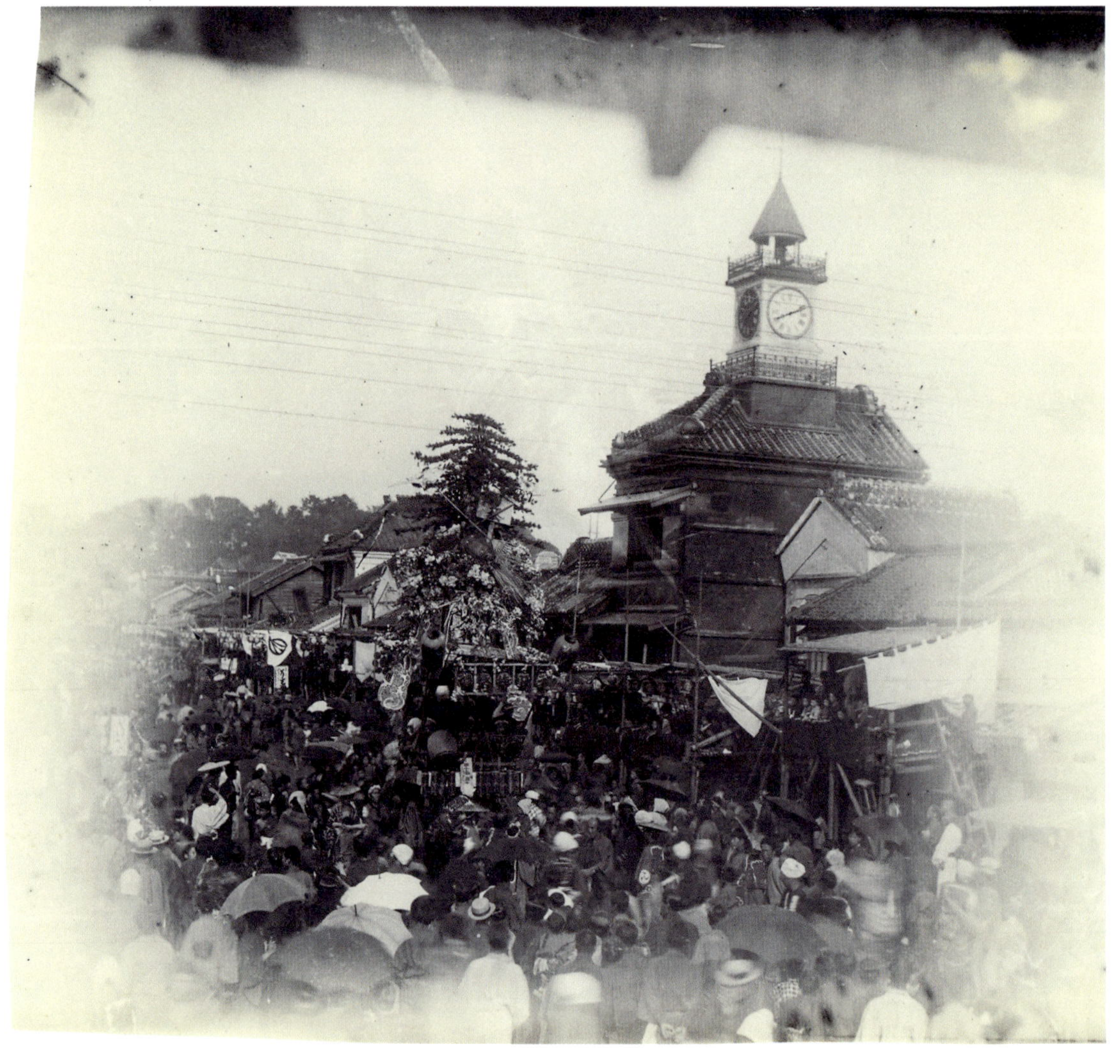

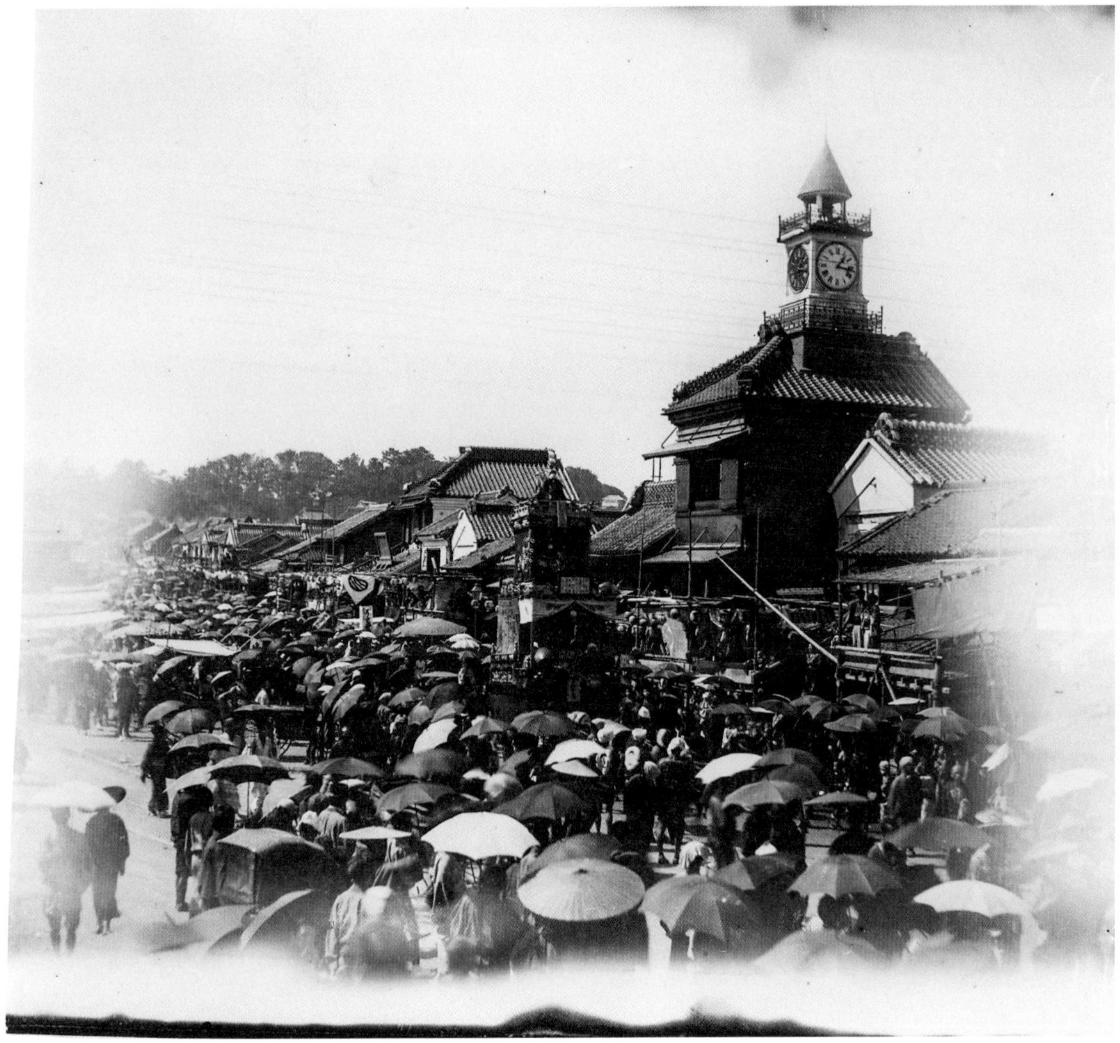

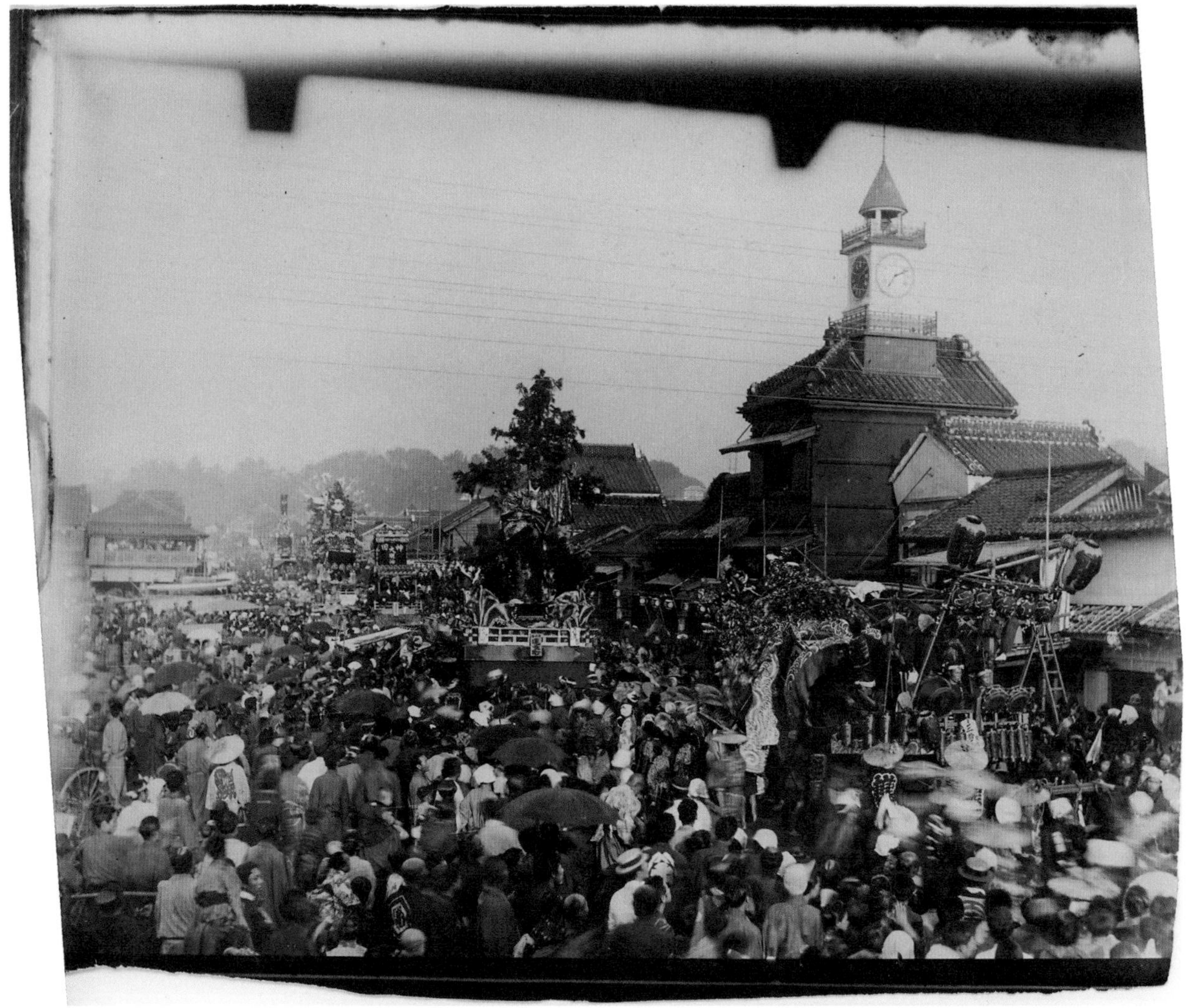

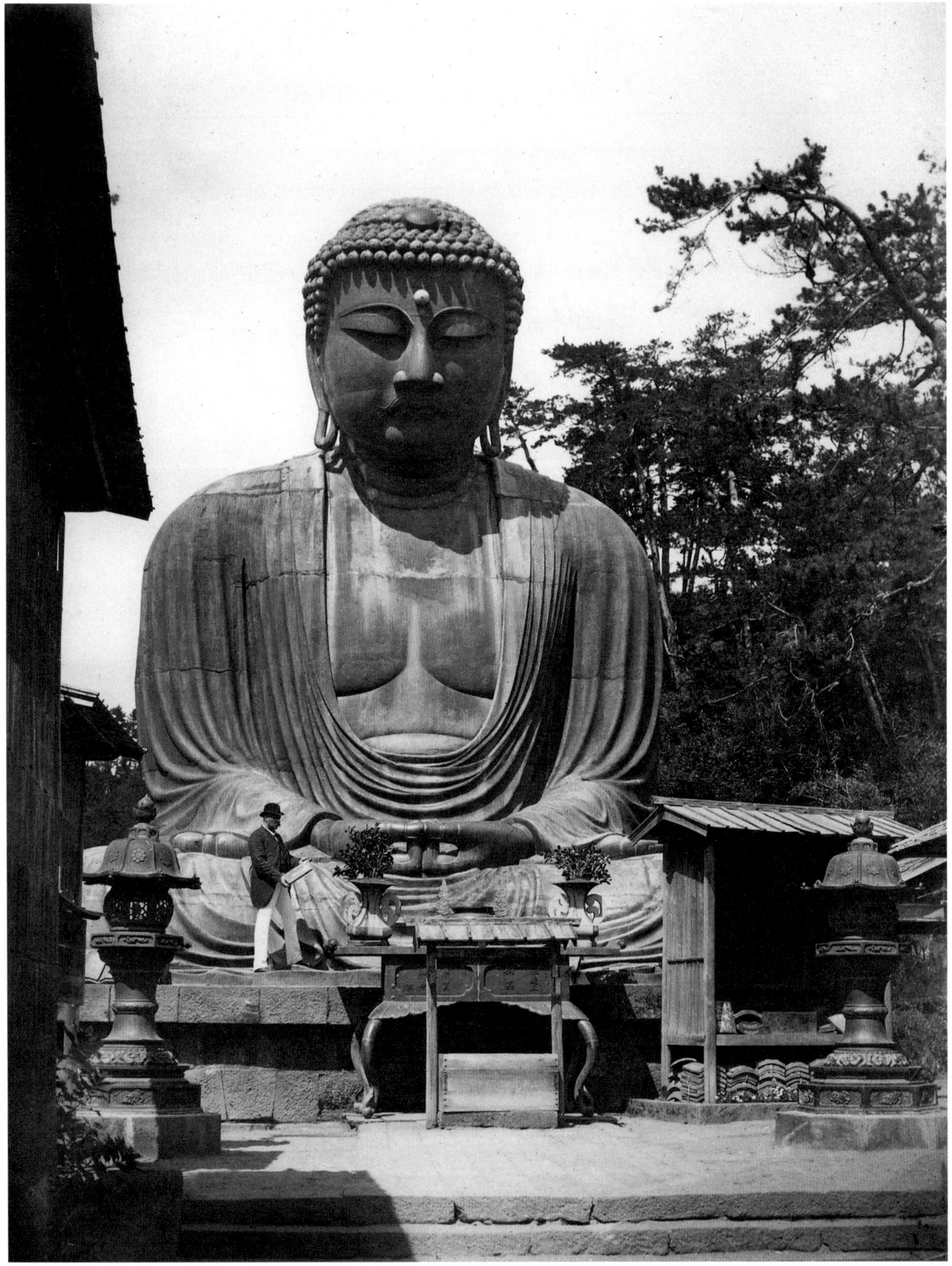

Previous pages: On 15 September each year, writes Stolpe, an endless procession passes through the city district of Kanda in Tokyo. Those being paid tribute in this popular festival are various mythological persons and gods placed on giant floats: 'The floats bearing the gods were drawn by bulls, and they were all surrounded and preceded by men in festive garb.' The event attracted a huge number of onlookers, and Ekholm and Stolpe had the good fortune to experience the procession in September 1884. Hjalmar found it 'incomparable', and appears to have really motivated his photographer Oscar, who lavished no fewer than 14 expensive glass plates on the event. Thanks to the clock face on the clock tower above the procession we know almost exactly when he took the photographs. The first plate was exposed at 7.20am, after which the hands of the tower clock show that the exposures were made at 7.50, 8.05, 8.11, 13.10, 13.15, 13.17, 14.35, 15.35 and so on. Sometimes it is indeed hard to understand how Oscar managed to reload so quickly. An excited Hjalmar was doubtless standing beside him, egging him on. With regard to the procession he told his wife that he 'will describe it in detail with the aid of the photographs'. He never did.

Left: In front of the giant bronze Buddha – a so-called Daibutsu – in the city of Kamakura, just south of Yokohama, Hjalmar is posing with an object in his hand. He sent the picture to Emmy, explaining: 'Roof tiles bearing the bestower's name were offered, and it is one of these that I'm holding in my hand. It is believed the statue was cast in 1252. Height 49.7 feet [approx. 11 metres]. Inside is a chapel with several altars.' A repository containing roof tiles is to be seen at the bottom right. Three of them were purchased, together with '1 string of prayer beads, 3 pairs of large straw goats (offerings), 4 towels from the washing cistern'.

Following pages: A purchased picture from the small island of Enoshima, not far from the city of Kamakura.

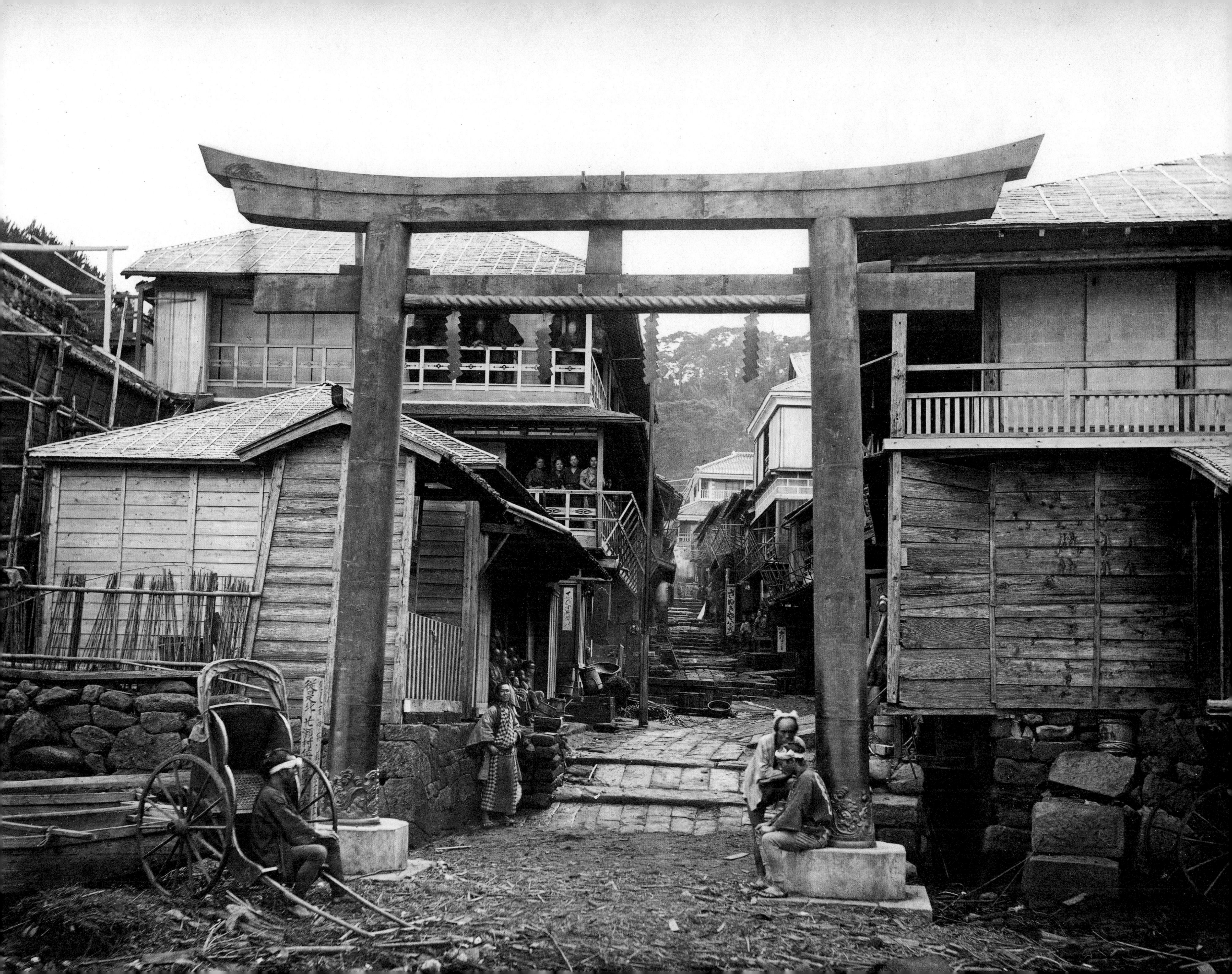

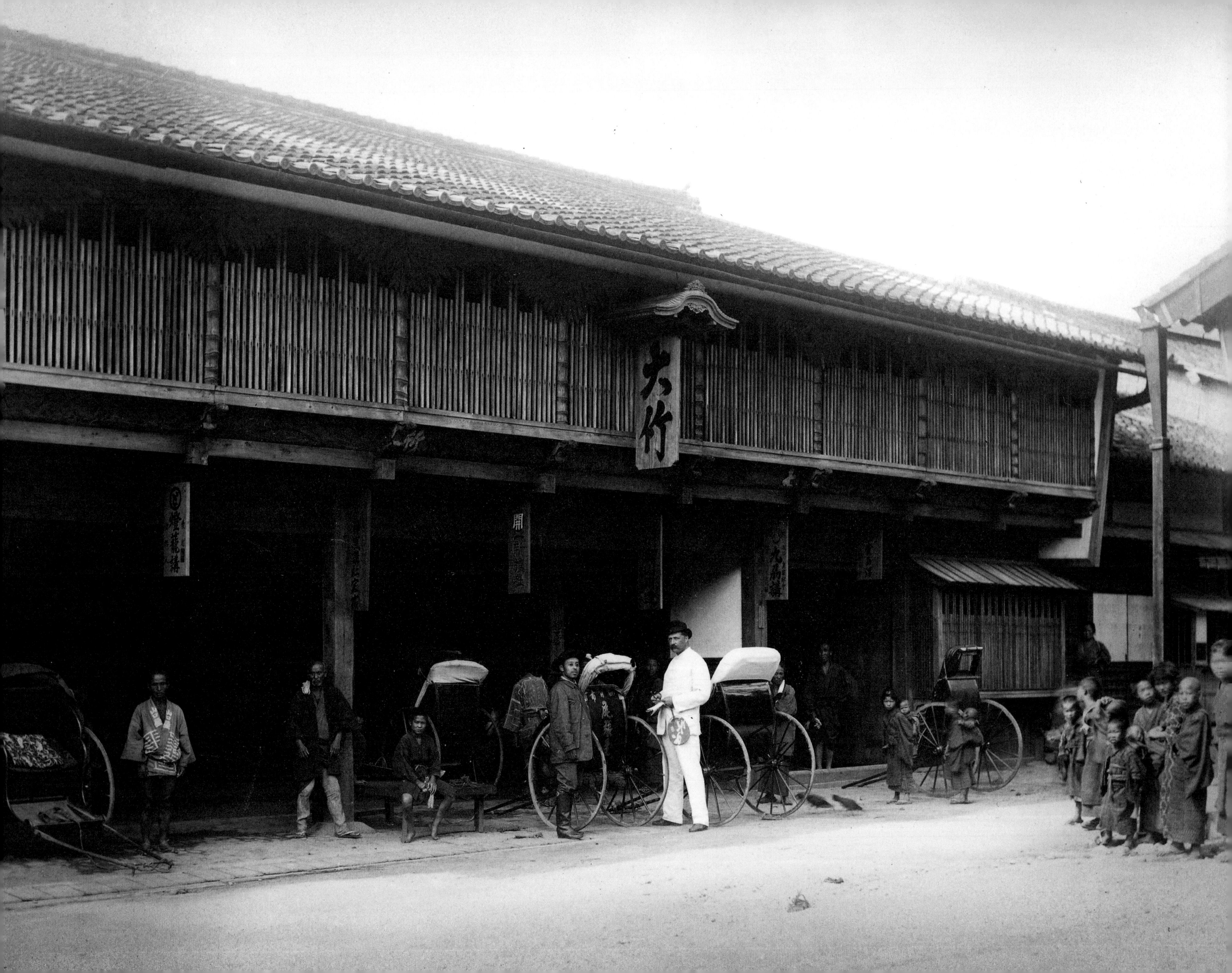
大竹

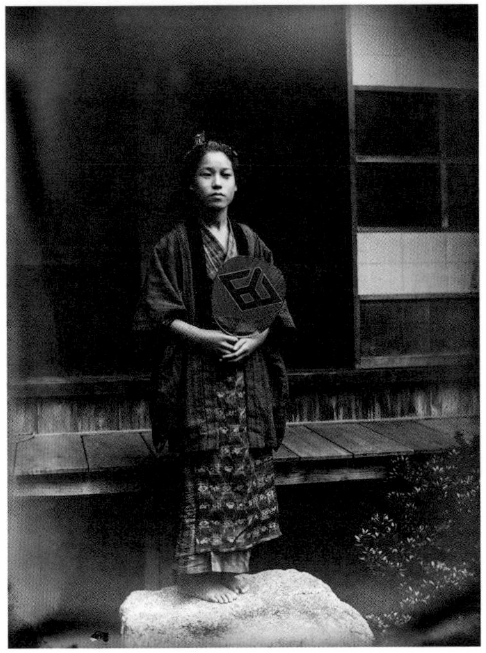

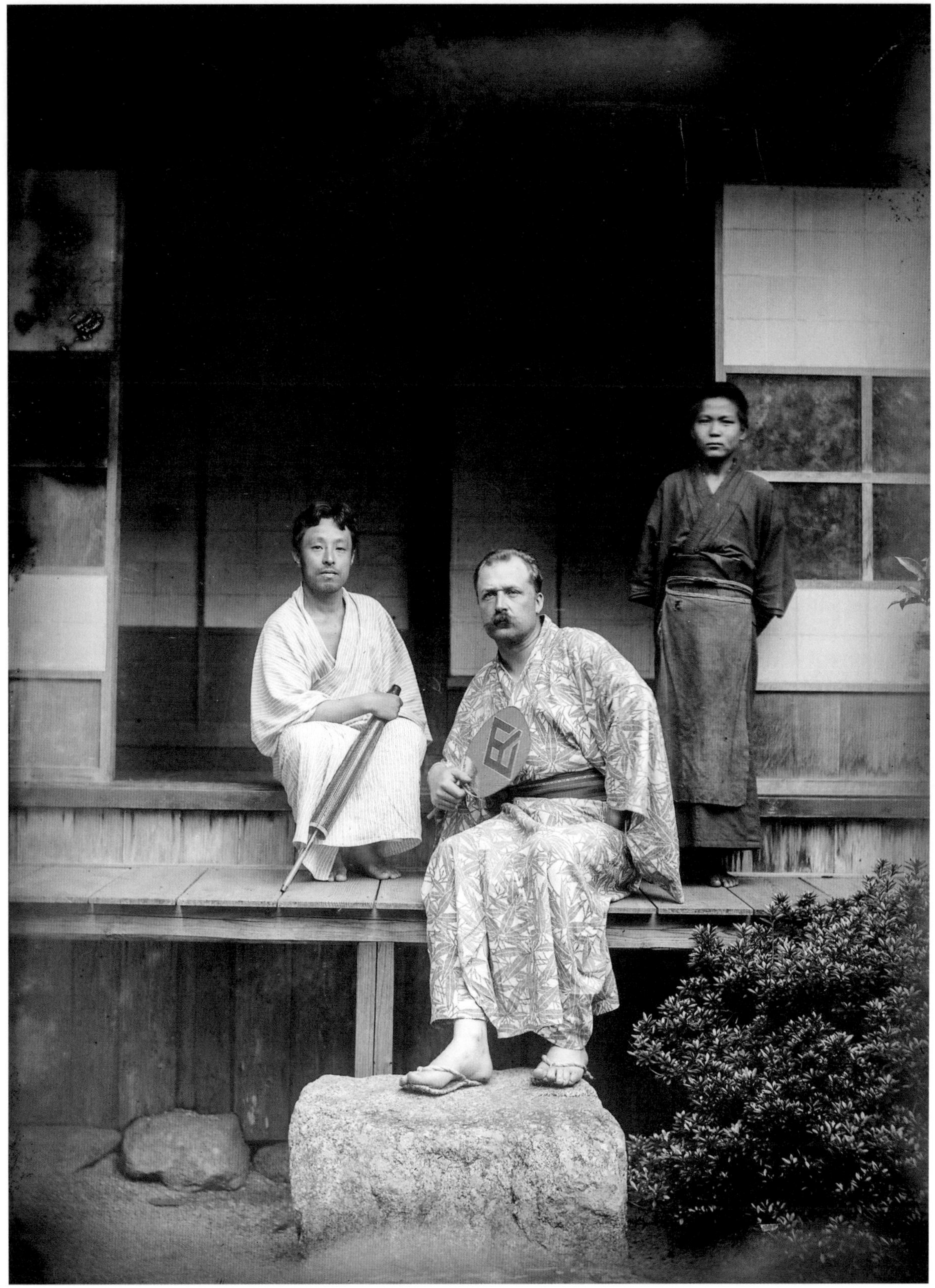

Above: One of the women who worked at the Odake ya teahouse in the village of Sakanoshita, where Oscar and Hjalmar stayed together with the Japanese interpreter Ymaïzoumi.

Opposite page: Outside the teahouse in Sakanoshita: 'You [can see] us on the street about to set off, and Ymaïzoumi is dressed for travel in big boots. On the far left my jinriksha man. Note the children on the right, each of them with a small child on their back. This is characteristic of Japan.' The small ethnographic expedition progressed in three rickshaws: Stolpe in a white suit, the interpreter in big boots and Ekholm with his heavy photographic equipment. Plus suitcases.

Right: 'Ymaïzoumi and I in the village of Sakanoshita, Odake ya teahouse, morning of 24/IX. The boy Suyekitchi, usually referred to as Suye, age 16.' Together with the Japanese interpreter and Ekholm, Stolpe now made the 'overland journey', which extended from the east coast to Nagasaki in the far south. During what was just over a month in Japan, the trio were almost incomprehensibly busy.

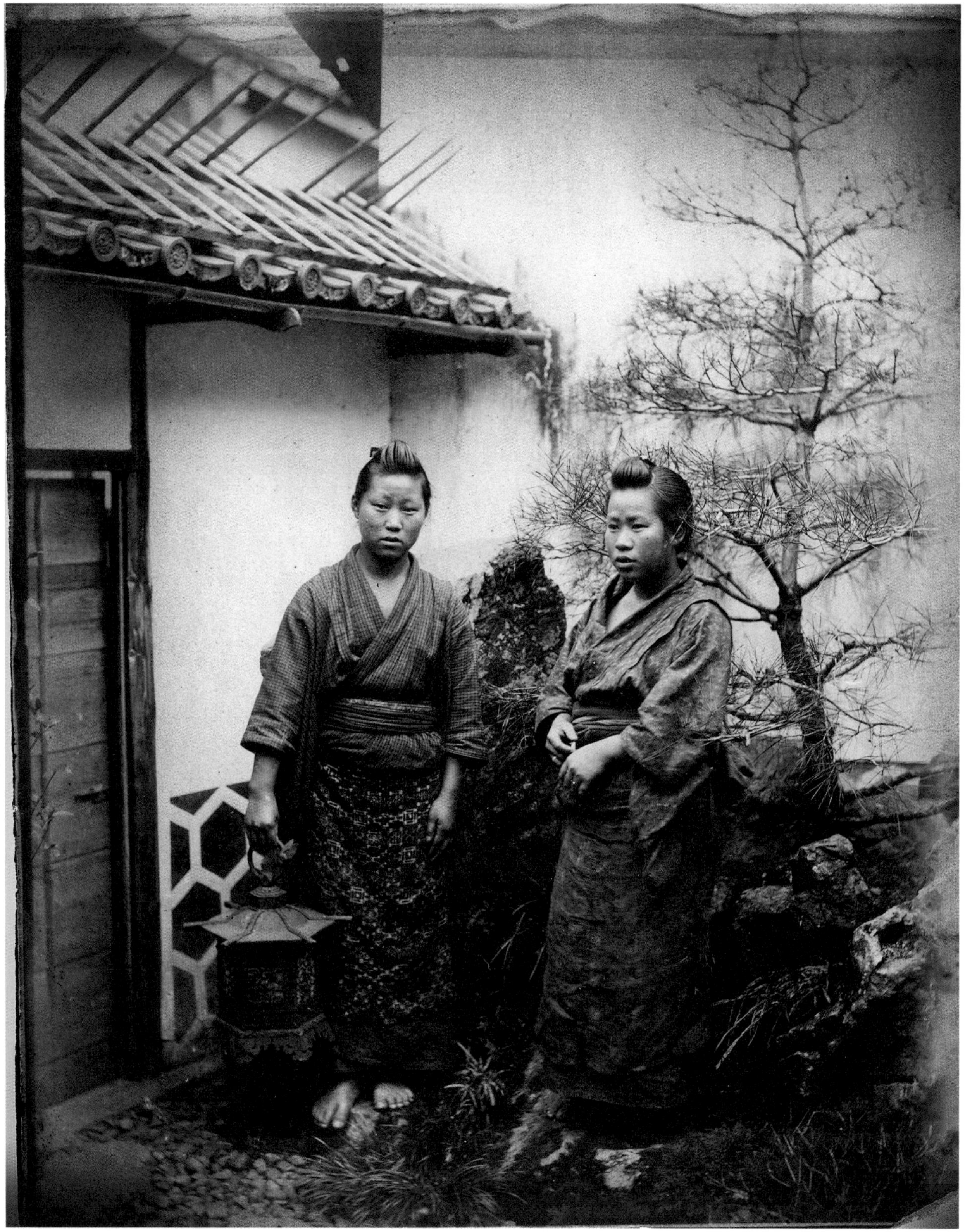

Left: A teahouse in Isibe and '2 girls in a courtyard'.

'Layman and priest, taken on the street, and splendid', Hjalmar wrote on the back of this photograph, taken in Sakanoshita. The boys are aged 12 and 15 respectively – the younger one is said to be a pupil at a temple in the locality.

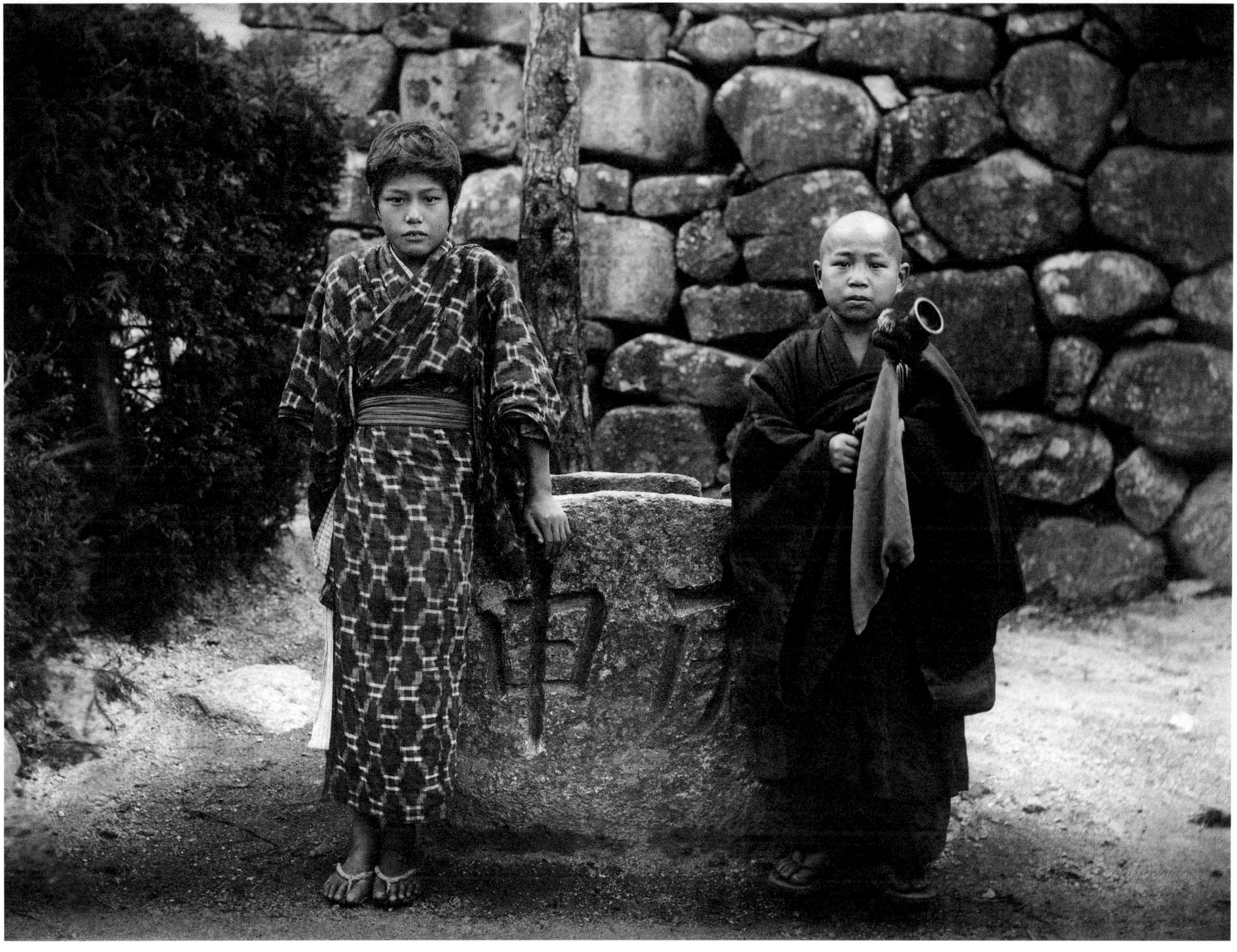

'Tamura myojin' – according to the text on the picture taken during the overland journey in September 1884. A list of photographs in Stolpe's archive states as follows: 'Wooden horse behind a grating with slips of paper by Takakura-no jinji.' On the slips one wrote down wishes, which one expected to be granted. He also describes the small and carefully painted boxes 'containing peas with which to feed the holy white horse standing saddled at many temples at the behest of the gods. It's not always the case that the horse is living – it can also be a white-painted wooden horse, tethered by its manger, but it will in any case eat peas.'

Left: 'Treasure tower at Ishiyamadera and a young Buddhist priest.'

Opposite page: 'Village on the road between Sakanoshita and Mina Kuchi.' Hjalmar, in a white suit and bowler hat, is standing at the bend in the road. The aim of the 'overland journey' – which took three days – was to find ethnographic objects and image motifs from rural Japan. Up to that point they had mostly stayed in the cities of Tokyo and Yokohama. Purchases were made in Mina Kuchi, Kamimangari, Kusatsu and Otsu.

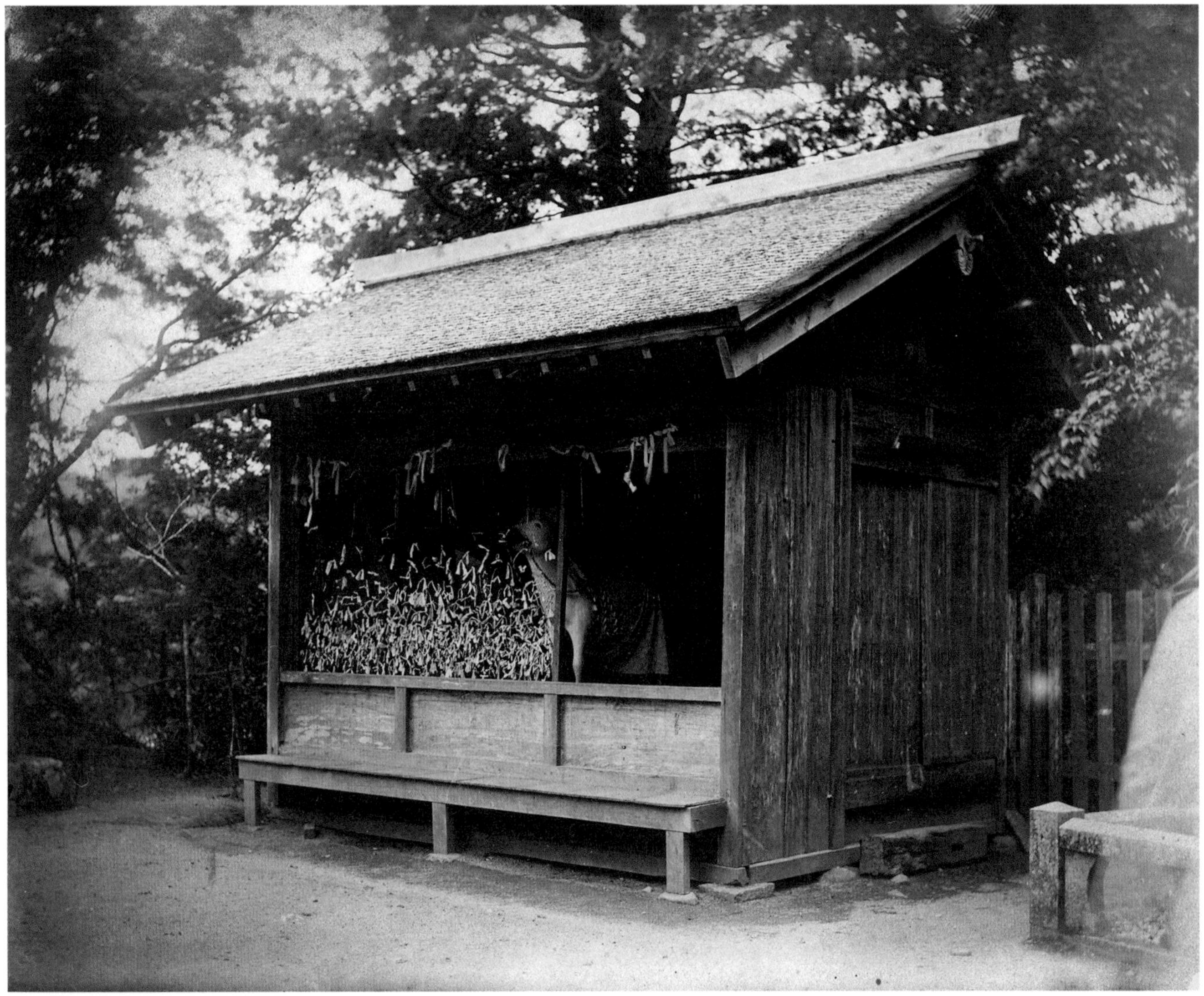

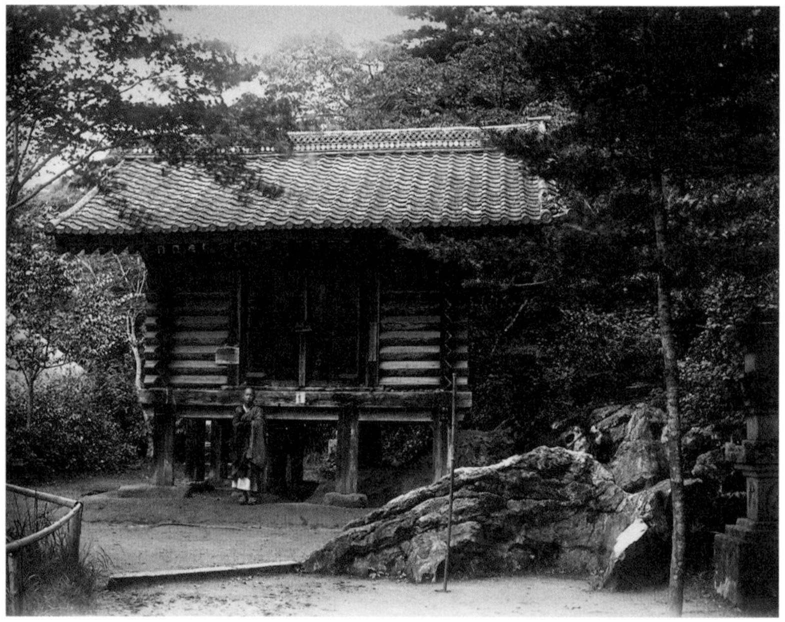

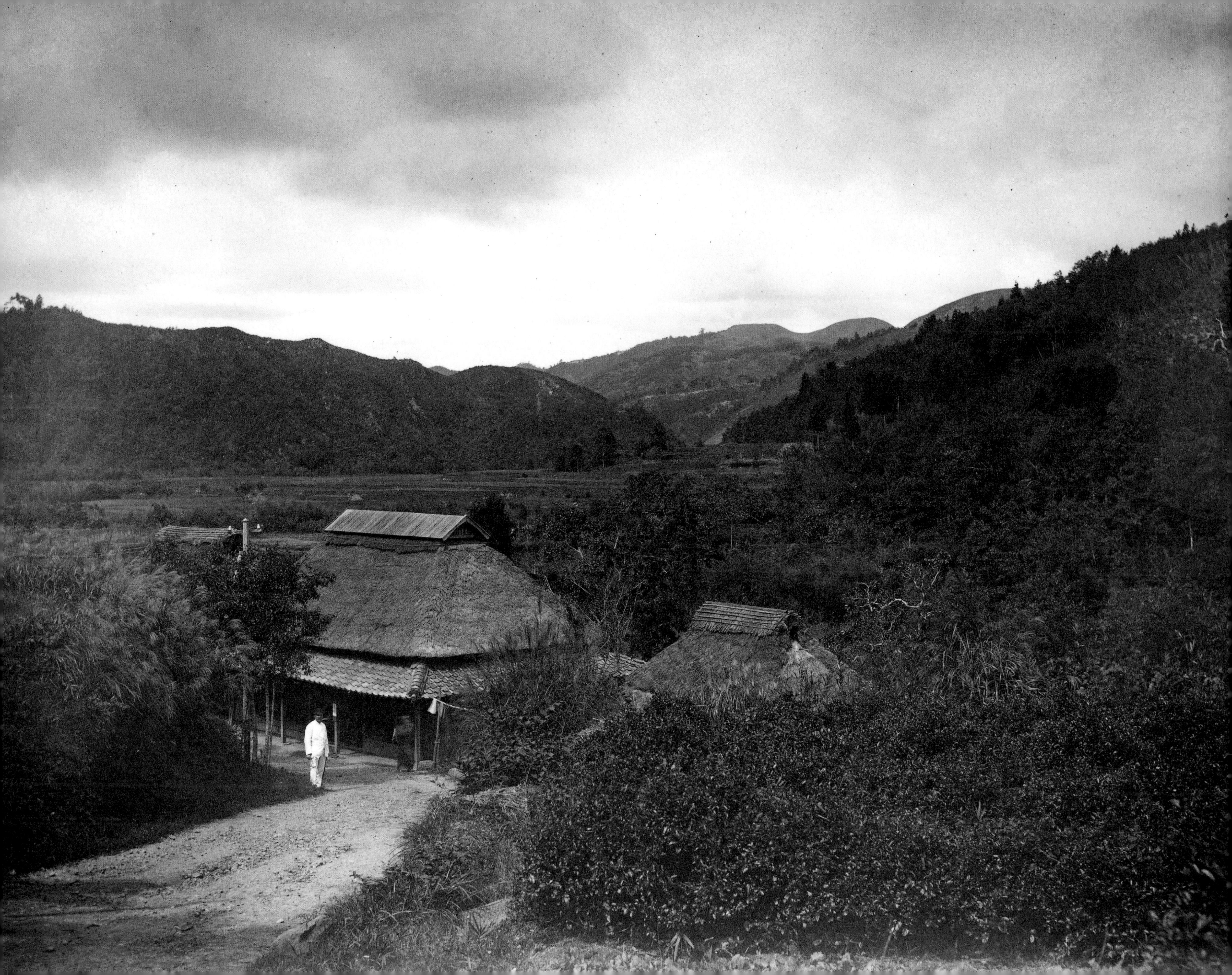

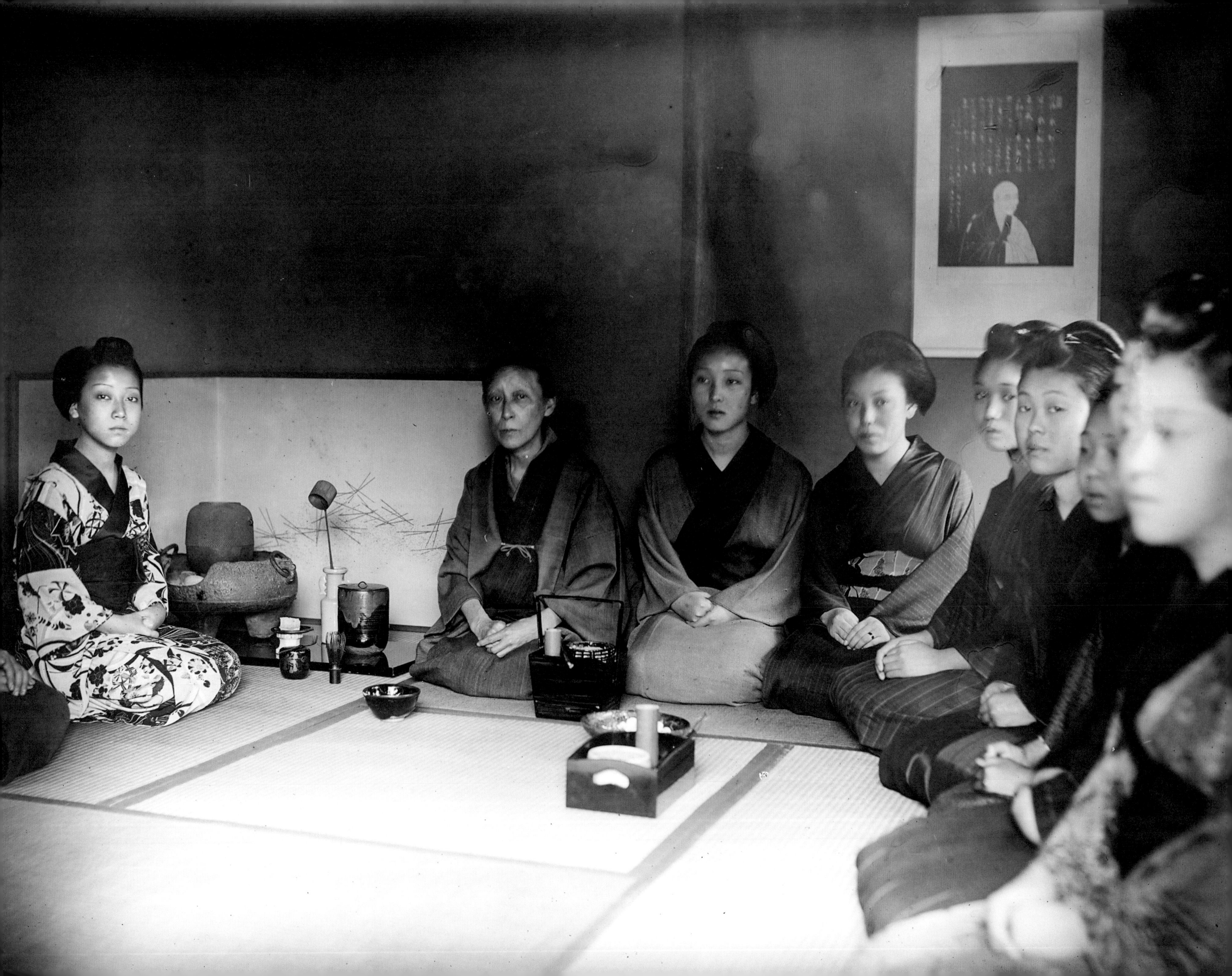

Opposite page: At a girls' school in Kyoto, Stolpe got the chance to study the Japanese tea ceremony, which has its origins in China nearly a thousand years ago. What is consumed is simple: green tea ground to a powder and whisked in hot water plus something sweet. But the ceremony and the equipment are all the more complicated, and Hjalmar was very proud of the 150 objects he collected in connection with the tea ceremony. He noted: 'There is scarcely anything in Japanese life that is better suited to showing us the mood of the people in all its strangeness and that has at the same time had a big effect on the development of the people ...'

Left: 'Tree-felling' during the overland journey somewhere between Suzuka-tage and Tamura-nava.

Below: The Ainu are an ethnic group on the island of Hokkaido in northern Japan. This indigenous people had long interested Stolpe, and he bought a series of photographs documenting the life of the Ainu people. After centuries of conflict and colonisation, the ethnic group was subjugated by the Japanese in the 19th century. What remain are just a few 'museum villages' that attract tourists. The picture shows some of their simple temple buildings. The Ainu language has now almost died out.

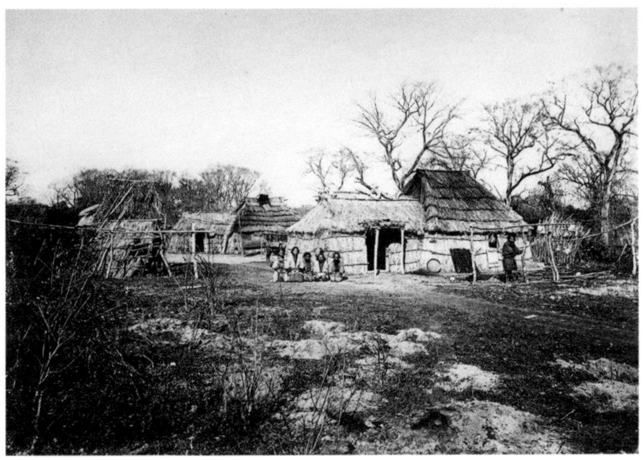

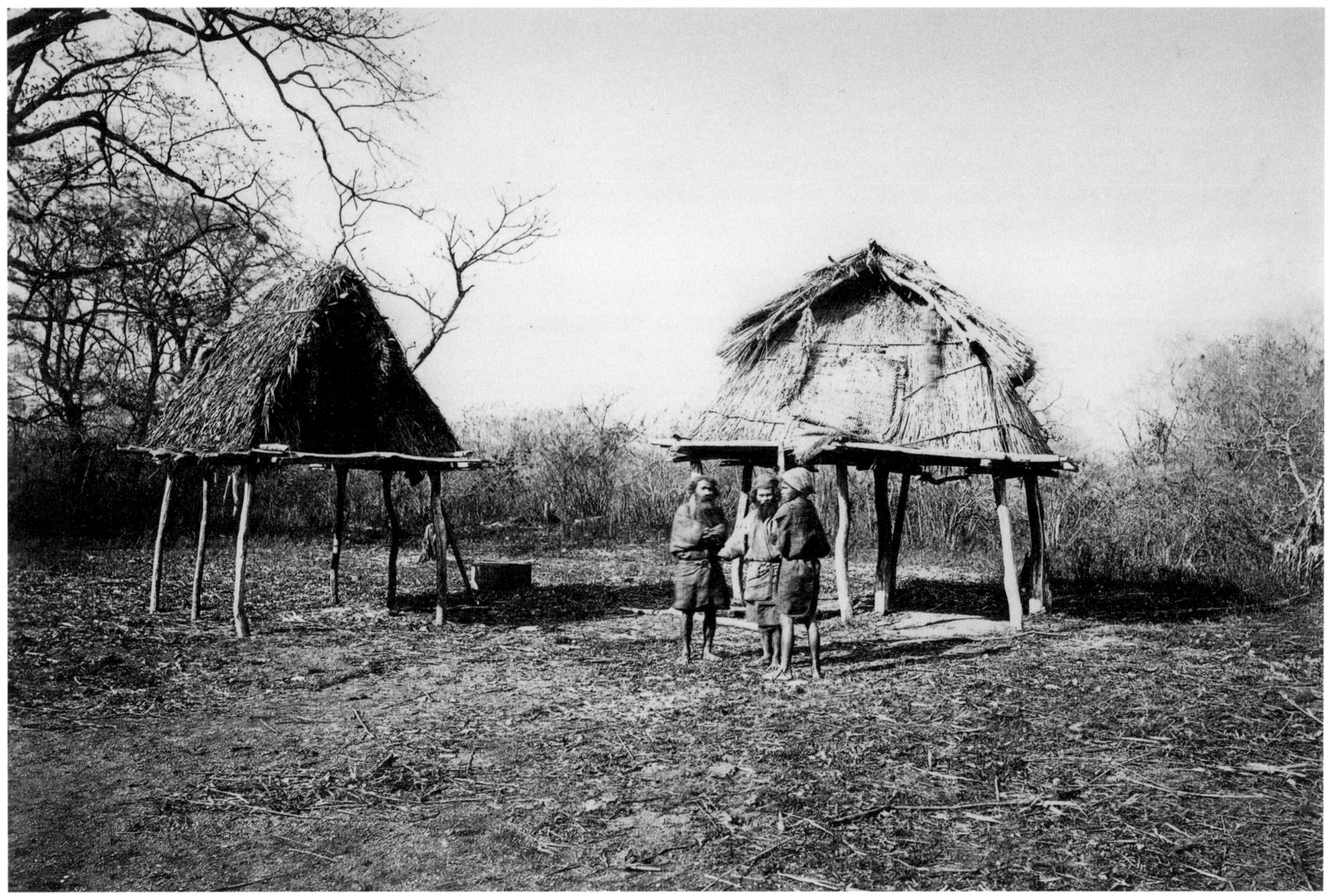

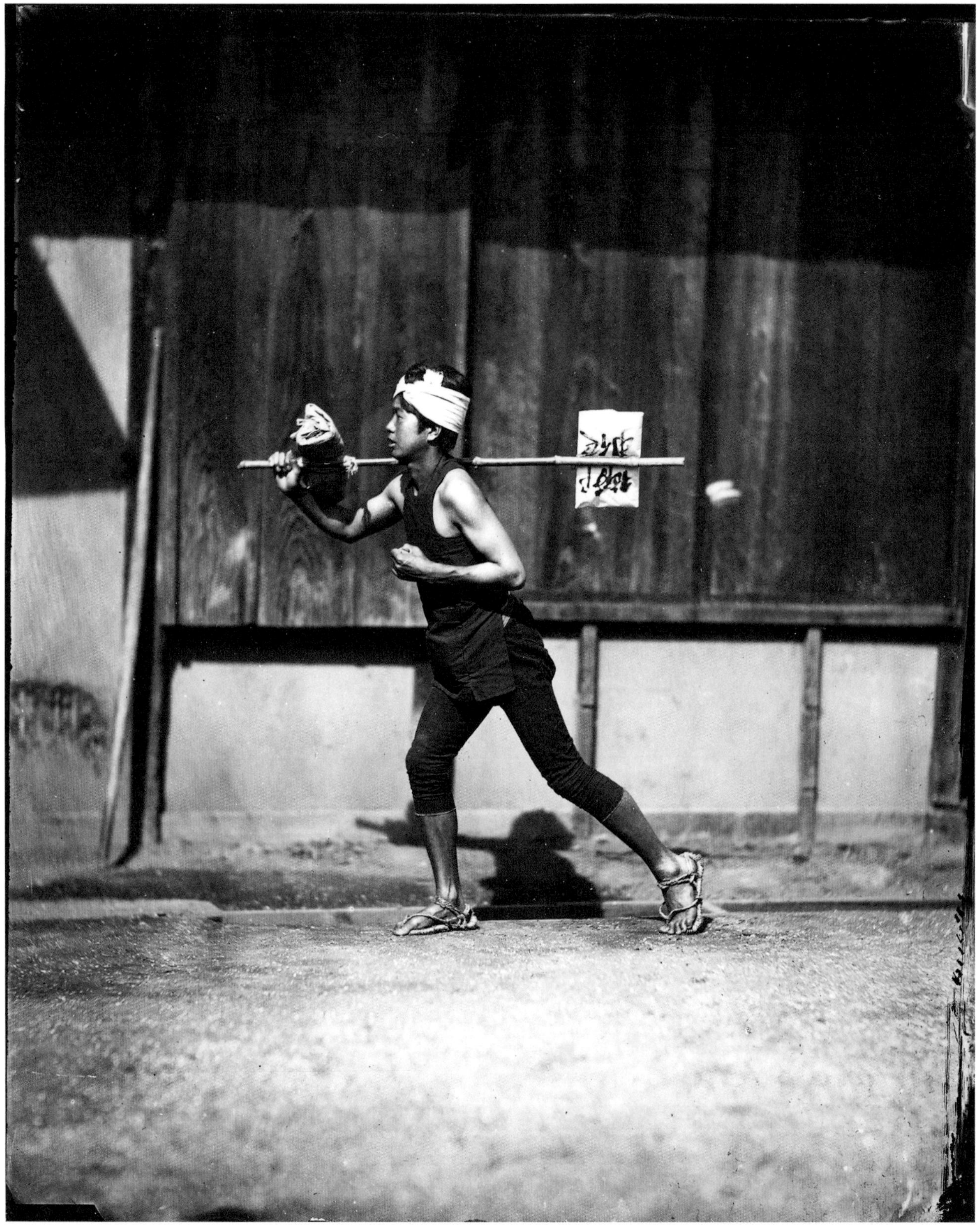

'A Japanese postman.'

Opposite page: 'Shop selling drums.'

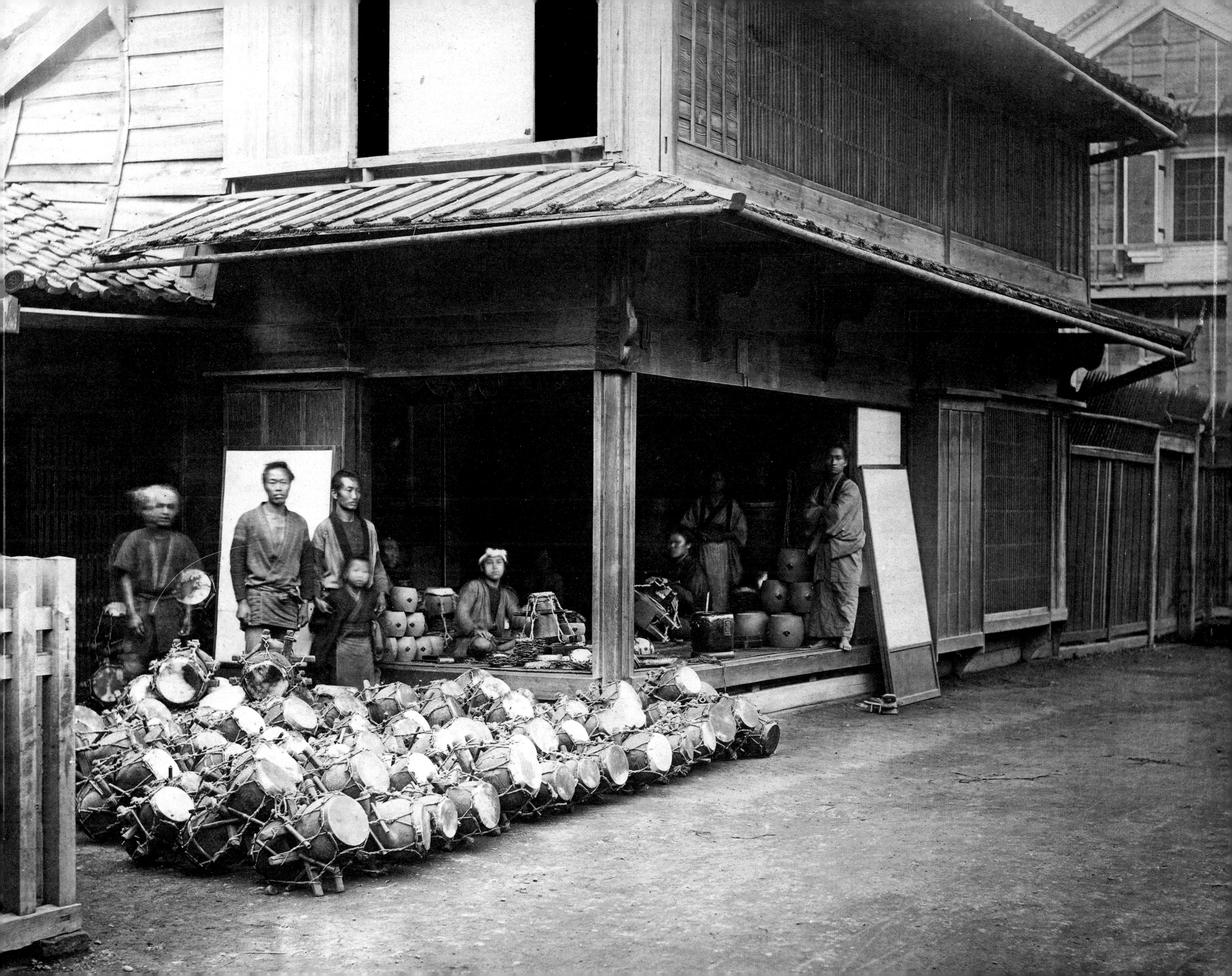

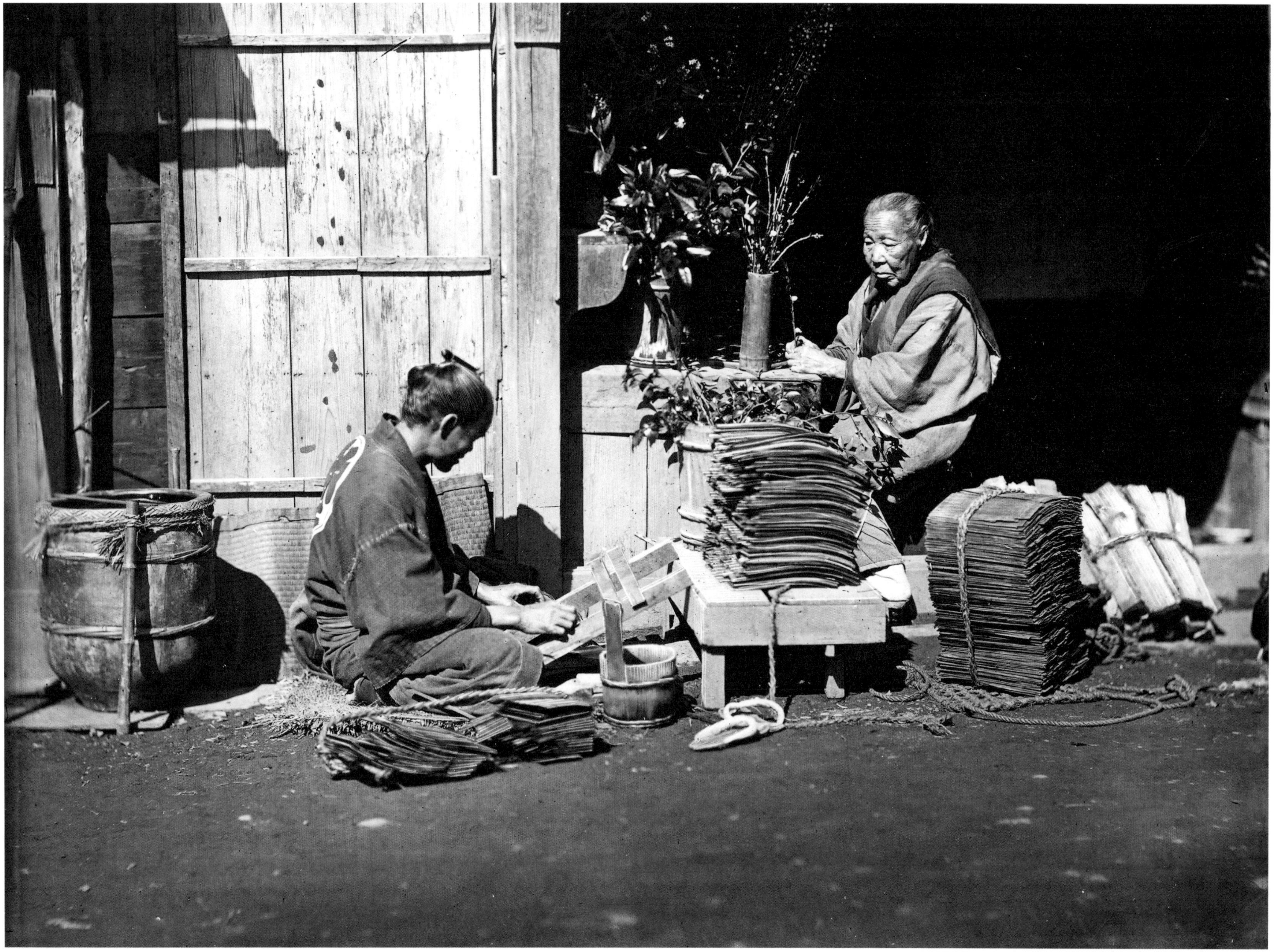

Opposite page: This photograph from Stolpe's purchased collection bears the German text 'Schindelmacher und dessen Mutter', i.e. showing a 'Shingler and his mother' at work.

'Fencing men.'

'A hairdressing salon in Tokyo.'

This photograph from Stolpe's collection of purchased pictures bears the German title 'Kämpfer', i.e. 'Fighter(s)'.

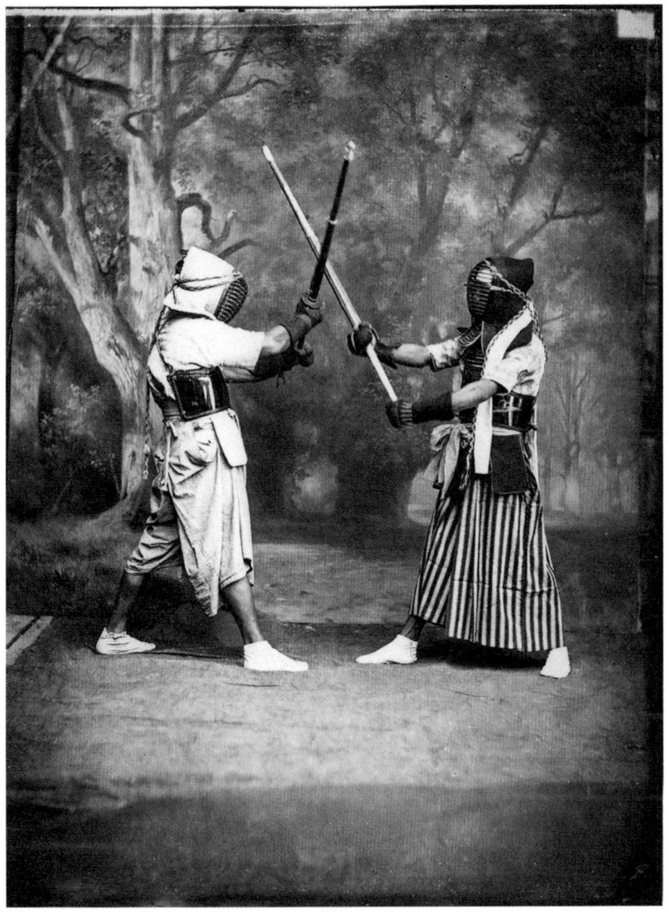

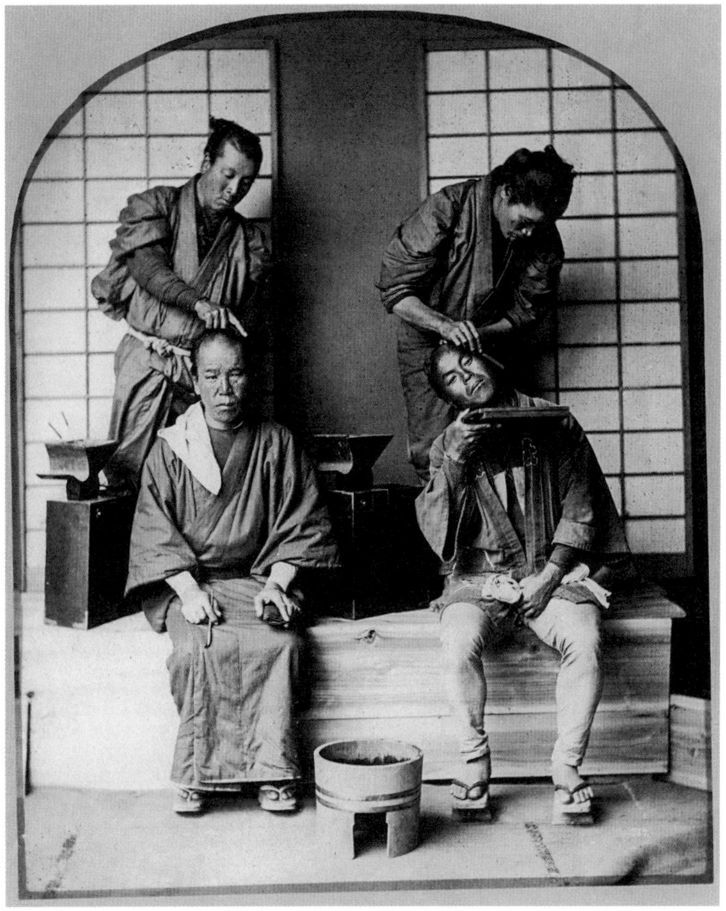

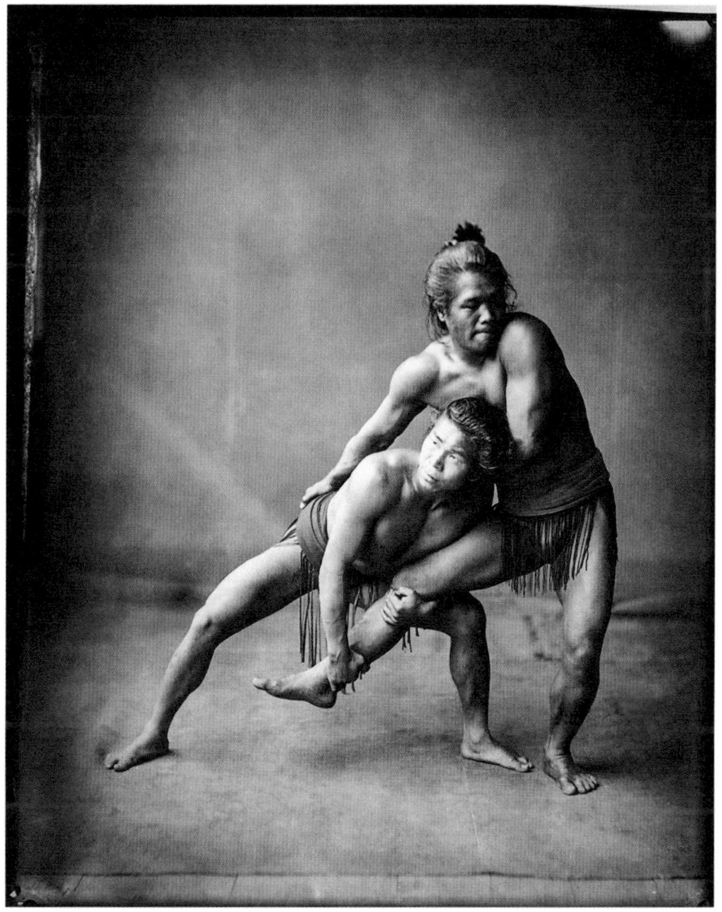

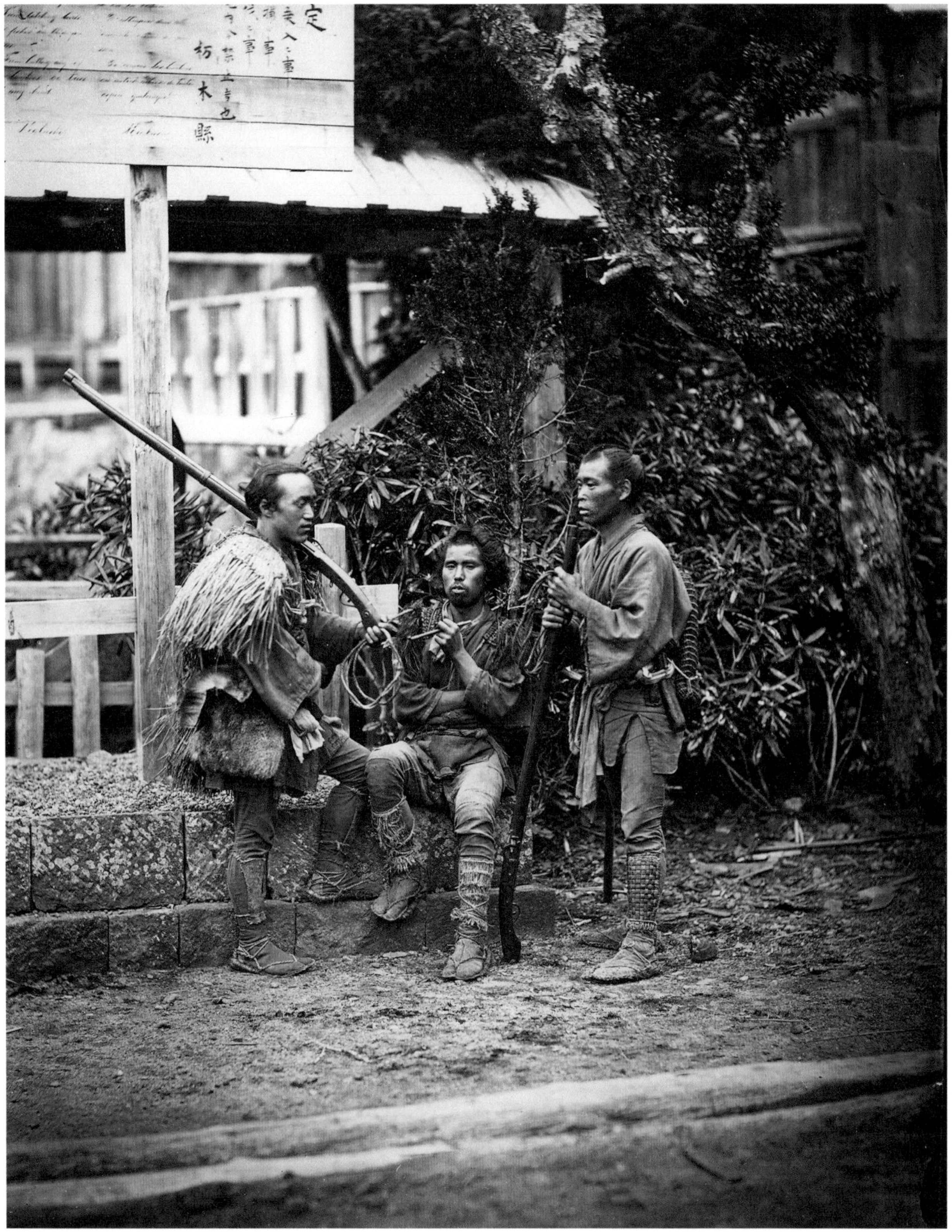

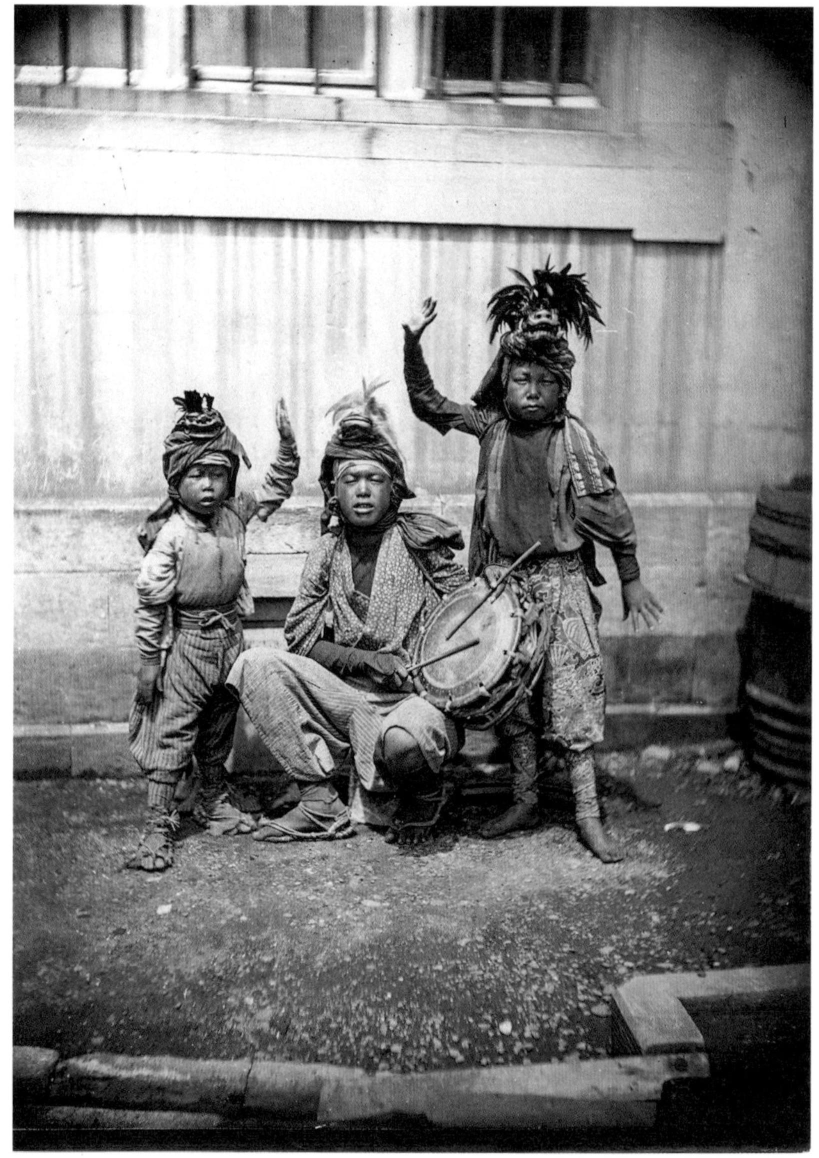

Left: 'A group of hunters.'

Above: In one of Stolpe's German lists of pictures the motif is described as 'Strassen Akrobaten', i.e. 'Street acrobats'.

Opposite page: 'Portrait of an old man.'

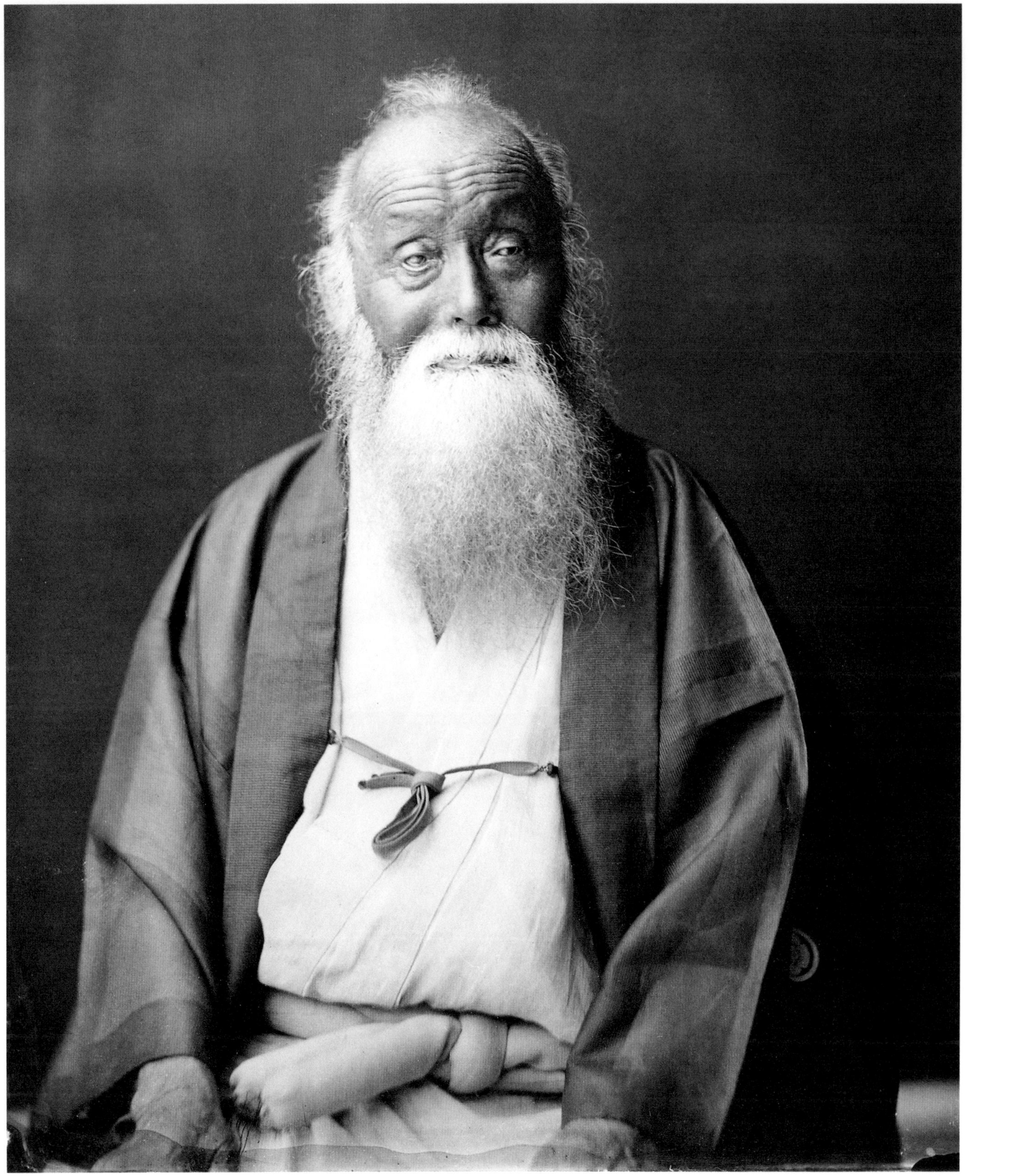

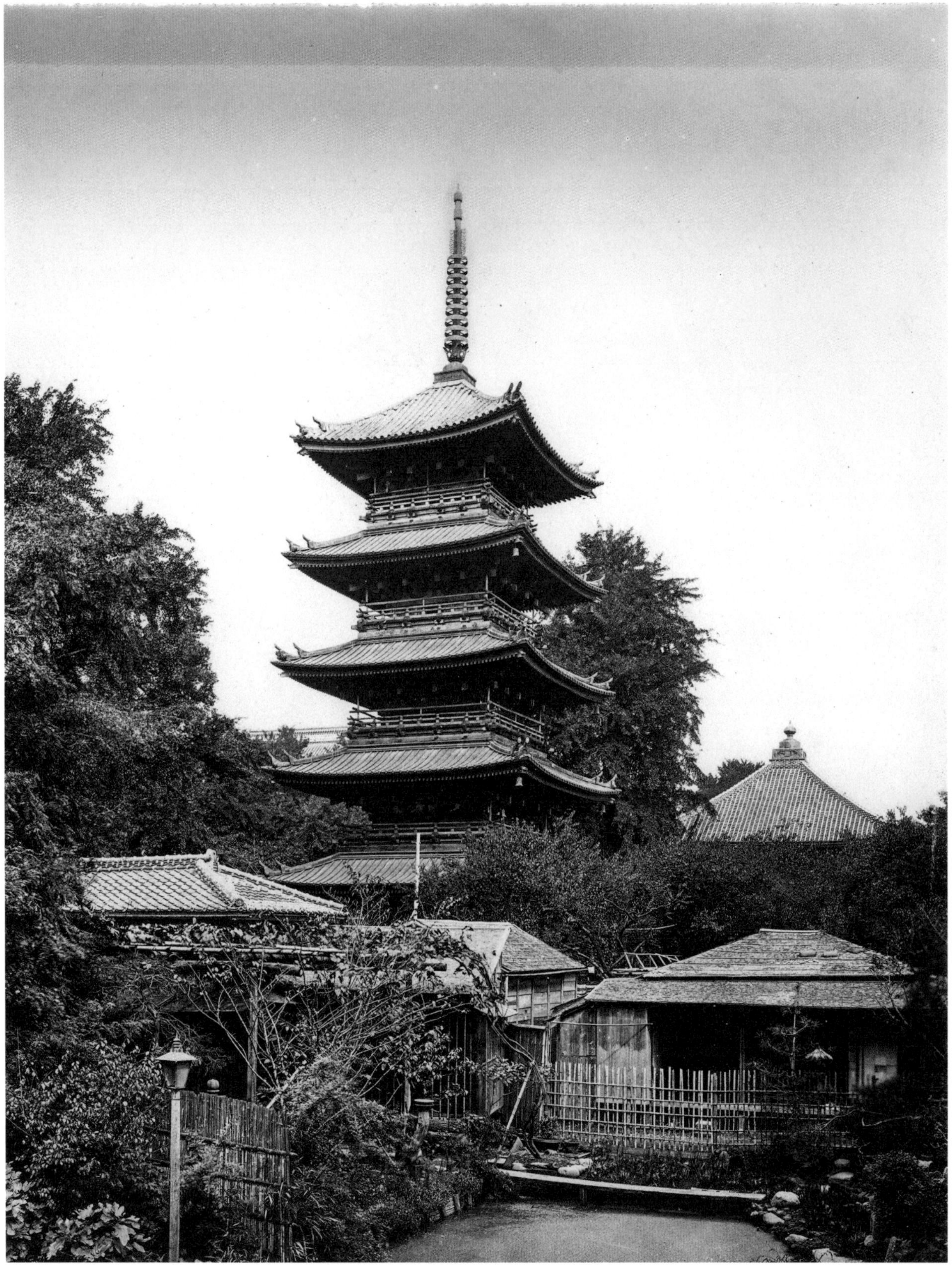

Left: Asakusa in Tokyo is the location of this pagoda, i.e. a temple building with the typical projecting eaves.

Below: 'Warriors wearing cuirass.'

Opposite page: The picture showing the two statues of gods is from the city district of Asakusa in Tokyo, the location of the city's oldest Buddhist temple, Sensoji, which is devoted to Kannon. On the back of the purchased picture Stolpe has written 'Kannon' with regard to the idol on the left and 'Seishi' with regard to the other one.

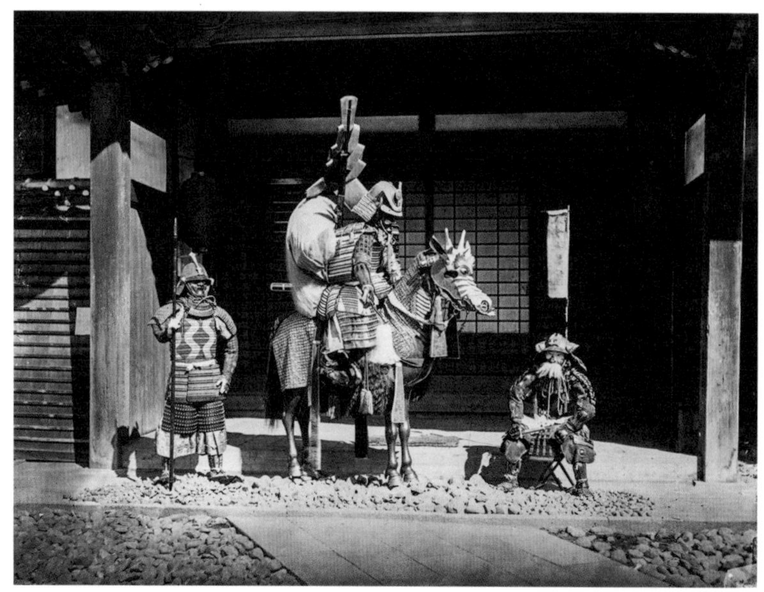

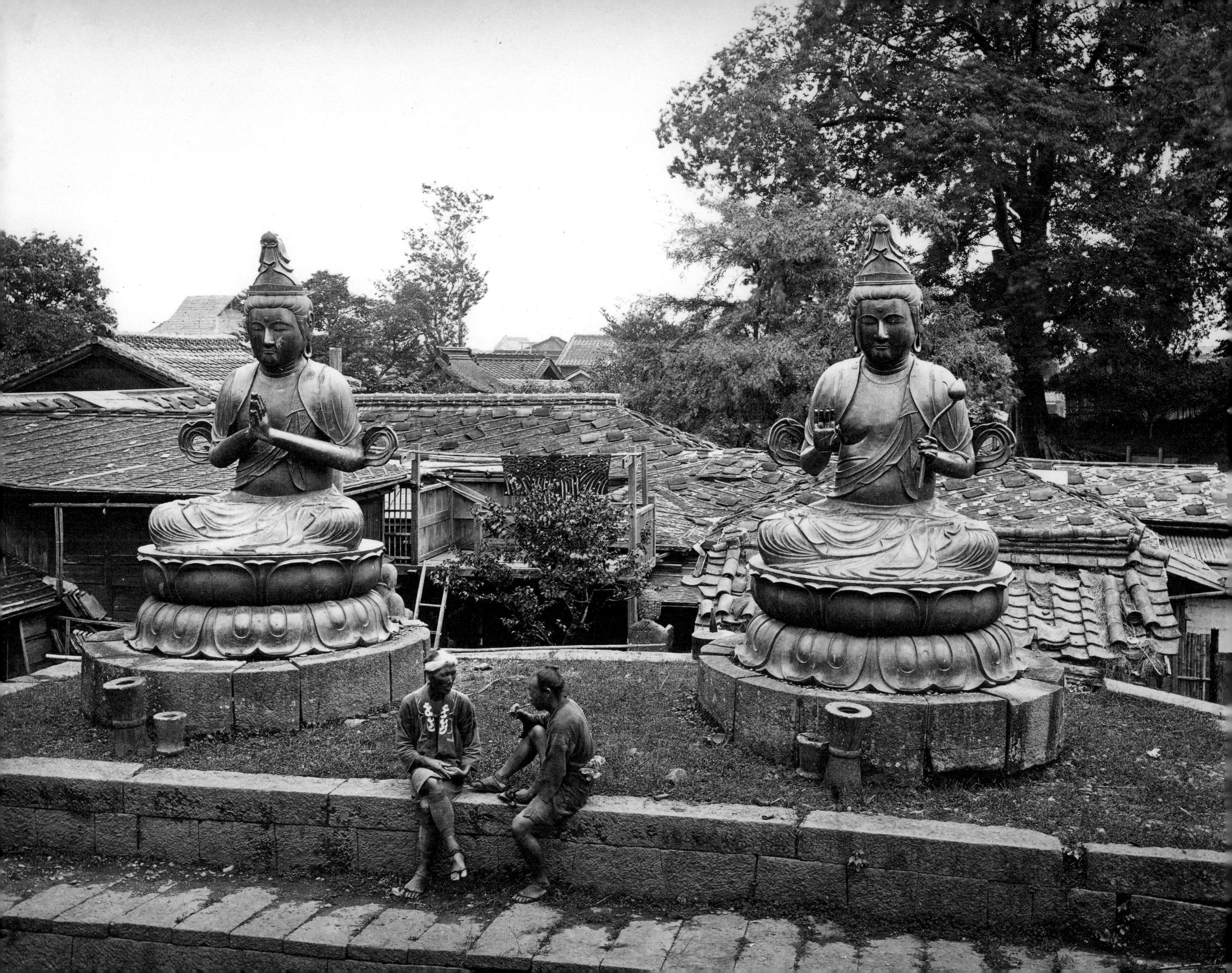

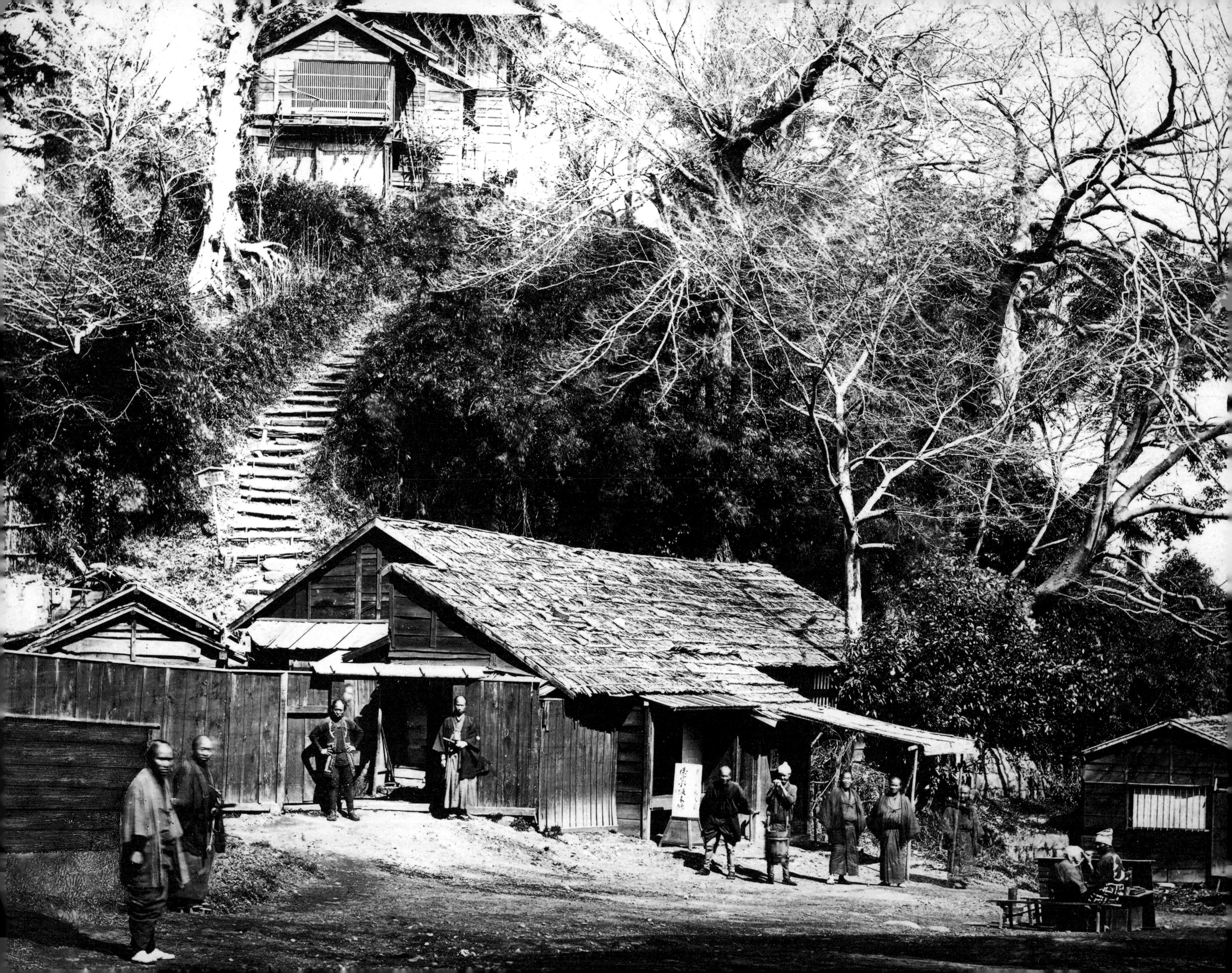

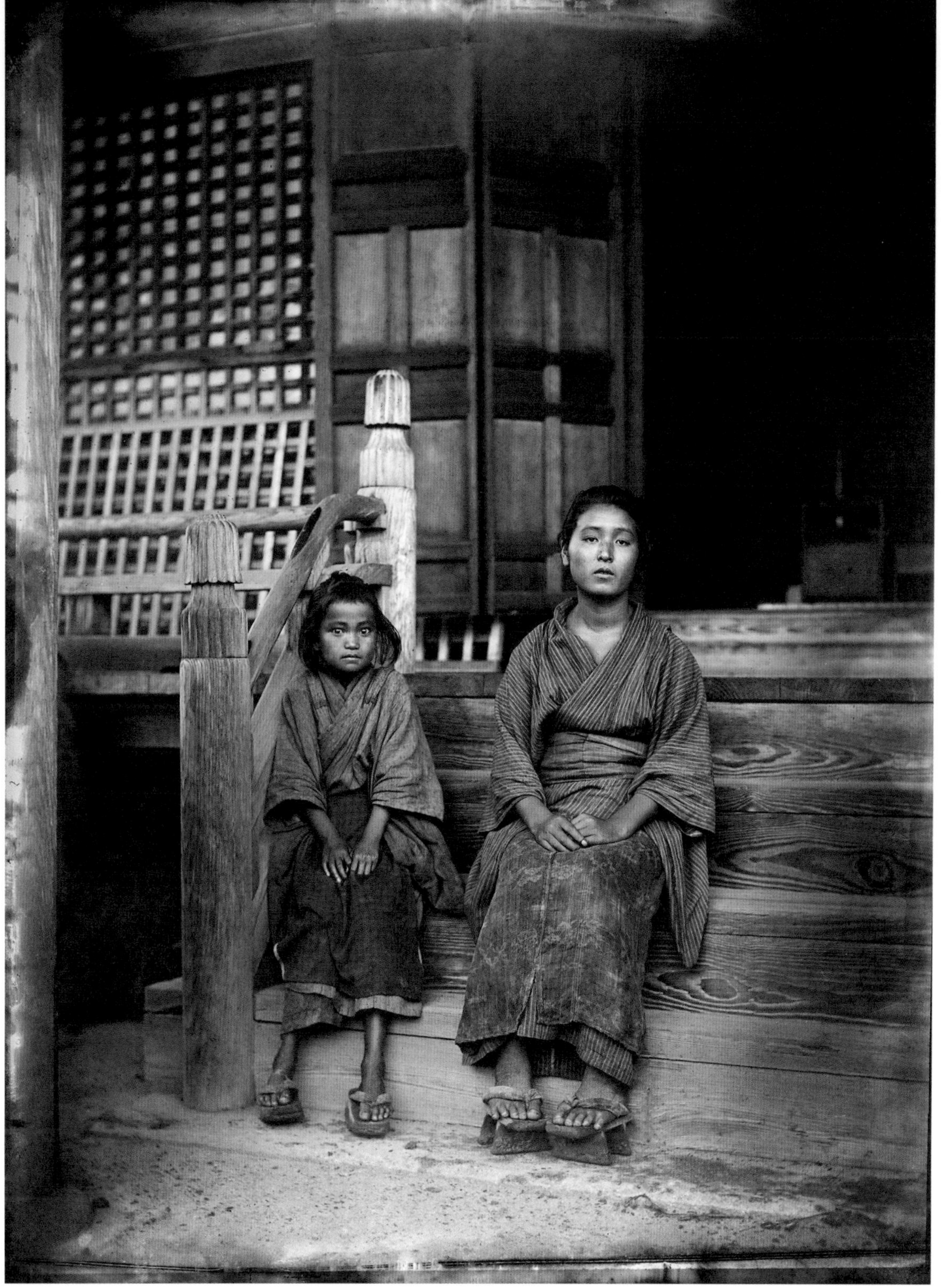

Opposite page: 'Fukiage, an imperial park' in Tokyo.

'Kiku with sister on the temple steps in Kusatsu'. This information is to be found in the list of Ekholm's photographs kept at Roesler's studio in Stockholm, where Stolpe placed a large collection of glass plates, to facilitate the processing of orders for copies.

Right: Nagasaki was the last Japanese port the Vanadis visited. Hjalmar and Oscar were invited to stay for several days with the Dane, Julius V Petersen, an official of the 'Great Northern Telegraph Company'. The little Petersen family is to be seen on the left, with the Japanese servants on the right. The view from the villa of the city and the harbour was magnificent.

Following pages: On 8 October 1884, the Vanadis anchored in Nagasaki harbour. Using two overlapping photographs, Oscar Ekholm created a panoramic image of the city and the roadstead with its ships, the Vanadis being recognisable on the left by virtue of her so-called 'Nelson chequer', a white belt extending along the vessel's sides and framing the black-painted gunports.

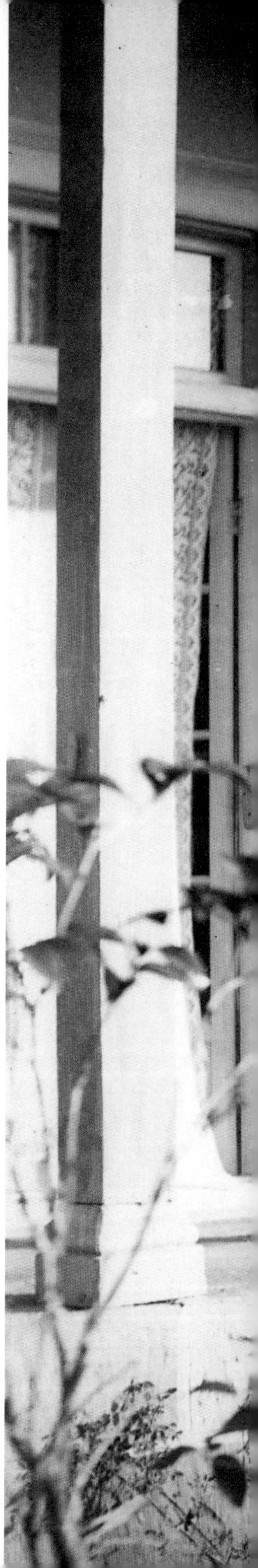

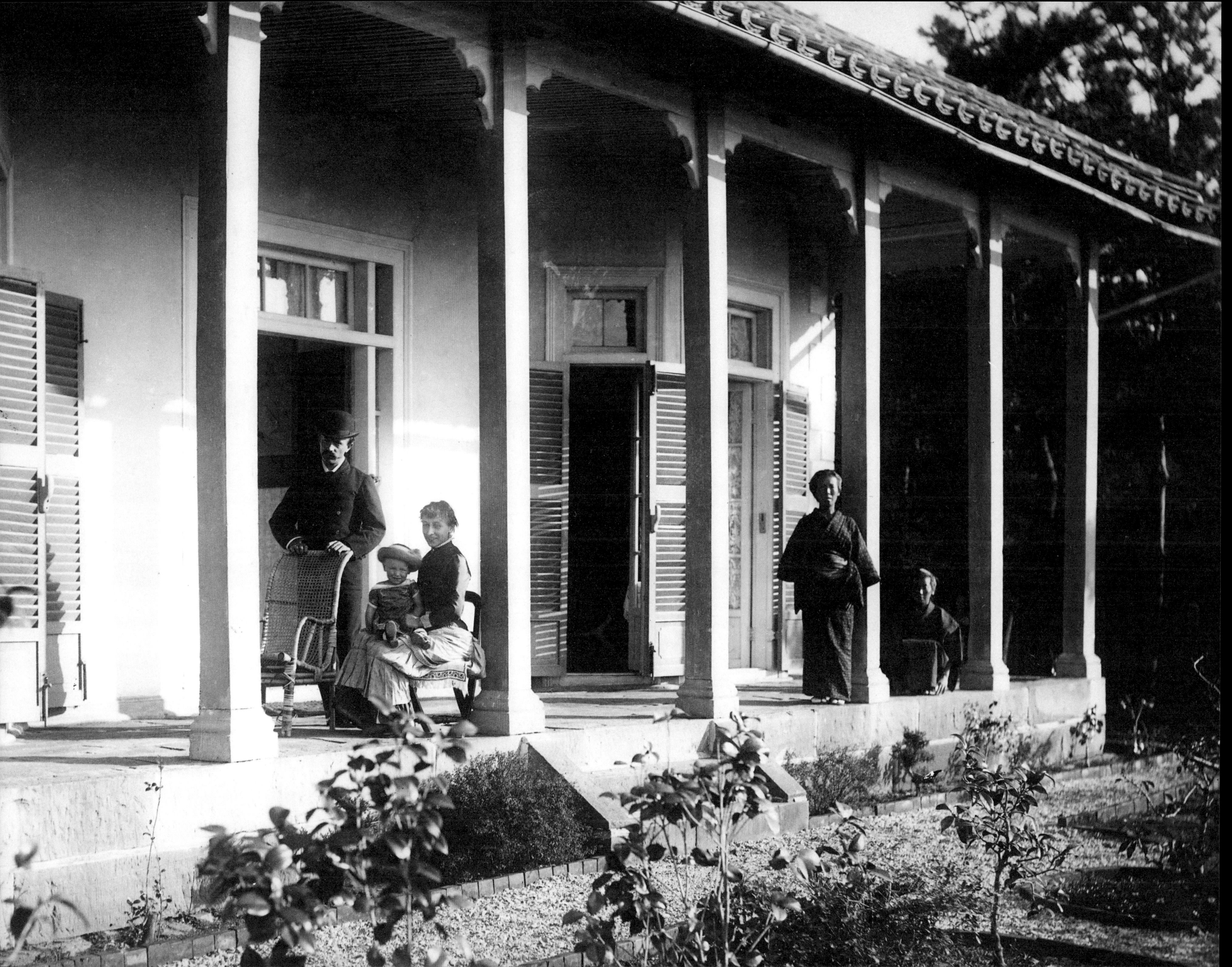

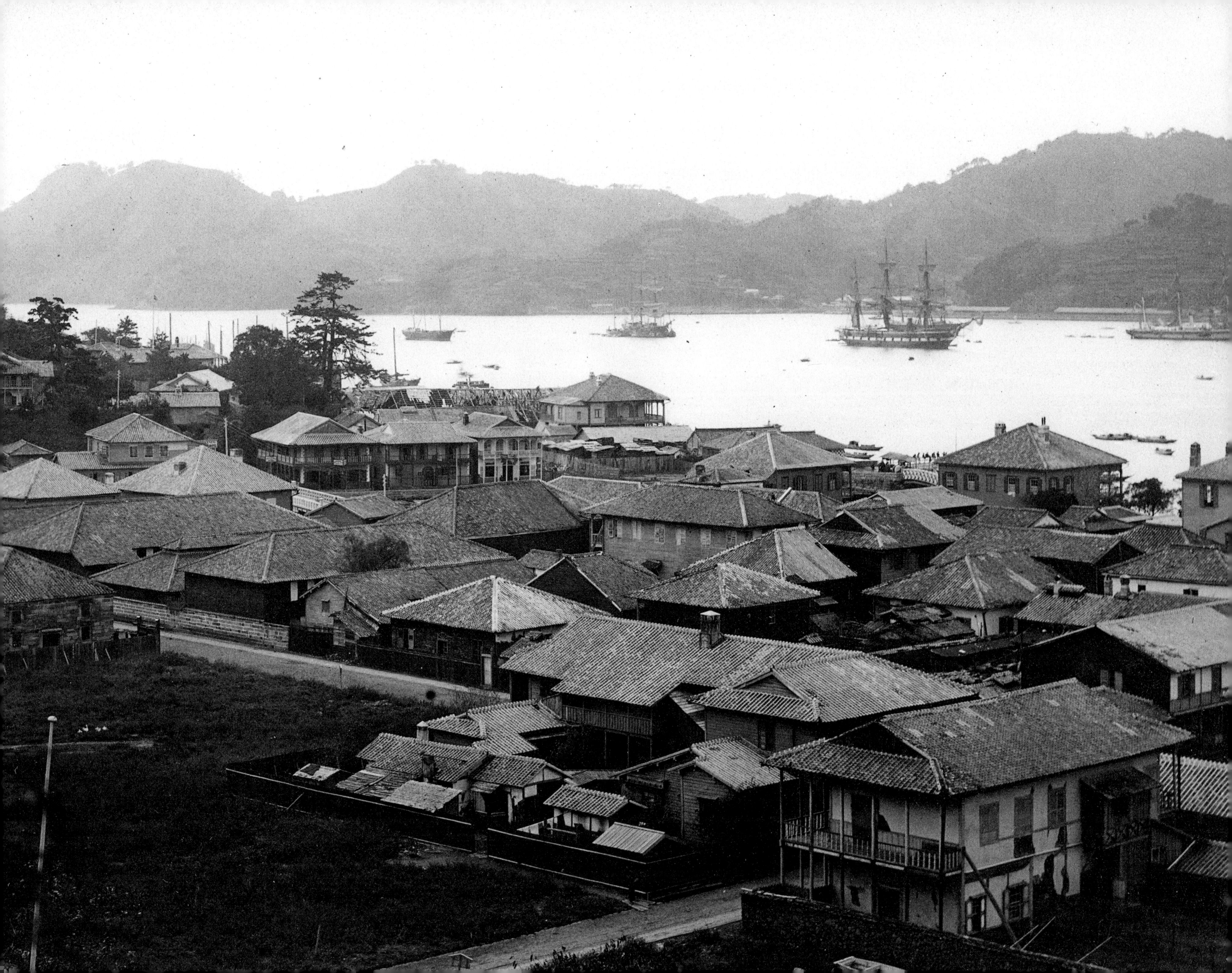

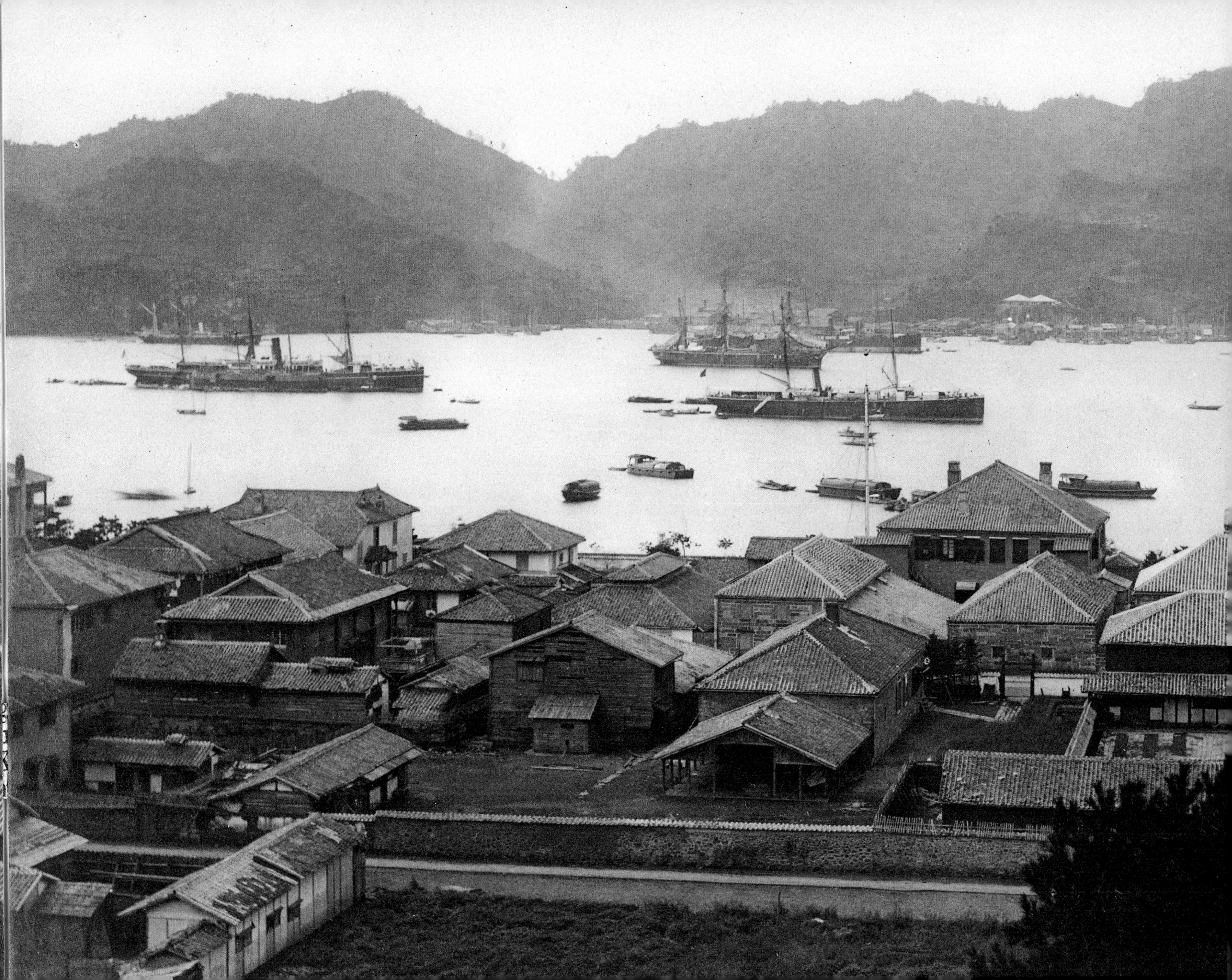

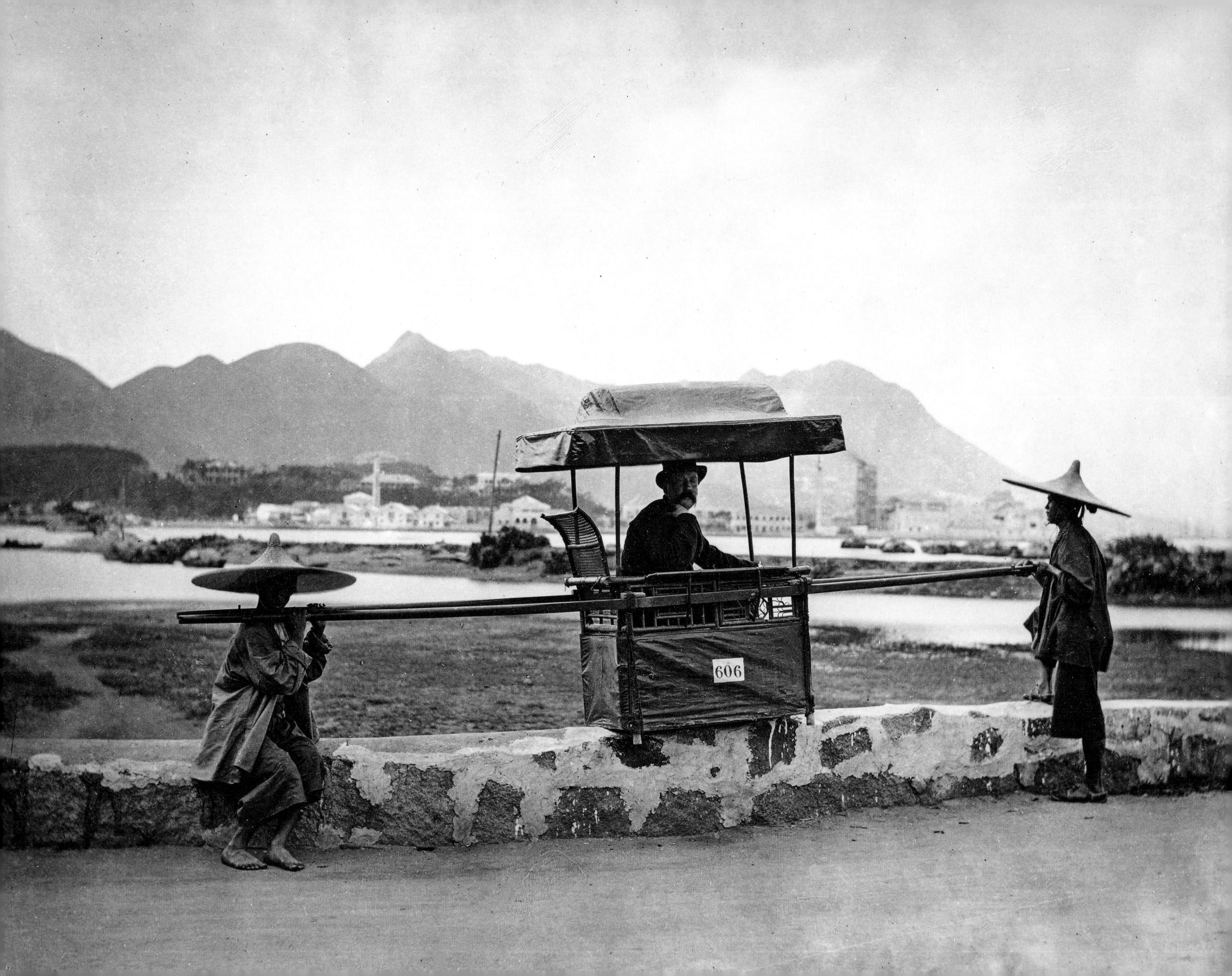
606

THE PHILIPPINES AND THAILAND

On 1 November the Vanadis anchored in Manila harbour. The main remit on the Philippines was to study the indigenous population, the so-called 'Negritos', and for that reason Stolpe rented a small steamer with a particularly mixed crew. This pursuit of 'the Negritos' was in the process of ending in catastrophe, but it took a slightly farcical turn, as Stolpe recorded in his notebook:

> I was to cross the bay to Mariveles in search of Negritos. Rented a steam sloop for 2 days. My companion was Mr Au, a young pharmacist, formerly a senior lecturer in pharmacology in Breslau who was now engaged in scientific research trips on behalf of the ethnographic museums in Berlin and Leipzig. An excellent chap for me. Two others (1 pharmacist, 1 old clockmaker) attached themselves and made the journey less pleasant than otherwise.
>
> I didn't get the boat until 10 in the evening. Glorious moonlight. The journey took 4 hours. I'd dropped off in the cabin when Au came and woke me, asking me to come up for safety's sake. The mechanic, a Tagal, had overheated the boiler and the machine was out of order, so the preheater was not being fed with water, the boiler was thus almost empty of water and Au was fearful of an explosion. We were about 1/2 an hour from the nearest land. Au had immediately put out the fire and prevented the Tagal from *flinging cold water into the boiler* (!), which he was on the very point of doing when Au woke up and grasped the situation! We were that near to being scalded.

They anchored, called for help and fired off revolver shots. The mechanic swam ashore and got hold of a canoe that could get the expedition leader to the shore dry-shod. The steam sloop had managed to get close to the village of Cabcaben in the Mariveles mountains, and luckily enough there was a group of savages close by:

> People were sent out into the forest to call out, and within 4 hours around 20 Negritos – men, women and children – had assembled. The photography now started. I cannot say they were anywhere near as ape-like as they are depicted as being. There were on the contrary intelligent faces. The heat was *dreadful*. Hard to work. Everyone wanted to sleep. Ekholm and I meanwhile walked around photographing buildings etc. We were constantly plagued by the heat. Better towards evening. A Spanish priest helped us buy Negrito things. I got 3 bows, 8 arrows, 2 bone ornaments and 2 combs, and this was deemed to be *very* rich pickings for one day. Also bought a load of Tagal stuff.

The notebook shows that despite the murderous heat Stolpe also carried out measurements of natives and copied their tattoos. Ekholm's yield was 25 photographs. It was finally time to return to Manila:

> The clockmaker repaired the machine sufficiently for us to be able to use it to get us home, but the Tagal [the mechanic] went out on a test trip and completely destroyed it. Spent half the night endeavouring to get hold of a sailing canoe. Finally succeeded, and cruised home in absolutely glorious weather. Arrived at about 10, worn out but content with our catch.
>
> Oh my dear friend! There's now less than 1/2 a year remaining. Then I'll be with you and Eterli [a nickname for Stolpe's daughter] again! Bought some beautiful summer hats for you, and hopefully they'll fit my Sparrows! The kisses will be applied in Bangkok so they're fresh.

Regarding the fresh kisses – Hjalmar always ended his letters by kissing the writing paper six times for his wife and six times for his daughter, Estrid. Around the kisses he drew big round rings.

The Philippines had for centuries been a Spanish possession, and the Spanish governor made a great effort to pay homage to Prince Oscar, to the extent that in his diary the latter wondered: 'What is the motive for all this friendliness?' Dinners and lunches, fireworks, pigeon-shooting, cockfighting and a 'monster concert' involving 400 musicians were laid on. And it was there that they went on a steamer excursion to Santa Cruz, by the famous giant lagoon on the main island of Luzon. The surprised prince recounted:

> The alcalde mayor (governor) immediately collected us in a magnificent boat powered by 10 dark-skinned rowers, and we were taken ashore accompanied by the usual letting-off of rockets, shooting and music-making. A special landing stage several thousand feet long had been set up in the shallow water to facilitate landing.
>
> Flags, decorations and gates of honour embellished the shore and the road as far as

During the Vanadis's visit to the British colony of Hong Kong in October 1884 Hjalmar decided to be lazy and occasionally allow himself to be carried. But he was very heavy, weighing a hundred kilos, and the bearers had to rest and put him on a stone wall. The collecting in Japan had been hectic – almost 24/7.

> the eye could see. Four-wheeled carriages with a coachman and servant in the Swedish colours took us up to our accommodation at the alcalde mayor's residence. Several thousand people greeted me with resounding hurrahs, and 23 bands played at various places on the way, where the carriages pulled up, escorted by natives on horseback and militia. From the festively decorated windows flowers were thrown by the town's young ladies.
>
> Our host, the alcalde mayor Don Iriate, was an extremely peculiar type – the very sort one sometimes encounters in a remote corner of the world. Dressed in the red uniform of the Knights of Jerusalem with several brilliant-adorned decorations and an épée set with precious stones, he also looked very impressive.
>
> It is by the way an observation that we seafarers often have cause to make, namely that in no other place is one received in such a hospitable, friendly and warm manner as on these very remote, secluded island groups. It's as if their population were seeking with a certain anxious enthusiasm to hold on to each point of contact with the big, outside, more powerful world.

The visit to the Philippines was thus a memorable experience for the prince and the officers. And Stolpe would much later come to deal with a dramatic episode in the history of these islands. The 1890s saw the start of the Philippine independence struggle against the Spaniards – a struggle in which the medical doctor and author José Rizal played a significant role. When he was executed in 1896, the tempo of the Philippine revolution increased, and the fate of the freedom fighter deeply affected Hjalmar Stolpe, who was to depict the story of his life.

THAILAND

The voyage from the Philippines to Thailand (Siam) took a good ten days, and in Bangkok Stolpe stayed with the Swedish/Norwegian consul Wilhelm Theodor Müller, whom he described as 'an extremely jovial German who lived in a decrepit hovel – a miserable place the like of which I've never beheld'. Stolpe described his stay as follows:

> The first days involved going to the palace every morning – all meals were to be taken there, and we were to be involved in all the attendances. The procedure on such occasions is moreover that you arrive in a group, move forward in turn and bow, then stand and watch for a while as the prince converses with the king, after which you can go. The only enjoyable part is seeing how the inconsequential people eager to be decorated push forward and seek to make themselves noticed, and hearing the clever comments that are made ...
>
> I was involved in attending on the *other king*, an old man of honour and the king's uncle, but I then found that I was wasting my time going around with the others, and I parted company. I had put the photographer up at a cheap hotel, where he lived vastly better than I did, and I went there for my meals instead of visiting the palace, and it was only now that I was able to do something of real benefit. I took with me a Chinaman obtained through the consul, and we strolled around the bazaars, returning every evening with a boatload of stuff.

Thus despite the short stay a bumper yield was achieved in Bangkok, involving over 400 objects illustrating the Siamese people's way of life. The everyday objects were complemented by 'a complete set of musical instruments', including drums, cymbals, wind and string instruments, Jew's harps and carillons. Hjalmar was satisfied with his collection, of which '26 images of God made of gilt bronze are the most splendid examples'. In Bangkok gifts were also contributed by persons with an interest in ethnography, e.g. the manager of the Siamese customs authority, Fidelis da Costa, and the commander-in-chief of the Siamese fleet, A Duplessis de Richelieu. The latter was originally a Danish naval officer who together with his brother had emigrated to Siam, where the pair enjoyed a military career.

The climax for the officers of the Vanadis was their meeting with the king of Siam, Chulalongkorn. Prince Oscar had been commissioned with presenting to him the Grand Cross of the Order of St Olav and conveying greetings from his father, the king, in the interests of promoting trade links. As usual, the reception was a lavish affair, involving parading troops, resounding salutes, huge bands playing the royal anthem *Ur svenska hjärtans djup* [From the depths of Swedish hearts] and tables loaded with food and drink.

The opulent grandeur and riches surrounding the 'king of the land of temples and elephants' made a strong impression on both the officers and the prince, who recorded the encounter with this incomprehensibly rich sovereign, newly married to a beautiful 19-year-old and with his enormous family clan assigned to all the kingdom's key posts. Prince Oscar wrote in his diary:

> The king, surrounded by a numerous and motley retinue, stood out on the stairway, where he amiably bade me welcome through his interpreter, secretary and brother, Prince Dewar. Moving hastily, he ascended the stairway into a hall, closely followed by myself and the horde of people. As a result of the great speed all the companions got left behind, and I gave my little speech, together with the greeting from my Father and King as well as assurances of his friendship, and discharged my assignment of

> presenting the Grand Cross and Chain of St Olav without my attendants and the secretary Lagerholm having managed to present the bright-red cushion bearing the decoration. The speech came first and then the cushion, after which the king presented me with the Siamese Order of the Crown.
>
> After I had presented the captain and the officers and had spoken for a while to the king, who is a heavily built man of average height with a particularly dignified and elegant appearance, he followed me in to meet the queen. We passed through large, magnificent throne halls and other beautiful rooms, and through a covered garden where water gushed forth in cascades from artificial caves, we entered the Siamese queen's reception room. I was pretty astonished and surprised to find myself face to face with a slim 19-year-old figure splendidly dressed in sort of page-boy clothes displaying refined taste, whom the king presented to me as his consort ... The companions who got to come were, like myself, entranced and delighted at everything we'd seen here, not least the extremely stylish queen.

Everyone was entranced – all, that is, except Hjalmar. As usual he was only moderately diverted by such affairs, and his reports home reflect this indifference. It's possible he had intended to reserve the depiction of Siam's enchanting 'One Thousand and One Nights' world for the Vanadis book. In which case he would have been able to present a signed portrait of the king's brother, Prince Diss Warakumar, which is still to be found in the Stolpe photo album. In conjunction with paying homage to the king, Ekholm took a group photo in which Prince Oscar is flanked by the king's brother and a bleakly staring Stolpe. Disregarding the royal glitter, Oscar Ekholm was moderately impressed by the country:

> The appearance of these people is somewhat reminiscent of the Japanese, but the former are much grubbier and live in wretched huts, which make up the entire town, the royal palace and temples alone being magnificent. Amongst the latter there is one that has been worked on for 300 years but was completed two years ago and cost the current king a total of 18 million kronor. It is gilt from the foundations to the steeple, and has cost 1 krona 80 öre per square inch.
>
> The unpleasantness of the Siamese people's appearance is greatly increased by their constant chewing of betel, which makes their lips red and black. If their mouths are for a single moment not full of betel they eat sugar cane, both of which are on sale on the street corners, so their chewing apparatus is in constant motion.

Ekholm was very busy in Bangkok, and photographed the temple area Wat Cheng and the big 'central pyramid' Pra Prang, which, Hjalmar wrote, is said to be 900 feet high (90 metres): 'One recognises a Pra Prang from the curious spire at the top, which I'm guessing is supposed to represent the tree of life, which is such a haunting part of India's ornamentation'. In this monumental building 'you can ascend to a great height on the many steep steps, and I sent the photographer up to take views of the town while I was studying the details', Hjalmar explained to his father Carl Johan back home in Norrköping. He added:

> Just as I was going to take a look, along came an old shaven priest in yellow silk attire and grasped my hand. His betel-filled mouth distorted itself into a well-meaning grin, and my Chinese friend presented to me the temple's high priest, Pra Si Sam Pot. He was hugely affable and followed me everywhere, but he was ignorant and unable to tell me anything. I photographed the old man for decency's sake, and will send him his portrait from here.
>
> The whole day was taken up with photography at Wat Cheng and a visit to an old missionary's widow, from whom I wanted and got a collection in the form of the so-called Bangkok Calendar, which had been published by her husband and contains a host of notices of use to me and my forthcoming book.

Much later Stolpe returned to the encounter with the high priest in Wat Cheng and the latter's reaction to the question as to whether he might buy some pictures in the temple:

> As evidence of the harshness of the punishment for contributing to taking holy images and objects out of the country, I recall the Wat Cheng high priest's descriptive gesture around his neck, as notwithstanding all his good-naturedness he was compelled to most firmly oppose my cautious intimation that I would like to buy a few small pictures of the temple. He was also careful to add that he would probably be made shorter to the tune of his beautiful shaved head if any of the pictures my greedy eyes were regarding were to disappear from the temple in some manner or other unknown to him (and naturally to me too).

Hjalmar Stolpe concluded his letter home to his father with a sigh: 'I've actually conveyed in brief everything I experienced in Bangkok, and I could weep tears of blood at this having been so little'.

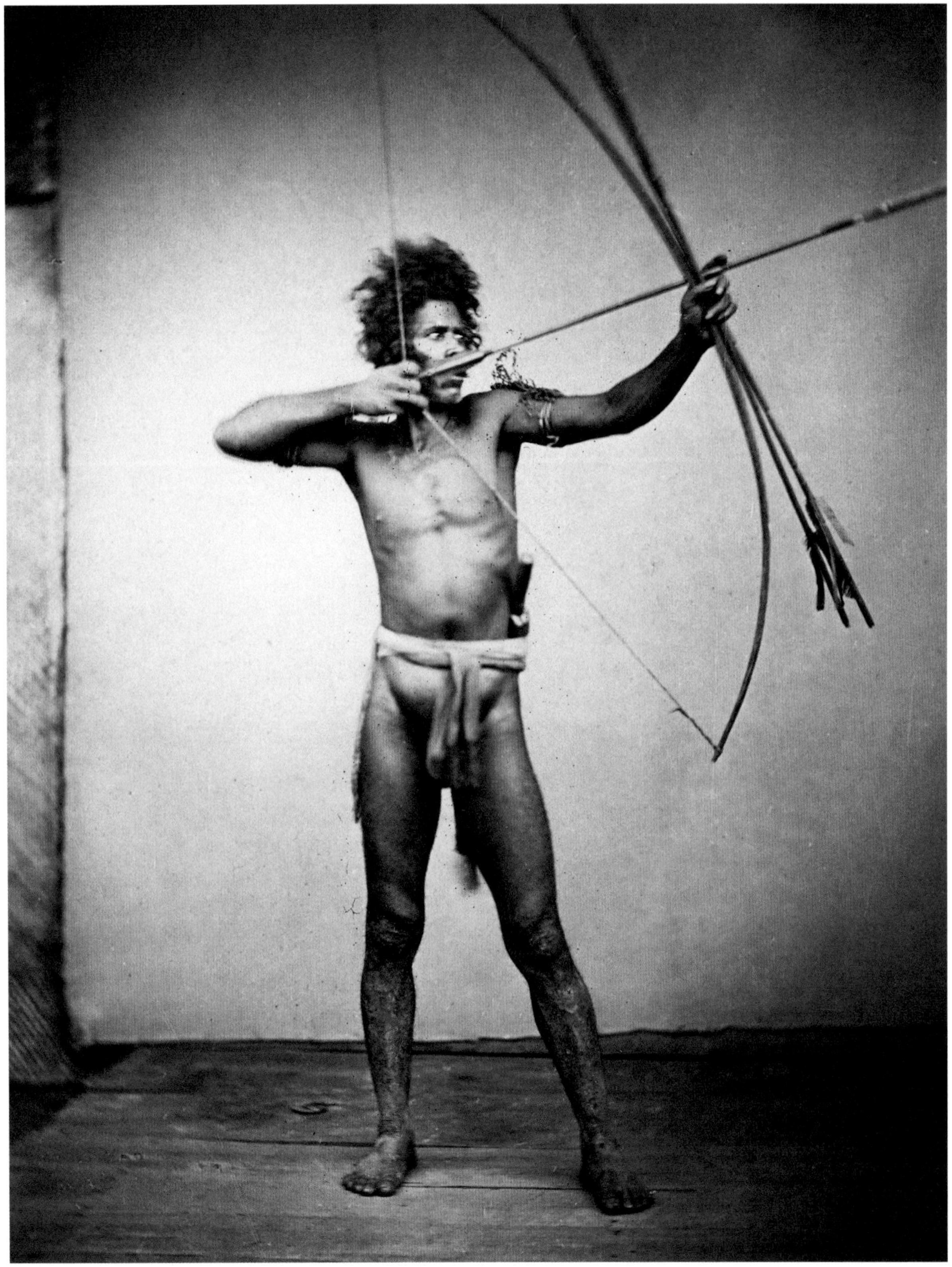

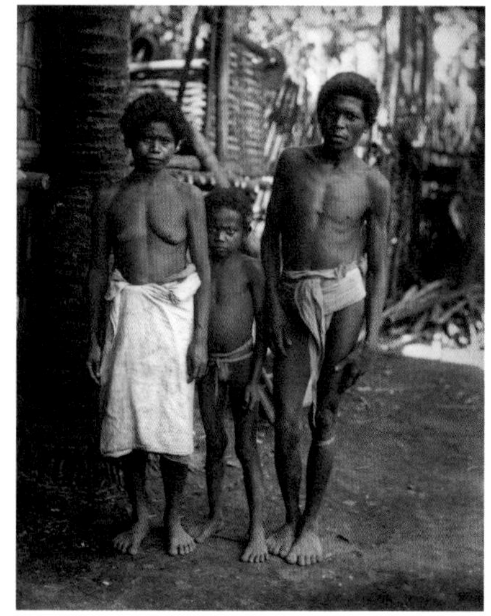

Above: 'Woman, boy, man with a band of raffia around his left leg (leaning). Man's height 150.2 cm; woman's 137.7 cm.' According to Stolpe's list of pictures the photograph is from Cabcaben on Luzon. In his notebook Stolpe depicted their tattoos in detail.

Left: A 'Negrito' from northern Luzon. In a lecture ten years after the Vanadis voyage Hjalmar Stolpe described how he visited 'the Negritos', and showed Ekholm's photographs. 'The actual name indicates the characteristic feature of the people, namely their dark, negroid skin colour, in which respect Negritos differ from the islands' main Malayan population. ... Culturally their status is very low; they subsist on hunting, constantly roam from place to place and have no dwellings other than a sort of canopy woven from branches, under which they seek protection during the night. ... Their weapons comprise bows and arrows and short, harpoon-like spears. They live monogamously, and never divorce, as far as is known. Their bodies are buried, and a shed is set up over the grave, but the whole area where the death took place may not subsequently be entered by any of the tribe's members, on pain of death. This is the chief reason for their constant change of domicile.'

Opposite page: 'Seven dancing Negritos and two Negritas'. They were photographed in the village of Cabcaben, according to Stolpe's list of pictures from the Philippines. In a museum catalogue he describes them as 'black, curly-haired Negritos *or* Aetas *of short stature remaining at the lowest stage of development, from Sierra de Mariveles' on the northern side of Manila Bay.*

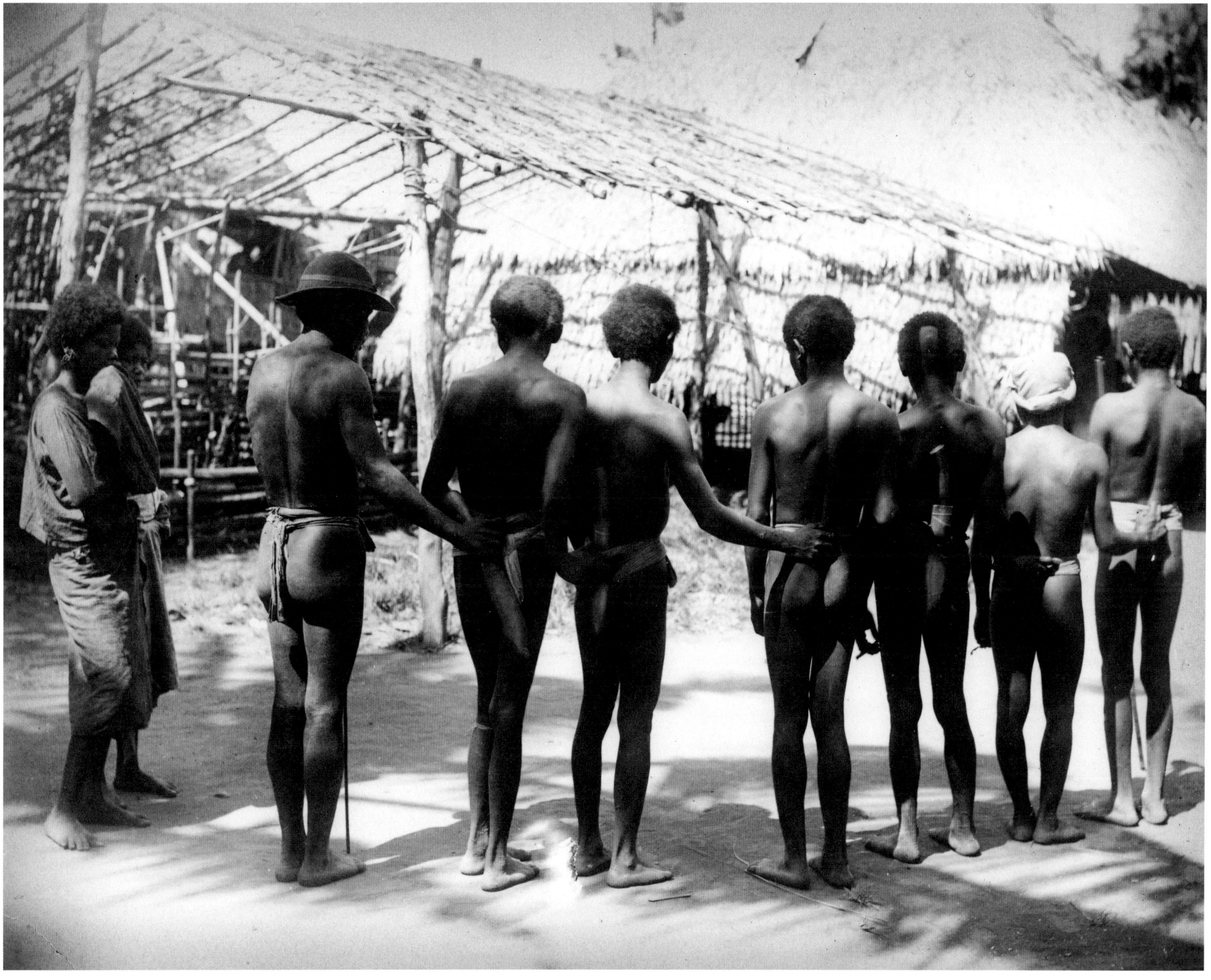

The indigenous population, whom the Spaniards called Negritos, had for millennia above all populated the main island, Luzon: 'They have thin lips, curly hair and a shaved parting on their neck, concealed by a large bamboo comb; they are willing to work, have no agriculture, hunt for wild boar and deer, kill fish with a bow and arrow and keep dogs as their only pets. ... The sole traces of religion that have been found are offerings of boar in the event of thunder, prayers to the rainbow and dances on moonlit nights.' This is the summary of Hjalmar Stolpe's lecture at the school Stockholms Borgarskola in March 1900, when he started a lecture series illustrated with 'numerous sciopticon images' based on Ekholm's photographs. Stolpe had bought 'Negrito objects', bows, leg decorations and neck combs, as well as boar's-bristle knee decorations that the men wore around their left ankle. The objects came from the Aeta Negrito tribe.

Opposite page: The picture shows 'Cabcaben, a Tagal village below Mount Mariveles' on Luzon, according to Stolpe's notebook. It was probably in this village that the Negritos gathered and were photographed. 'Tagals' was what Stolpe called the people in and around the capital city Manila, where Tagalog was spoken. They had immigrated to the Philippines far later than the Negritos.

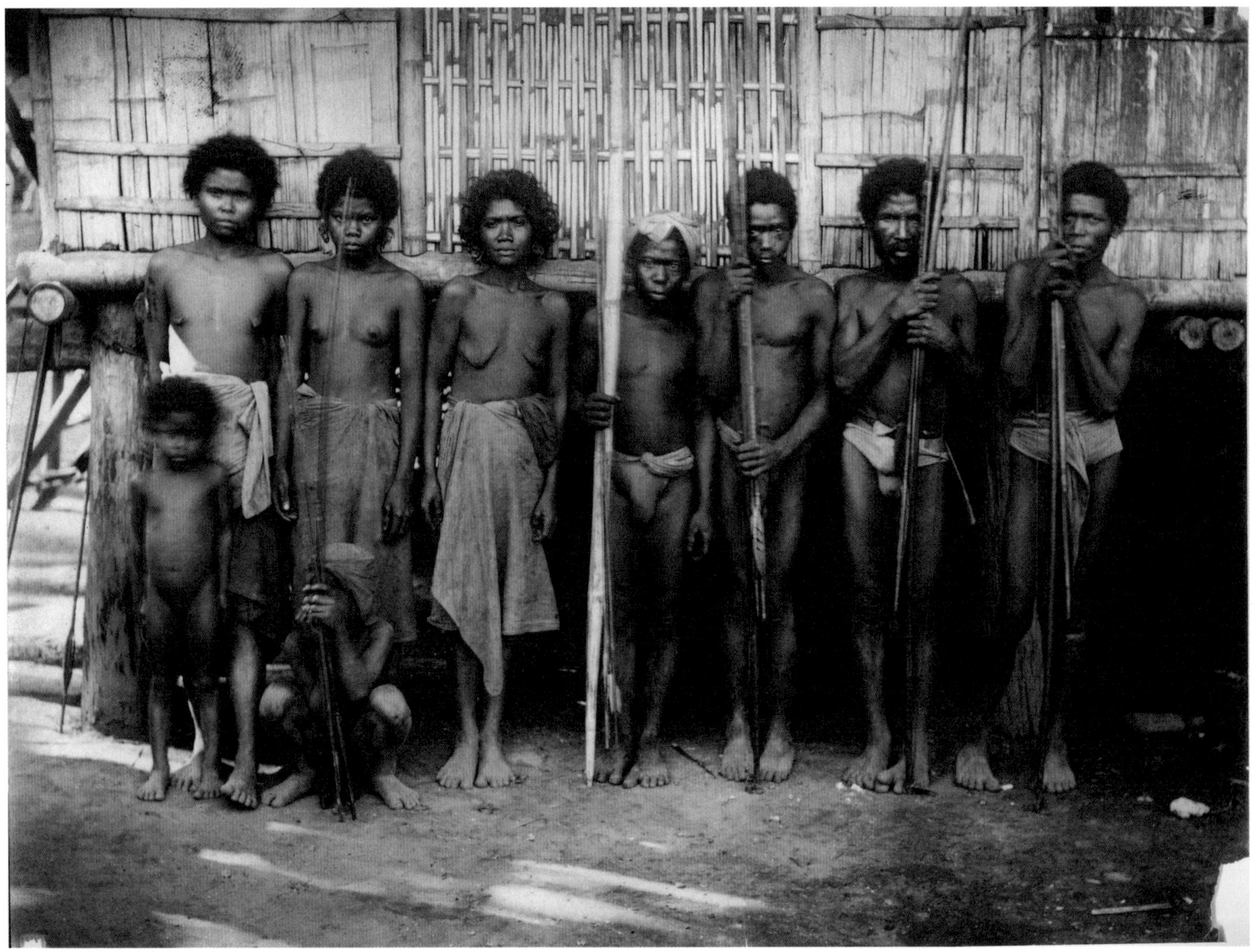

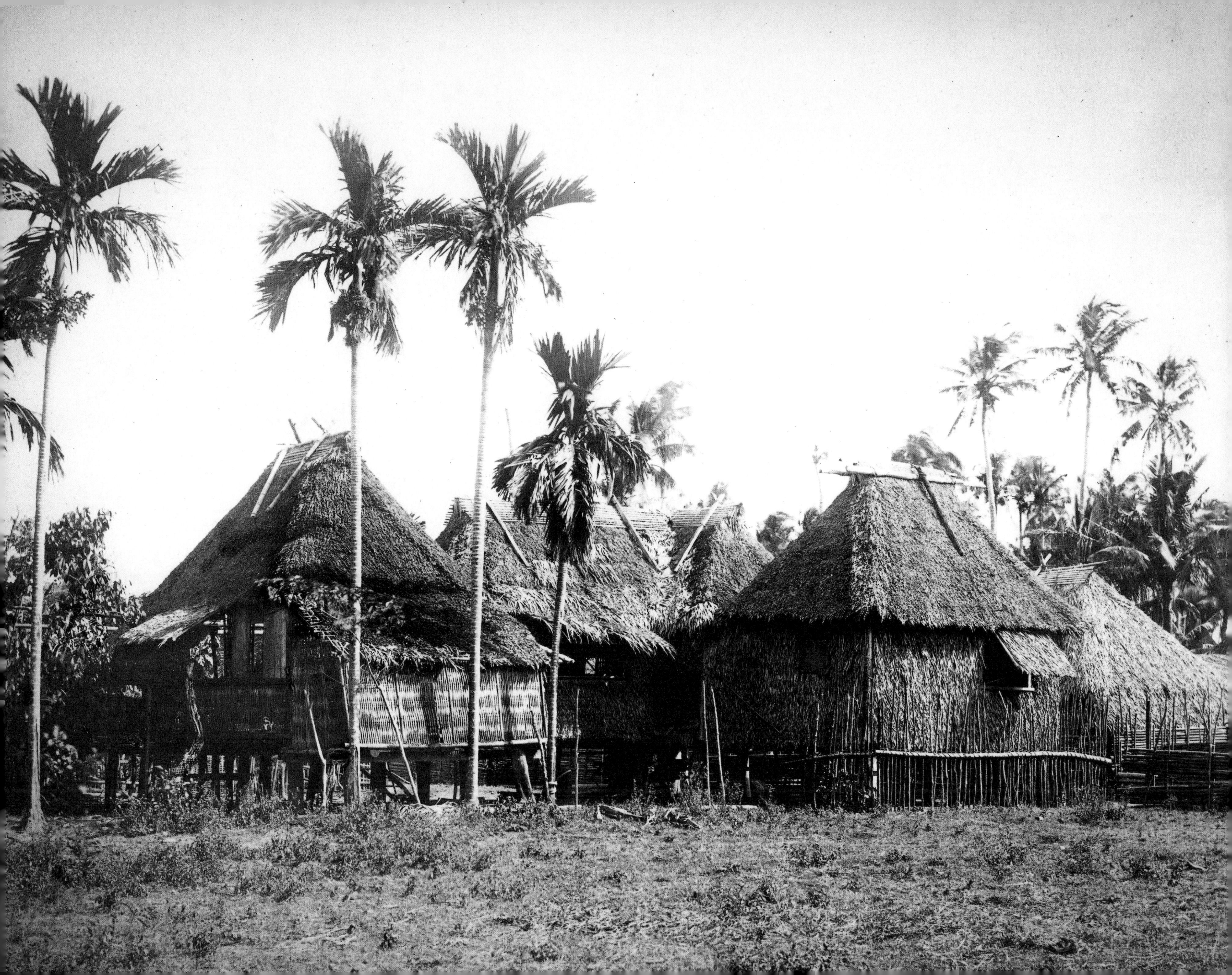

The King of Siam (Thailand) wanted his guest Prince Oscar and his Swedish/Siamese entourage to be photographed. Sarauromya Palace was placed at the disposal of the Swedes, who courteously allowed themselves to be dressed in festive Siamese trousers and stockings. How they ever got the commander-in-chief Lagerberg to don the garments is a mystery. Stolpe, who always refused to take part in 'spectacles', is missing from the festive gathering – it would have been impossible to get him to put on these clothes. Sitting, from left to right: Captain af Klintberg, unknown, Prince Oscar, Prince Diss Warakumar (the king's brother), Commander-in-Chief Lagerberg, Lieutenant Sundström, Consul Ehrenborg, Lieutenant Carlsson. Recumbent, from left to right: Lieutenant Lagercrantz, the assistant ship's doctor Holmer, Lieutenant Dyrssen, Second Lieutenant af Ugglas. Standing, from left to right: Siamese dignitaries, then the three Swedes: Second Lieutenant von Krusenstierna, the pastor Kiellman-Göranson and the purser Lagerholm.

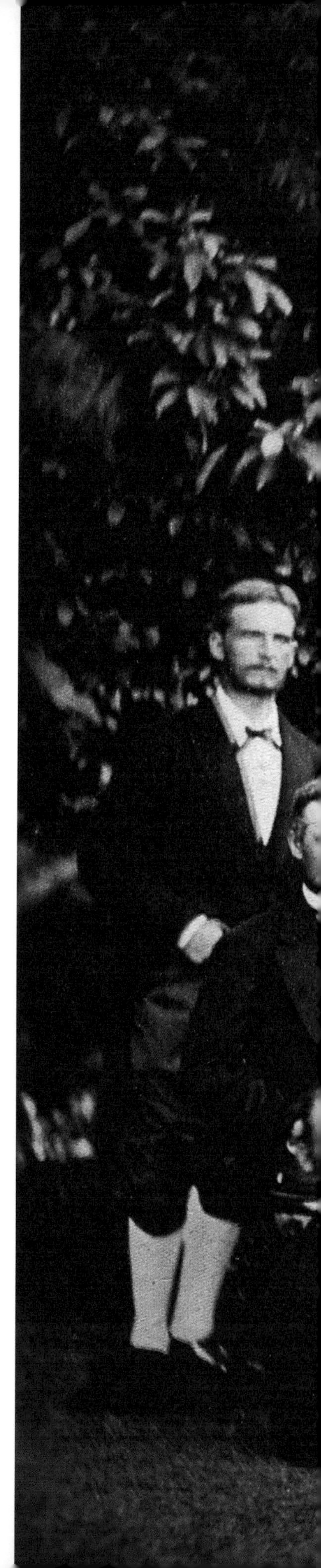

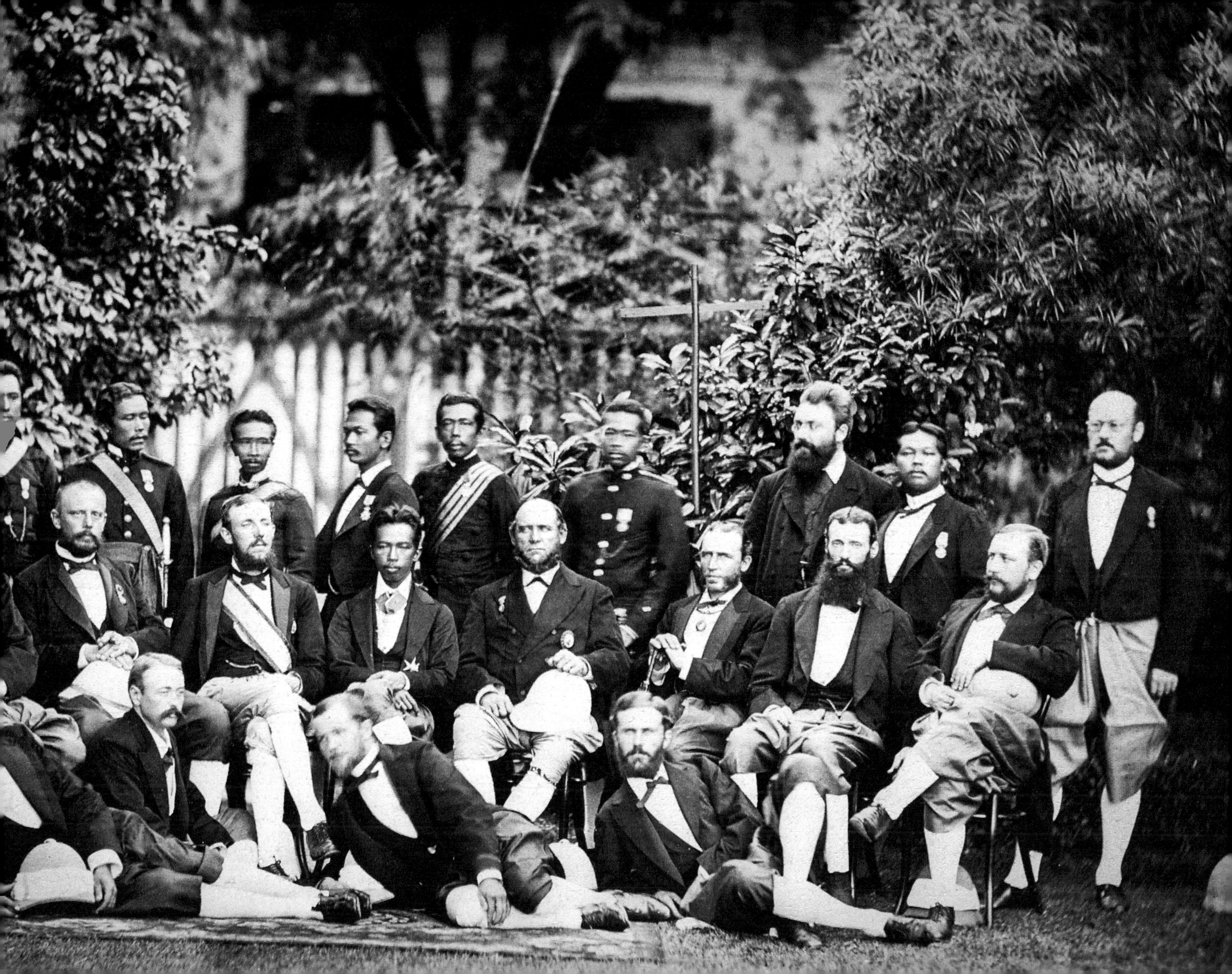

This page: In the lower part of the picture one can see the roof of a 'houseboat' moored in one of Bangkok's canals.

Opposite page: A view of Bangkok, with one of the city's 'canal streets'.

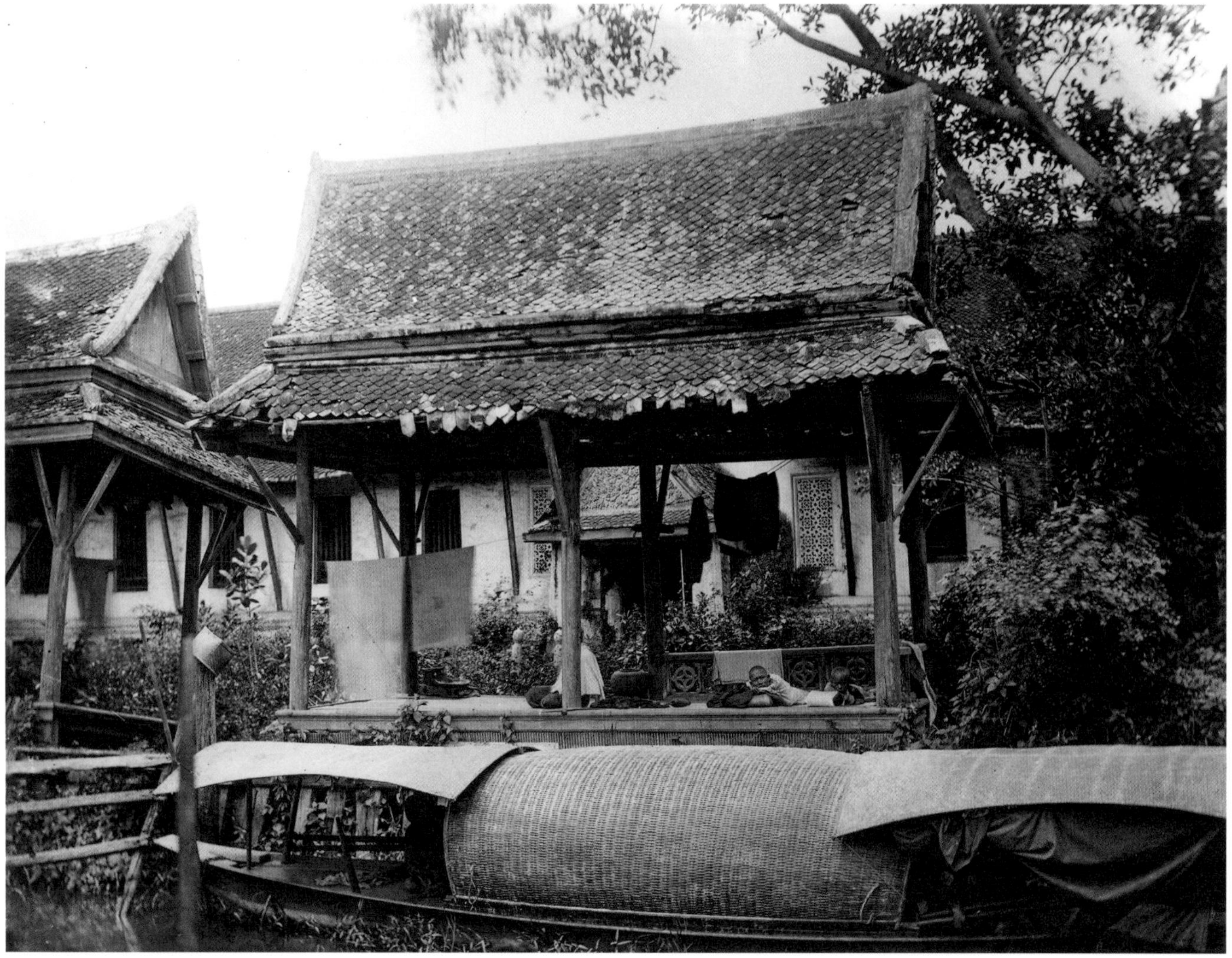

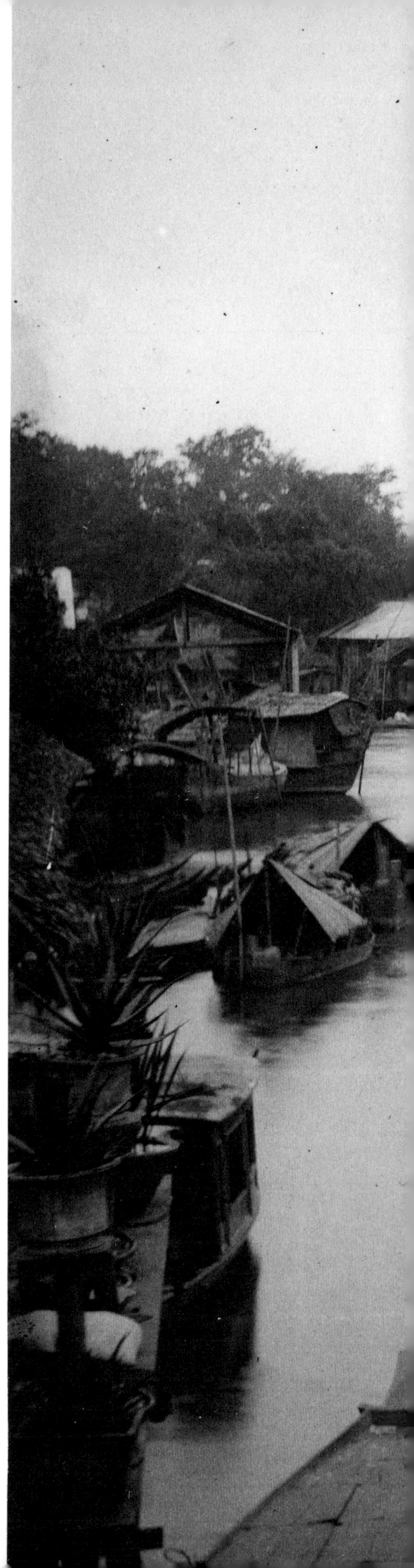

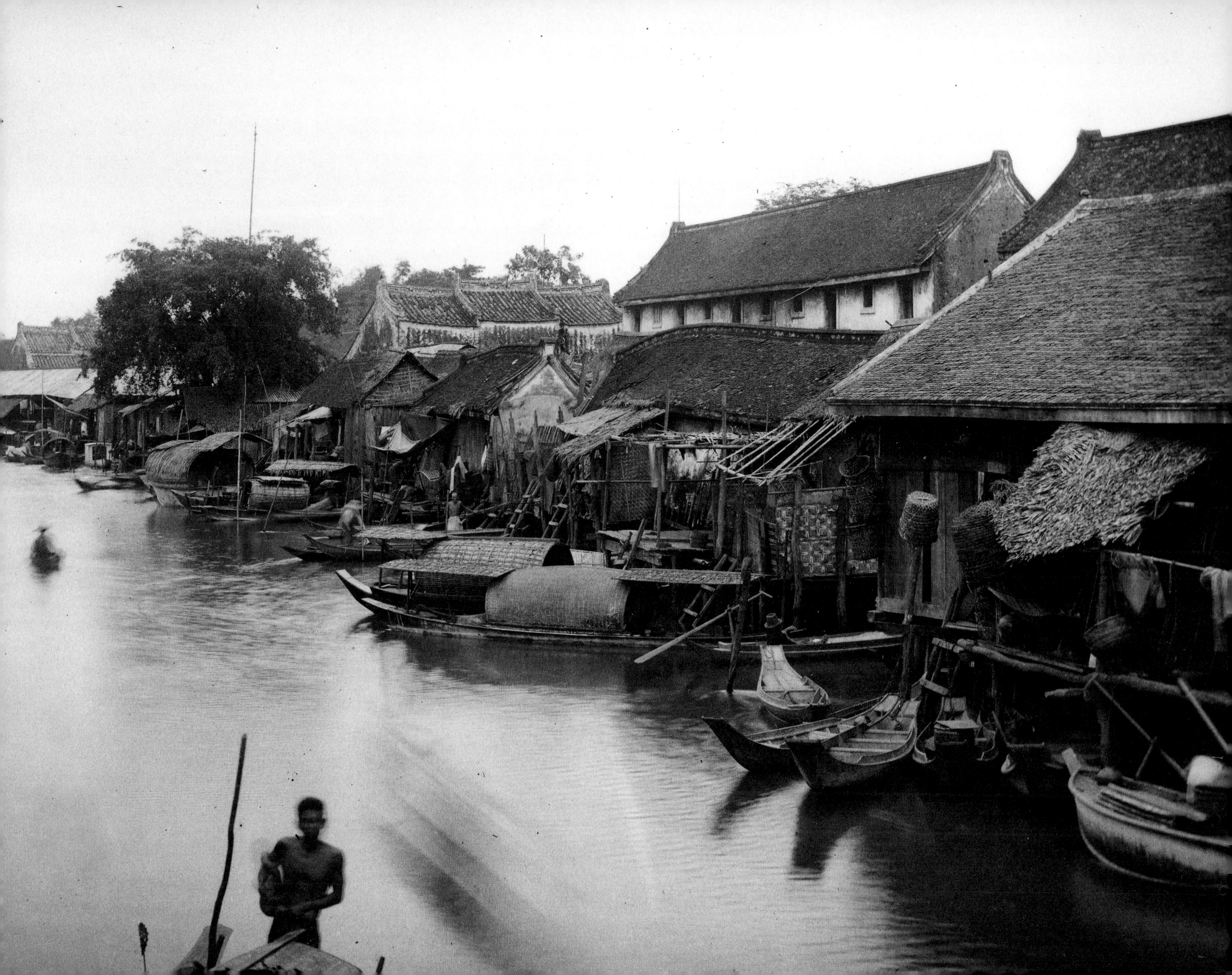

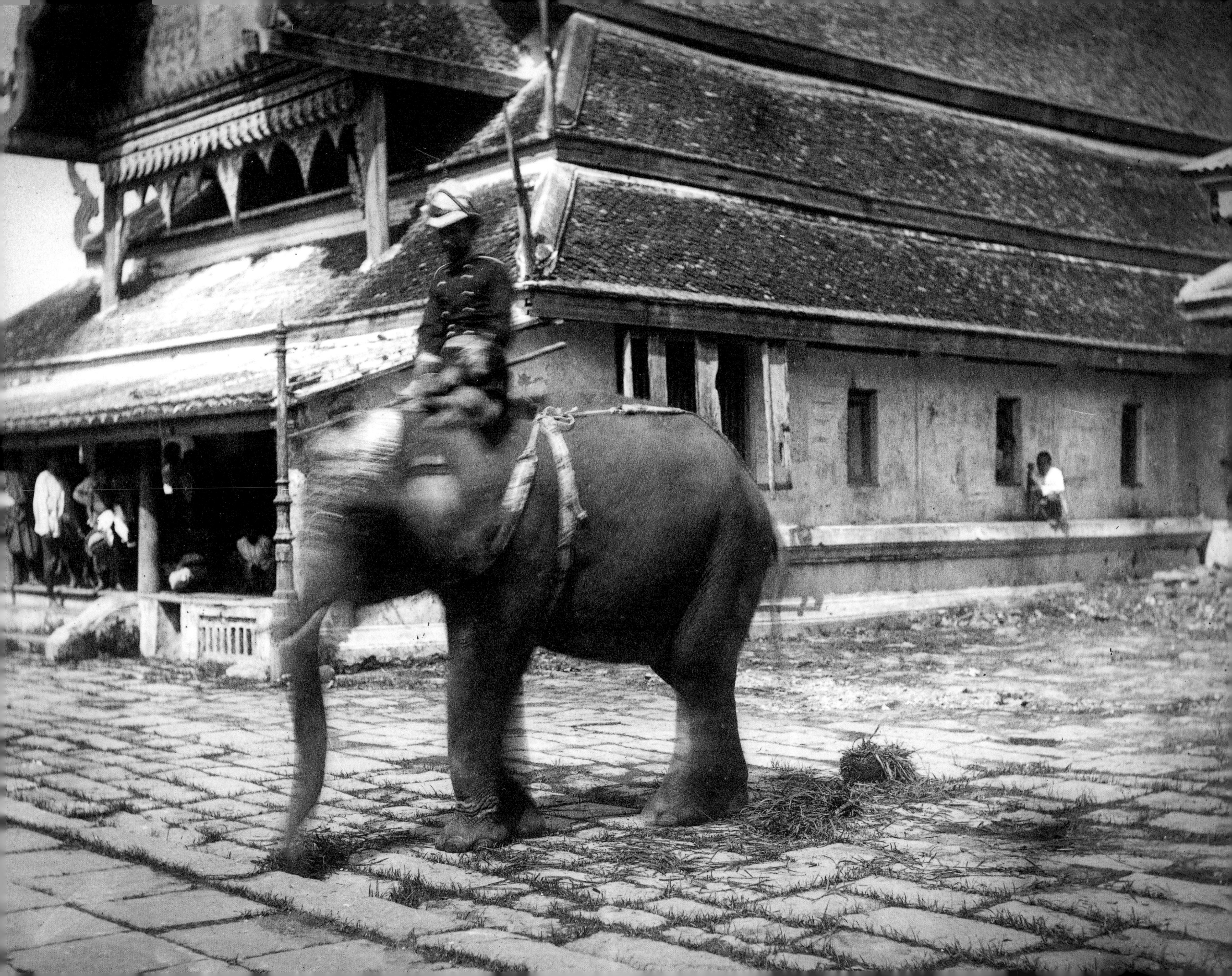

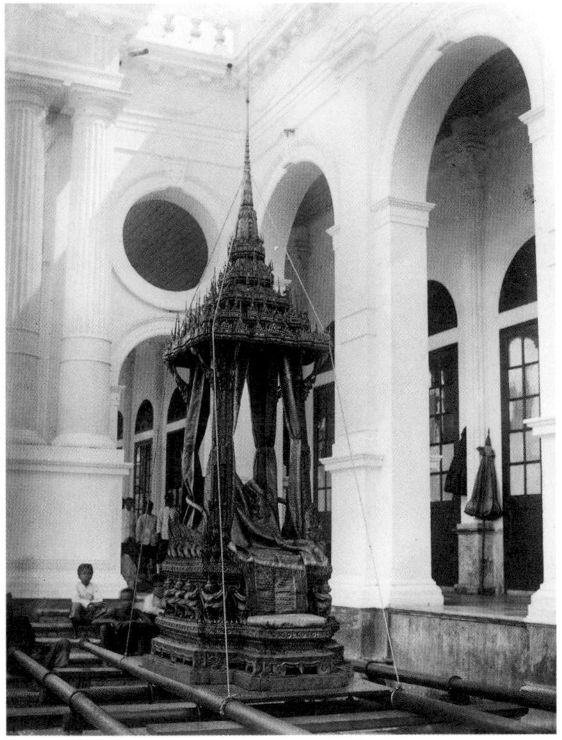

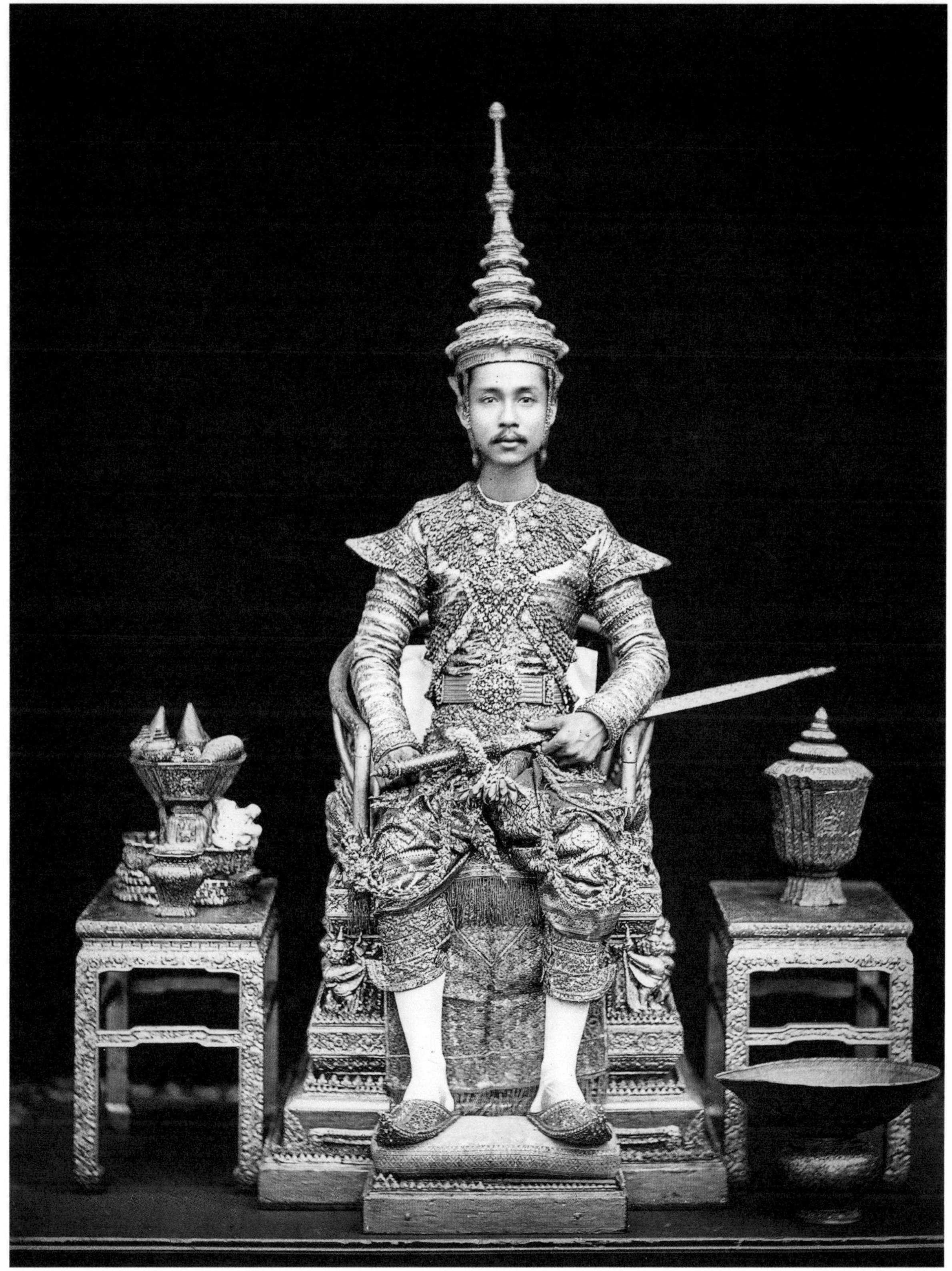

Opposite page: In Bangkok Oscar photographed this elephant, which to his irritation moved at the moment of exposure. The uniformed rider probably belongs to the police force.

Above: This is said to be the giant litter in which the King of Siam was borne on festive occasions.

Right: Hjalmar Stolpe took part in paying homage to the King of Siam, Chulalongkorn: 'What otherwise happens on such occasions is that you arrive in a group, proceed in turn and bow, then stand and watch for a while as the Prince [Oscar] converses with the King, after which you can go! The only enjoyable part is seeing lowly people desperate for decorations push their way forward and seek to make themselves noticed.' The king received the Grand Cross of the Order of St Olav, which was presented by Prince Oscar. Thanks to the presence of the latter, the men of the Vanadis were as usual given a regal reception, involving parading troops, resounding salutes and huge bands playing the Swedish royal anthem Ur svenska hjärtans djup [From the depths of Swedish hearts]. *The king liked Sweden, and in 1897 he visited the Stockholm exhibition and the region of Norrland. In the 1990s a Thai pavilion was erected to commemorate him in Ragunda, in the county of Jämtland.*

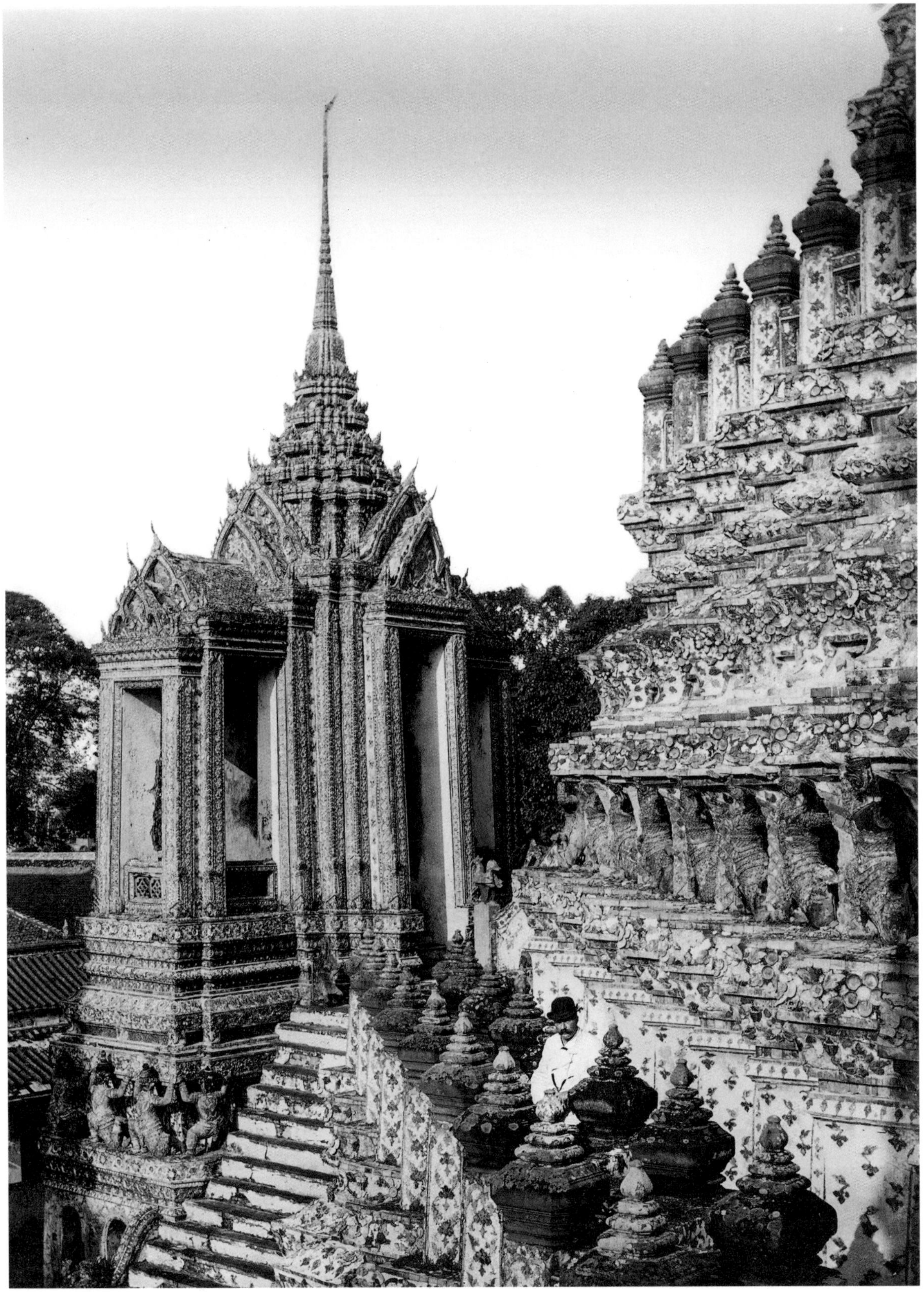

As so often with Ekholm's pictures, a small but elegantly dressed Stolpe pops up. With some difficulty one can pick him out in the lower part of the picture, wearing a bowler hat and a light-coloured suit. He is in Bangkok, and the main motif is the pagoda (temple) behind him. According to the text on the envelope holding the photograph: 'Pagoda of the Arun Rajvararam temple, height 81.85 m.'

Opposite page: During the visit to the Philippines, Stolpe added 80 or so photographs to his collection – over half of them purchased and the rest in the form of a 'gift of photos' from his friend the German ethnographer Adolf Meyer. The collection includes portraits, landscapes and settlements around Manila and Batangas. The text about the group on the beach includes the word 'Mariguina'.

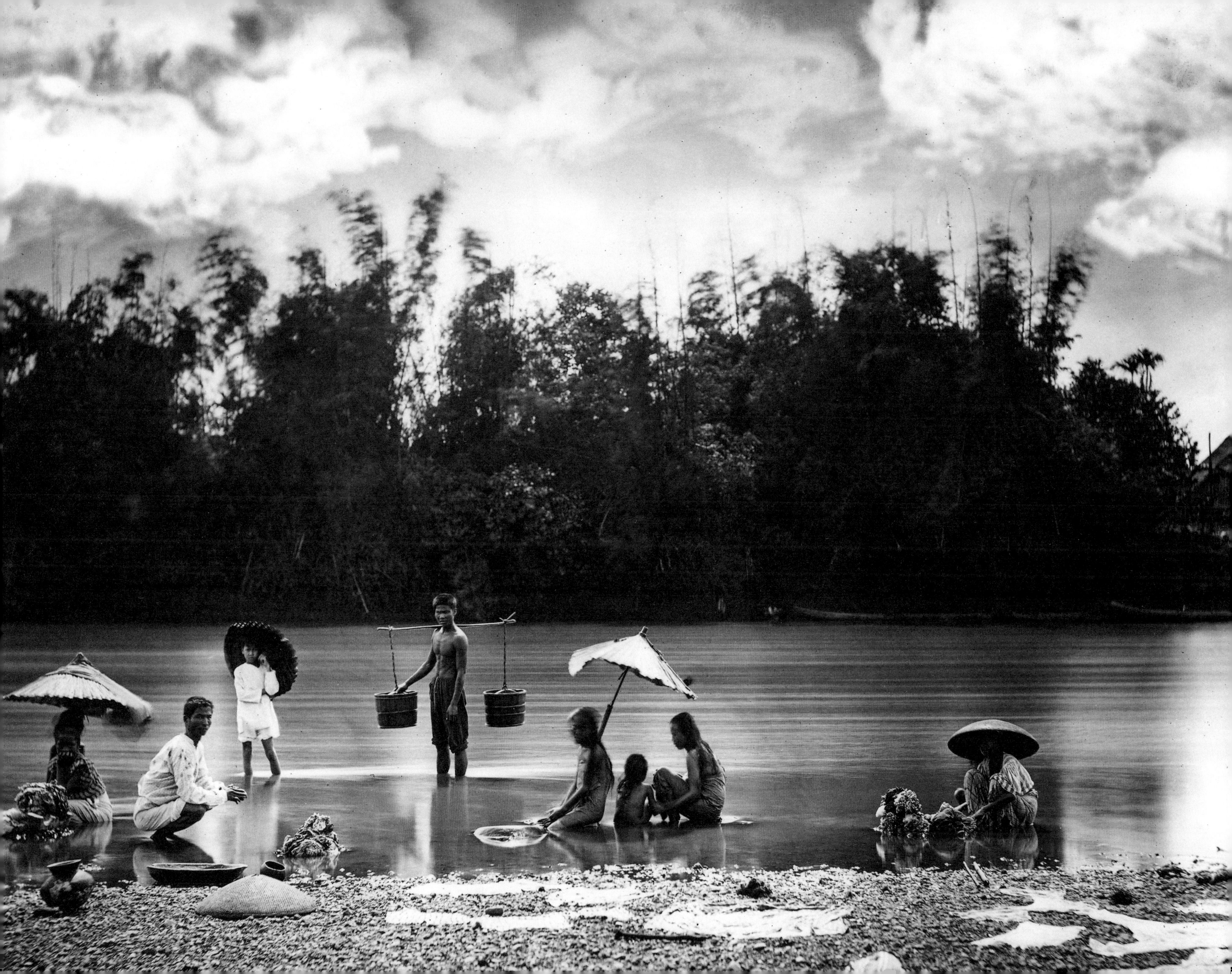

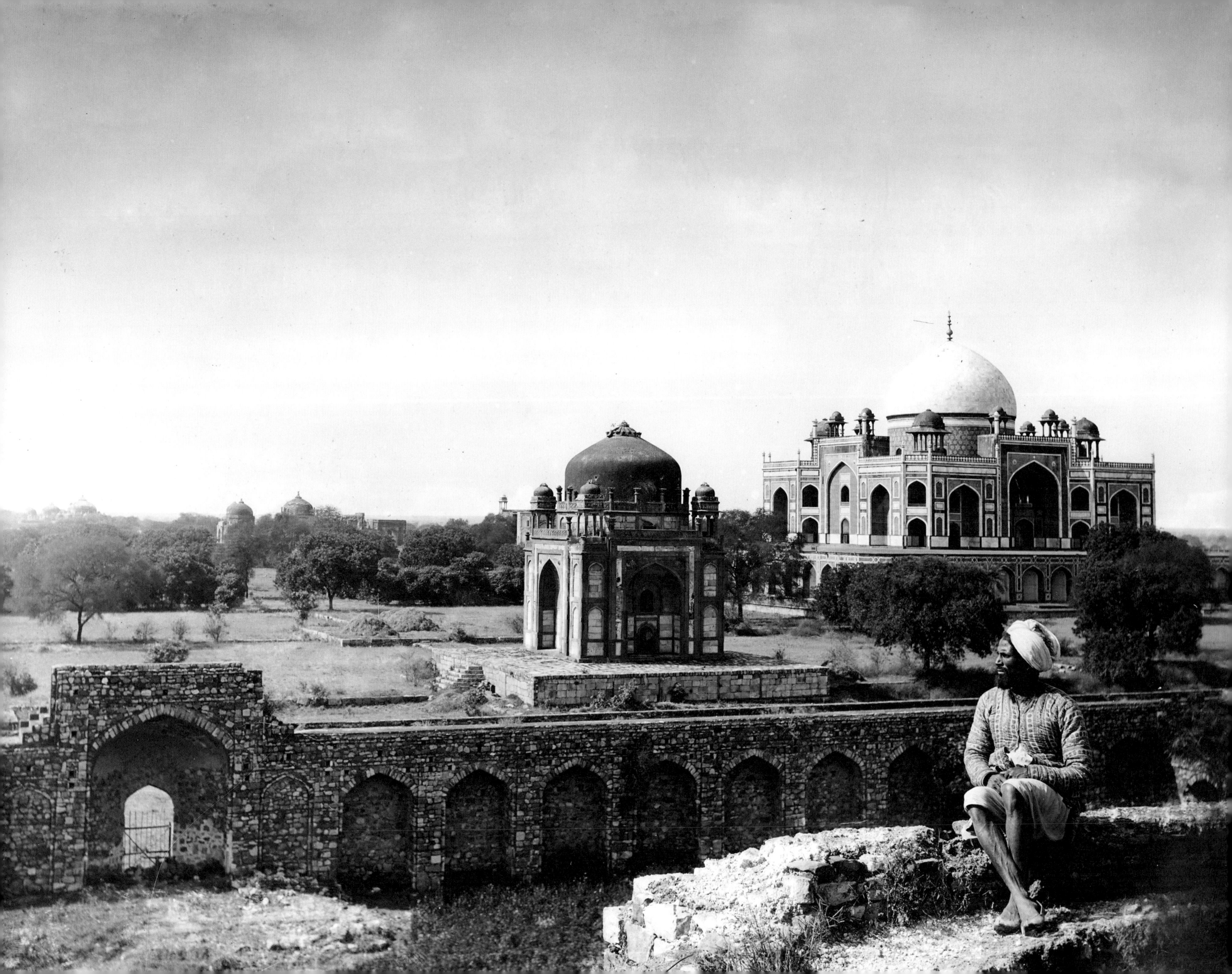

INDIA

When the Vanadis reached Calcutta on 19 December 1884, the journey around the world had been in progress for nearly 400 days. Ekholm and Stolpe had spent over half their time on board the ship. Hjalmar in particular had now reached a limit in terms of what he could put up with, both privately and professionally. Letters from Emmy enabled him to shut himself in his cabin on the frigate, where he 'enjoyed reading that everything was in order and seeing that I am cherished. It does you good, when you are for so long amongst indifferent or simply averse people!' His longing for his wife seems to have overwhelmed him, resulting in passionate words of love in cipher. The night before Christmas Eve, he wrote: 'Tomorrow at 7 o'clock Norrköping time I'll be drinking a toast to you – I maybe won't be able to do that, as service will have finished here by then, unless I take a bottle of beer into my cabin. I'll do that!'

Hjalmar Stolpe then made his dramatic decision. He abruptly broke off the journey round the world and left the Vanadis together with Oscar Ekholm. The ethnographic expedition was to continue on terra firma, and on his own terms. Behind the decision lay a general dissatisfaction with the conditions and the people on board, and the enforced idleness of life at sea: 'The sedentary life on board must have engendered the utter melancholy from which you see me suffering! My isolation and reticence on board are probably also contributing'. But the foremost reasons for his decision were a feeling of powerlessness and the inability to discharge the ethnographic remit of the expedition. With the exception of that full month in Japan, Stolpe deemed that the time pressure had been disastrous for his collecting of objects and photography. And this stress was heightened by the need for all the assembled objects to be quickly packed up and left behind before the Vanadis steamed off to its next destination.

The break led to a new rift – now a financial one – between Stolpe and the ship's commander, who, as Stolpe wrote back to his civilian boss in Stockholm, Hans Hildebrand, had once again erupted 'in rage at my being included on the expedition in the first place. Can't the fleet be left in peace, etc. – the same old story I've heard all the way'.

Hjalmar Stolpe was happy about the separation. The great ethnographic adventure could now start – a 3,000-mile journey across the Indian peninsular, from Calcutta in the east via Kashmir in the far north to Bombay on the west coast. This journey was to take three months, the greater part by train, but also by boat and donkey cart, on horseback and riding unruly elephants. For the first month Oscar Ekholm was his fellow traveller, before having to return to the Vanadis for the trip home, for financial reasons. Together they dragged along equipment and purchased objects, which were sometimes smuggled into the trains' toilets, to save spending money on transportation. In his travel invoice to the Royal Swedish Academy of Sciences, Stolpe later confessed that he had, on his conscience, 'the fact that I have as often as possible smuggled large or small boxes into the Indian railway carriages' very spacious toilet cubicles – an economy for which I quite rightly once had to pay a moderate fine'.

During the 90 days in India the yield of objects was enormous. Stolpe's initial listing recorded 2,700 objects, which tallied up to 3,000 after the cataloguing back in Stockholm. In the hotel rooms he filled wooden boxes with purchases embedded in straw, and then nailed them shut himself. A total of 40 boxes were sent to the Swedish consulates in Calcutta and Bombay for onward transportation to Sweden via London. The collections predominantly included those items used by the lower echelons of the population, such as household utensils, ornaments and clothing, and icons and objects of religious significance. And then there was what Hjalmar found especially interesting: objects made of brass and other metals, decorated with exciting ornamentation, plus tools for working these metal objects. Stolpe told his father Carl Johan – a retired juridical burgomaster in Norrköping – that he had absolutely no time at his disposal for sightseeing:

> I'm stuck in shops all day, and my notes frequently contain absolutely nothing but details of what I've bought. I'll fret over it henceforth – I'm now under the sway of a monomania for just collecting, and suffer when I cannot find enough to do in that respect. I hope it will one day be said of me that I worked like a man, and that I laid good foundations for a museum – on which others will in future be able to build.

Hjalmar's manic collecting in India is reflected in four small notebooks, which were evidently intended to be used in the planned – but never to be realised – book on the Vanadis voyage. The long letters home to his wife Emmy were also to serve as documentation for this book. What emerges is a picture of the trials and tribulations of collecting and travelling – with moments of happiness and a great sense of freedom. He called himself the most fortunate man in India. But there were also percipient insights into the political and social situation in this huge country, which

Not far from the Old Fort (Purana Kila) in Delhi are two of the city's best-known funerary monuments. The monument with the white cupola was built in 1569–70 for the Mughal Emperor Humayun. The small building in the foreground was built for his father Babur.

had become a British crown colony a good quarter of a century before the Vanadis arrived. Hjalmar became ever more critical of the Englishmen's advance, and this criticism grew into abhorrence at their arrogance and brutality and their senseless assumption of superiority. He predicted the collapse of the British empire:

> How blind they are! I'm convinced we'll see the day when England's Indian possessions will fall away and Australia will declare itself a republic. What will then become of proud England? A small and insignificant state. This is thanks to their policy of exploitation and their ruthlessness and perfidy. They rule here with lies, perfidy and, in extremis – poison! They've never kept a promise to a native state, and as they've not otherwise been able to get the better of the regent he's been swept out of the way! I've found out more on this journey than is commonly known in Europe.

The day before New Year's Eve, 1884, Hjalmar wrote to his wife from Calcutta:

> I'm still here, and it will be a week before I set off on my long journey. Yes, I've burnt my boats (if I could only get to tootle on our old boat in real life!) and taken my leave of the frigate ... I could now say a good deal about Calcutta. I've strolled through pretty much the whole of the city, through all the main streets and a number of labyrinths between the mud huts. I've walked amongst the most wretched holes. Everywhere I've seen something for sale I've made a note of it, but have not yet bought anything, as you first need to seek out the best, and the city is big – very big.
>
> Early tomorrow morning I'll start buying, accompanied by my slave, a Hindu, who constantly stands behind my chair and furthermore goes in and out of my chamber, looking as if he thinks he's doing me a service. I'm staying at the Great Eastern Hotel, right at the top in a little cabin, or rather a room built on the actual roof ... The dining hall is very long, with a long table along the middle and small ones along both sides, and with all the places fully occupied. If you just imagine at least 120 or so people eating at the same time, and each and every one of them with their turban-clad slave behind their chair, or fighting with their companions about the food on behalf of their masters, you will get an idea of the crowd of people gathered here. About the city itself there's little to be said. The Englishmen have all of a sudden tried to make a little London out of it, but are nowhere near finished with that. Colossal open areas are to be found, with giant palaces in the English Gothic or Renaissance styles, but in the midst of them is an entire city of mud huts ...

Ekholm and Stolpe remained in Calcutta for three weeks. The city's museum had an ethnographic department, and amongst other things Ekholm photographed 'a number of old stone sculptures'. Stolpe was also in contact with the German trading house Ernsthausen & Co in Calcutta, which had offices in London.

In Hong Kong, Oscar Ekholm had apologised for the absence of letters from him to his main correspondent: 'Dearest Brother! My writing muse has never been as silent as right now'. But in Calcutta off he went:

> Allow me to escort you on a walk here. The first thing we have to do here is turn a deaf ear to everyone offering their wares on the streets, which in my opinion they do noisily. Then we need to make sure we don't trip up over the praying Hindus lying with their noses to the ground facing the sun – or in the case of Arabs with their noses facing Mecca.
>
> Upon your next step you will come upon someone being shaved, with both the person being shaved and whoever is doing the shaving sitting in the middle of the pavement. If you cast your eye on the building frontage you will find glassblowers with their furnaces, type foundries, theatres for which the street is the salon etc. And finally incineration, which is extremely reminiscent of when mother used to fry herring on the embers.
>
> As well as this, there is the people's dark complexion, white turbans and light attire, which is the same for men and women (graciously throw a curtain around yourself – that is their clothing). Low, wretched houses, dirty streets over which thousands of ravens and birds of prey hover. That's Calcutta!

BENARES

The first major stop after Calcutta was Benares, on the Ganges, 3,000 years old and today one of the world's oldest extant cities, now going under the name Varanasi. It is held to be the god Shiva's earthly place of abode, is home to 1,500 temples and is one of the Hindus' holiest places. Pilgrims flock there to wash away their sins in the holy river, often at the final stage of their life, and to die on its banks. Ekholm and Stolpe arrived from Calcutta by train, the ethnographer noting:

> The following morning we got to the river in good time and took a big rowing boat, on the cabin roof of which we sat, observing the 1,000s of bathers along the shores of the Ganges. We rowed slowly right through the city. Everywhere jetties teeming with the pious, who were to wash in the holy river. You could not imagine a more beautiful backdrop: tall palaces and temples.
>
> We also photographed 2 cremation sites on the banks of the river, amongst all the goings-on. No enclosure – merely a pile of wood right at the water's edge. On one of them, which was splendid, lay the body of a woman on a bamboo bier, half in the river, and a man went and splashed water over her with a coal barrel. She was to have her ablutions before her immolation ... All the broad stone steps leading down to the river

are full of people of all ages, sexes, castes and statuses, dressed in all the colours of the rainbow – mostly red, white and yellow. From the stone steps, on which a host of religious symbols (Mahadeo or Linga-Yoni) are to be found, narrow wooden jetties extend, at the end of which a devout worshipper usually sits, staring motionless at the water while mumbling prayers or attending to his ablutions.

While the boat glides past I make notes on how everyone is behaving, and the photographer has his work cut out. No-one appears disturbed, and we put to amidst the bathers, and ascend the steps to see temples or photograph ethnic types. The people freely comply with my wishes. I sometimes find it hard to stop too many of them getting into the photograph.

One of the most interesting visits was that to the so-called Nepali Temple – a small temple devoted to Mahadeo [the god Shiva], made of tiles edged with magnificent woodcarving (door and window edgings in a style that should not be studied in the company of women, unless with you!!! Preferably put into practice!!!).

Ekholm was extremely busy during the five days in Benares. In total Stolpe recorded 40 or so motifs in his notebook, including a series of photographs from the excursion to Sarnath, a centre of Buddhist pilgrimage. In one of the photos Hjalmar Stolpe can be seen as he often appears during the Indian journey – a research traveller and gentleman typical of the period in a suit and bowler hat. In the letters of recommendation he had brought with him he is presented as a professor, with the additional information that he 'came out with the Swedish Princes'. Oscar has placed him in front of a huge tower known as a stupa, inside which there is normally a Buddhist reliquary. It is probably the so-called Dhamekh stupa – over 2,000 years old, and now a ruin rising above Sarnath's old temple area.

The number of objects purchased in Benares was impressive, and included objects made of brass. Hjalmar stated contentedly that three boxes had been sent to the Swedish/Norwegian Consul in Calcutta, but he was not as happy when they were being prepared for transportation:

> Packing collections in India is an abominable task. You can't get any decent assistance, and if something is needed (straw, paper, nails etc.) and you tell a Hindu servant about it, he will immediately reply 'Very good, sir!' and will hastily go off, but after ½ an hour you'll find him standing there enjoying the interesting spectacle of me sweating, packing and cursing about the straw not having arrived. I will immediately accost him and ask where the straw has got to, and will get the reply that he'll go and see if there's any in the hotel's stable, otherwise he'll take a carriage and ride to a bazaar! The most inferior people are dreadfully indolent.

In another hotel room, in Lucknow, things nearly came to blows when Hjalmar was trying to pack up his objects. A quarrel developed with:

> an impudent young Englishman who came into my room and forbade me to do any nailing, as he was sitting outside on the veranda, drinking on his own. I of course chucked him out but he came back in, and I had to call for the landlord, who took the wretch away. It would have been a nasty punch-up had I not controlled myself, as the fellow was big and strong. It was because of the surprise that I was able to throw him out the first time. Oh how I hate these Englishmen! I've never encountered a more ill-mannered rabble! I'll cheer the day they're driven out of India. They've frankly earned it through their thoughtless way of treating people.

ALLAHABAD

The next stage of the ethnographic journey was to Allahabad, situated at the confluence of the Ganges and the Yamuna (Jumna). Like Benares, Allahabad was and is an important place of pilgrimage for Hindus, who bathe in the waters of the river in their hordes. Hjalmar walked along endless alleys with bazaars, noting:

> My best work there was a whole day at the confluence of the Jumna and the Ganges outside the city, amidst the 1,000s of pilgrims who go there in January and February to bathe. Actually, 3 rivers meet there – the aforementioned and one invisible from above. The surroundings were different – just a bare shore with Akbar's Fort in the background on one side. The large number of people had attracted bazaars to set up there; there was a full market, and I bought a lot of excellent things, which were admired by collectors when I got home. I was particularly successful in terms of old brass, a speciality that I'd fallen in love with and that is one of India's most remarkable fields of pursuit. The textiles are largely English, and I allowed myself to be atrociously swindled in certain instances. I took a lot of photographs in the fakir camp.
>
> I must say, the beggars were *gracious* to me. All attempts to photograph women failed. They moved, or started to cry at the critical moment. Another difficulty is that you're not allowed to touch them in order to put them in the position you want – you should just see me standing and posing to show how I want them to be!
>
> I also visited a couple of extremely eminent fakirs from Oudh (Ayodhya), who were being entertained with music, which I attentively listened to for a good 1/2 hour. A string instrument and a drum, but touchingly beautiful – the drum was barely noticeable, apart from its marking the beat. I had no interpreter, and was dependent on my own cunning. I thus got out my pencil and drew the two eminent gentlemen *pretty* successfully, and showed them.

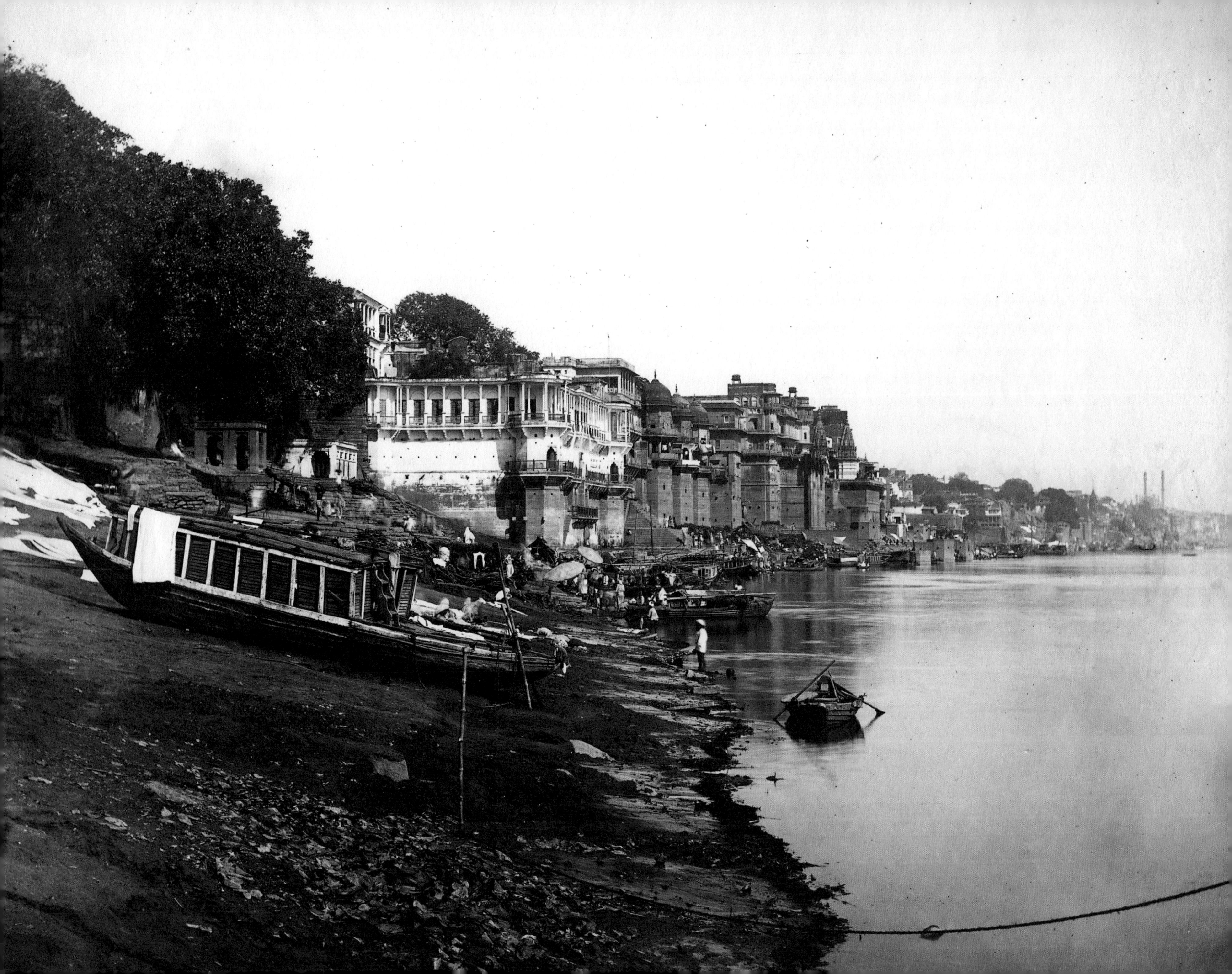

While I was occupied with one of them it was amusing to observe the other's jealousy. When I'd finished he looked worried, and his eyes followed the book and pen when I put them in my pocket. After a while I took them out again to draw the jealous one, and you should have seen his contented and proud expression! He adjusted his coat, sewn together using motley bits of cloth (rich penitents sew their clothes using all kinds of gaudy pieces of cloth, so as to thereby imitate poverty) ... Both of the men I sketched had the mark of the Shiva sect garishly painted all over their forehead.

When I'd duly kindled the vanity of the most eminent one (the one last depicted) by giving him to understand through mime that I would show him to the 'King', I got him to move out of the tent to be photographed, though he insisted on sitting under a canopy. I at last got him out into the sunshine, and at the final moment pulled away the parasol a servant was holding over his holy head.

But the major places of pilgrimage, Benares and Allahabad, were dominated by a completely different type of fakir. Stolpe encountered hordes of them, 'with ordinary fire tongs in one hand, sometimes holding a little brass bell'. He added:

Nothing is deemed to be as advantageous for tearing your thoughts away from this sordid world as sitting staring into the fire, and in the fakir camps one thus normally sees them seated, each with a small smoking fire in front of them. The poker is for tending this fire. In the crowd of people the holy men clap with the poker's prongs or beat them against the little bell, and the horde of people will immediately part and let them come forward. Their eyes are usually inflamed from all the staring in the smoke, and fakirs with their matted hair, angry eyes and ash-smeared bodies are not an attractive proposition.

Ekholm noted with some surprise the way that the people in Allahabad reacted to encountering the two foreigners:

On the streets one is avoided, and indeed many people look at us with repugnance, especially women, who take a really good look, pull their veil across and, if the street is too narrow to avoid us, stand with their nose pressed to the wall – the worst being young unmarried ones.

The young women are very attractive, fabulously big black eyes with long, dark lashes, which are artificially made even darker. But in the left nostril they always have a ring, sometimes up to three inches in diameter, and they wear big bracelets, finger and toe rings, and on their shins ones as big as bootlegs, all made of brass. The men usually have a red stripe across their forehead. Amongst the souls on the streets one finds magicians and snake charmers.

DELHI

On reaching Delhi, the two travel companions were to part company after spending 14 months together. Ekholm's finances did not permit him to continue. He took the train to Bombay to continue the journey home on board the Vanadis – and he did long for home:

I'm beginning more and more to count the days until we get home, and I welcome the fact that we're fast approaching Europe and Sweden ... On the outskirts of the city the trees are full of monkeys and parrots, and I bought two of the latter in Calcutta – I wonder if one of them will get home alive.

Before they went their separate ways, Hjalmar organised a final photograph from India, getting a picture taken in front of an old grave outside Delhi, and explaining to his wife:

I'm sending this letter with Ekholm, who's now going to Bombay. He's been instructed to attach these lines to the portrait, which he will copy between Bombay and Aden and send from the latter location. I might have wished for better staffage for this final photograph of the journey, but was unable to find anything. May it serve, however, to show that I was thinking of you whilst on the banks of the Jumna.

The long journey together with Oscar Ekholm on board the Vanadis and in India was thus now at an end. It was possible, in the long term, that being at such close quarters had had a slightly adverse effect on their relationship. And once they had taken their leave of each other, Stolpe was moved to note: 'I feel free and happy'. In Delhi a telegram from the government in Stockholm also reached him, formally allowing him to discontinue the circumnavigation. He thus slipped away from the Vanadis and exulted in this, writing the following in his notebook: 'I'm the happiest person in India'. One month later he wrote to his wife, in effusive tones: 'Best of all – I'm alone in India, travelling totally off my own bat, no longer with the Vanadis to think about, and am consequently the happiest man in India!'

But he was not totally without worry. In Delhi he had problems with his guides: No. 1 'was ignominiously hounded out for dishonesty', No. 2 'was an obliging blockhead', and No. 3 'is extremely useful, though he also deceives me'. He furthermore got into a row with a tailor from whom he had ordered clothes:

They were ready, but of course not in accordance with my order. 2 hours arguing once again about the price, which was agreed a month ago. I boxed the scoundrel's ears and

Hjalmar hired a boat on the Ganges so he could study Benares's beaches. Oscar took this picture from the roof of the cabin. The boat was possibly of the same type as is to be seen on the beach.

was on the verge of leaving everything, but then at least 4 days (that's how much time I'd lost in total because of the tailor) would have been wasted, so I had to take the clothes.

LAHORE

The next destination was Lahore, a cultural centre and the focal city in the Indian Punjab district, now belonging to Pakistan, and it certainly met with Stolpe's approval:

> I used to think Agra was the most interesting city after Benares, but now place Lahore ahead of it. Nowhere have I seen such beautiful streets as in Lahore. I'm really sorry not to have my photographer with me! But I couldn't afford to take him across Europe ...
>
> I can tell you, I've already got some beautiful things in Lahore! Old Kashmir copper jugs, Persian wine bowls with quotes from Hafiz [Hafez, 14th-century Persian poet], drinking vessels made of silver and gold, embroidered cloths – so there's enough for my little wife too. And everything's cheap, but it will be expensive in that we'll have to hold parties to show how resplendently oriental our little home is! My dearest little sparrow! But I don't buy a single thread without wondering what you'll say when you unpack it!

In Lahore Hjalmar studied the city museum under the guidance of the director John L Kipling, formerly professor of architectural sculpture at the art academy in Bombay.

> I got loads of casts of the magnificent stone sculptures from the Punjab, and diverse items of pottery ... I should get the best material for my book from India. But it's given me some work to do! I've been at the museum again today under Kipling's auspices, and I think I now understand the big question about the age and origin of the Punjabi sculptures.

The museum director's son, Rudyard, was working as a journalist in the city at the time, and ten years later would write *The Jungle Book*, about Mowgli, Baloo the bear and Bagheera the panther – with illustrations by his father. In 1907 Rudyard Kipling was awarded the Nobel Prize in Literature, on the occasion of which he visited Stockholm.

In Lahore, Stolpe observed, 'the Hindu carnival (Holi) is in progress, and everything's upside down in every city. The celebrations have finally ended today. They basically consist in spraying water at each other – coloured water – or pelting red and yellow powder over each other'.

Hjalmar had now been out and about for 16 months, and his longing for his wife Emmy became ever greater. For a long time her letters went astray, whilst he wrote assiduously and in detail about his movements and activities. And, during his final weeks in India, he enthusiastically added to his stock of presents:

> I've had lovely garments made here. It will be fun to dress you up in them! Amongst other things, I have a complete Mohammedan bridal costume for farming folk. I today got hold of various silver adornments from Benares, which I bought for you. I was again really surprised upon seeing the adornments. I can dress you *solely* in silver!

However, he added, by way of confession: 'Enjoyable though it is, I'm tired of this roving around, buying and – packing. I'll already be broke by the time I get to Bombay! The consul will have to make provision for me if necessary'.

SIALKOT

In Sialkot at the foot of the Himalayas, Hjalmar stopped over for a few extra days. The reason for this was that he came across 'the most zealous guide I've ever had in India, and decided to exploit this excellent opportunity to get everything I could out of him, and supplement and enhance my previous notes'. Sirdar Kehar Singh was not only knowledgeable but also honest – a tax collector by profession, one-eyed and fanatically religious, and a merciless haggler with everyone except his fellow believers, the Sikhs. In negotiations with them he said:

> I'm delighted to have found an honest man; give me a piece of paper and let's see. The aim was for the 'honest man', who without any qualms initially lied about the price, to write it on a slip of paper. And sure enough – without any introductory haggling he straight away wrote, for example, 4½ rupees for something for which he had *verbally* asked 7! A Sikh cannot lie in writing!

With the help of Sirdar Kehar Singh, Stolpe made 147 purchases in Sialkot. Sirdar, according to Hjalmar, translates as gentleman, meaning that his guide was therefore unable to charge for his services. One day he even told Hjalmar that he had 'fallen in a kind of love with me'.

KASHMIR

Hjalmar Stolpe got from Sialkot to Kashmir and its winter capital, Jammu, after an arduous trip by donkey carriage, though he rode on an elephant for the final section. It was the seat of the ruler, the Maharaja of Jammu and Kashmir, and his court. But in practice the alpine country was a British vassal state, acting as a buffer against Afghanistan and the expansive Russians in the north. The exotic landscape made a strong impression on Hjalmar, who had with him a letter of recommendation to the princely state's prime minister, Anant Ram, and was to be the grand prince's guest: 'A telegraph has been sent to the Maharaja of Kashmir, saying I was coming.

Nobody gets into the country on other terms'. In the letter to his wife, Hjalmar gave extra detail about the donkey trip:

> What I've experienced today is already sufficient material for a book, or at least a chapter. The journey from Sialkot was comical to start with. The donkey on the right brayed, and the one on the left thrashed about with its rump. I had a carriage in which I sat as if under arrest, and was unable to see anything other than to the sides – the Himalayas being to the front. The country was flat, everything washed down from the Himalayas, acacia avenues all the way, and every hour a little village with a Shiva temple, which I was not allowed to enter. At one place the priest was kind enough to give me a handful of small raisins and a few bits of dry coconut, which assuaged my hunger.
>
> I crossed the river, coolies carrying my things, sauntered up a paved hill to a city gate and entered a city where everyone watched me with extreme mistrust. Eventually a man came up and greeted me in the usual manner by raising his palm to his forehead and repeating several times 'Eska nami' [what's the name], whereupon I finally gave him a card. He looked embarrassed, as he couldn't read, but he went. He was a policeman, and should have known who I was.

In Jammu, Stolpe was presented to the Maharaja:

> An extremely handsome chap with a white beard, attired in yellow garb adorned with red Kashmir embroidery and a silver, white and red turban ... We sat out on a big open veranda. Before us lay the open plain towards Punjab, and in the foreground were some temples – one with a gilt tower. The view was glorious in the evening light. While we were talking everyone who was not part of the retinue had sat down at the far end of the veranda, where a lively discussion was in progress. It was the very day people gathered for debate. We now all went there, and sat down on the mat cross-legged, with me to the right of the maharaja. The discussion was to be conducted in Persian, and was to concern the nature of God ... My legs started to tremble and ache, but it didn't matter. The sun set during the discourse, and two torch-bearers then came and lit up the scene – the strangest thing I'd seen since King Kabua sat and danced on Jaluit by torchlight.

But after three days Hjalmar Stolpe left Jammu in an angry mood. The reason for this outburst was that the city's merchants, whom the Maharaja had instructed to come to the palace to display their wares, had done their best to fleece him. He drove them out, and instead visited bazaars and temples: 'They were busy building a tomb, and I did my best deal in Kashmir, buying a full set of all the masonry tools. On my way back I got an almost complete set of blacksmith's paraphernalia, and learnt how to shoe horses the Kashmiri way. There are no limits to an ethnographer's work'.

When Stolpe left the city he was offered a free elephant ride by the Maharaja, and took the opportunity to acquire the implements with which the mahout controls this enormous animal. He later produced it at the Vanadis exhibition in Stockholm, describing it thus:

> When riding an elephant the mahout, who sits on the animal's neck, wields an instrument (*ankush*) in the form of a short iron boathook, with which he strikes the elephant on the forehead if the colossus shows any signs of not wanting to subjugate itself to the weak human's will. One of the examples exhibited here saw service when early one morning in the spring of 1885 I was leaving the winter city in Kashmir – Jammu – and had to cross the surging River Tavi.
>
> The elephant – one of the Maharaja's biggest, and about which I'd previously complained on account of its jerky gait but which I was now once more allocated for my departure – was in a bad mood and on no account wanted to go out into the water. It trumpeted tremendously loudly and looked as if it wanted to gallop home again, but when the mahout eventually managed to powerfully strike it several times in the middle of its forehead, so the tip of the hook penetrated its skin and went right down into the bone, it became compliant for a while. Having reached the middle of the river, however, it displayed an inclination to lie down, at which juncture the experiment had to be repeated – applying even more force, if that were possible.
>
> Another way of achieving obedience is to drive the implement's pointed tip down into a small wound that is always kept open behind one ear. If an elephant is to be made to go faster, the driver walking behind it jabs it in the hind legs with a short stick furnished with an iron spike at one end – this, too, a memory of the last elephant ride in Kashmir.

When Hjalmar and the trumpeting elephant had crossed the river they were met by a carriage – but it 'of course broke. No carriage can cope with the way people drive here. It's gallop, gallop all the way. Several times I was sure it would overturn. I finally sat on the box so I could see something'.

JAIPUR – AJMER

In Jaipur (Jeypore) the contact was Doctor Hendley, the director of the city's museum. Hjalmar undertook an excursion that he describes as one of the most enjoyable of his whole trip. The destination was the fort in the principality's old capital Amber, about seven miles from Jaipur and in ruins:

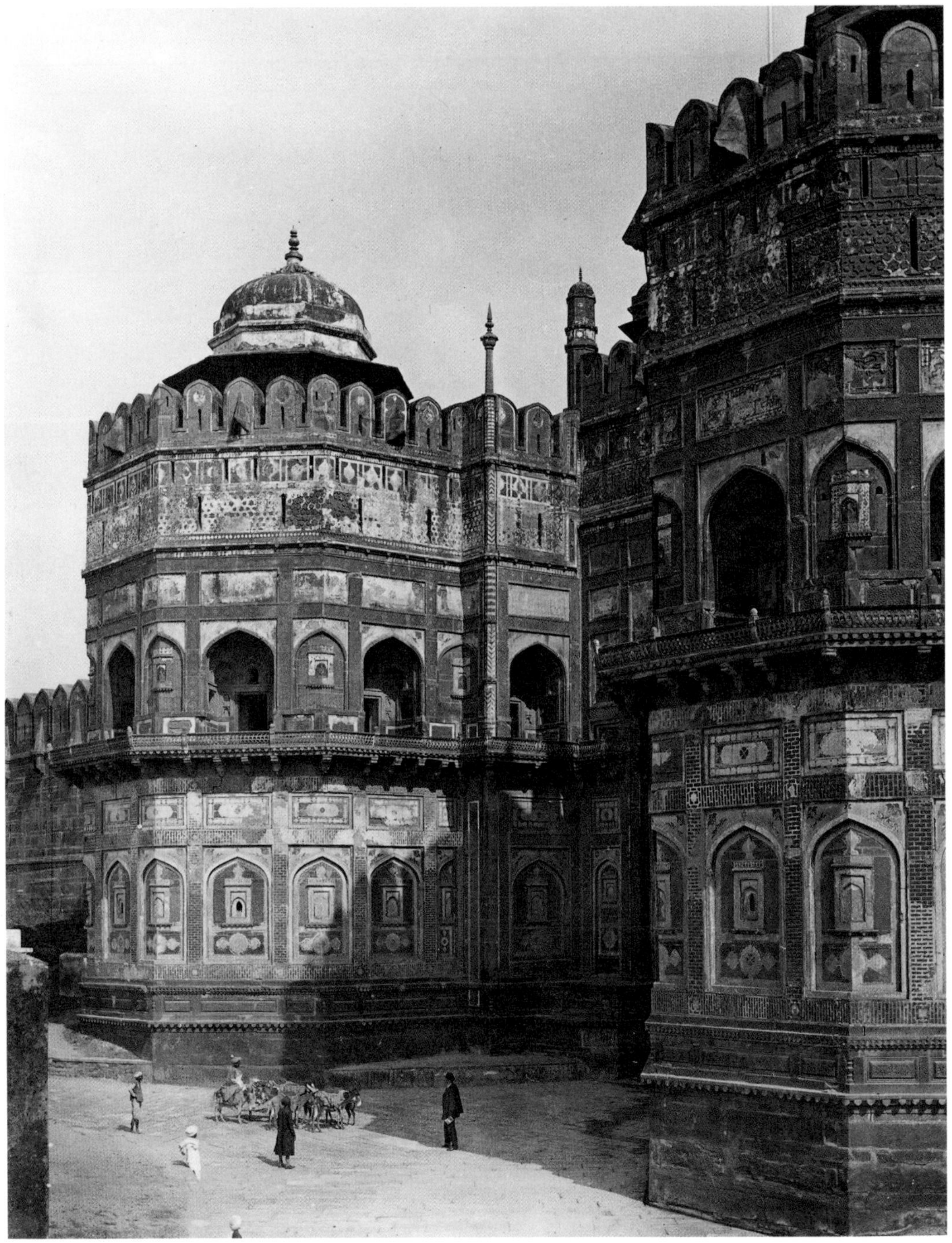

> I was to see the old Amber Fort. I did not set off until half past six in the morning, travelling by carriage through attractive burial sites (the excellent marble kiosks erected over the ash urns), gardens and palace in ruins in the Amber valley, surrounded by high mountains covered with cacti, which had just started to flower, finally reaching a place where the road became too steep for the conveyance. You always encounter elephants there, and I once again got to practice that art. It was a colossal beast. You never see such tremendous animals in Europe. The landscape is extremely beautiful, despite the scant vegetation. It is the colour that does it, and the remains of buildings. On the ridges there are walls, and here and there towers.
>
> The valley finally opened up, and Amber Fort was to be seen in front of me, rising high above a small lake, in which it is reflected. Oh if only I had my photographer with me. I thoroughly studied the palace, did drawings and made descriptions, so I have something for my book. It's a long time since I've been so interested in a building. Down by the lake were two gardens. I went through one of them. The lake was teeming with crocodiles. I found it hard to leave this place, but had to climb up onto my elephant again and get to the city before about 2 pm ...

The next place Hjalmar visited was Ajmer, from where he went on an excursion to the holy Pushkar Lake and the Temple of the Sun God, which he drew in his notebook, attaching an amusing note. For some reason he wanted to fish something out of the green sludge in the holy lake, which was teeming with small fish:

> But how was one to reach it? Anywhere in our fine country you could doubtless find a *pole*, but in India you *never* find what you need. In desperation I dashed up a tree, but the branches were too tough and brittle. I hung onto one (after having indicated in vain that someone should give me a knife), the branch cracked as a result of my Nordic weight, and there now commenced a twirling, twisting and running-around, whilst the holy gentlemen gaped in surprise. I ultimately expected that my crazed behaviour would cause them to deem me to be holy and smear my body as if I were a fakir. The branch at last broke off, and I went down to the lake and fished up the muck. And it was now of course assumed I was taking it as a relic.
>
> At a jetty below a temple there were a number of bathers in the holy lake. Bathing on the part of an infidel should of course be prohibited there, but before I had grasped this a chap offered a bath for 20 rupees and finally, as I was on the verge of accepting 1, he shouted 'Ek rupia' in my ears until I lost my patience and made as if to strike out, whereupon not just he but 50–60 others took flight. In normal circumstances Hindus are cowardly. A modicum of courage and off they run.

This picture shows the old fort in Agra. Hjalmar is standing outside one of the four mighty gates – the western one, completed in 1568 when the Mughal Emperor Akbar had conquered the area: 'Just as we were about to take a picture of the entrance a donkey driver came along, but I succeeded in getting the whole lot of them to stop. It was an excellent picture. I'm also to be seen amongst the other asses.'

In the notebook there is also a chatty text under the heading 'Lying in India and elsewhere', which appears to be intended for the planned Vanadis book. He summarises his impressions, noting that 'a love for the truth is not exactly writ large in Japan'. He continues:

> But lying there seems to me to be of a different nature than in India. In Japan people lie in order to be agreeable; they guess the answer you want, and that's the answer you get. Nothing bad is actually meant by it. For reasons of civility there is a wish not to contradict the opinion the other person has formed.
>
> But in India lying is of a more unattractive nature – people lie there for their own gain, openly and without the slightest attempt at dissimulation. Audacious and blatant lying straight to my face has so often caused my right hand to quiver in anger! I would draw the distinction that people 'fib' in Japan but *lie* in India.
>
> I remember asking the bookseller Babu Sushan Das in Benares whether he could provide me with someone reliable to go round the bazaars with me. 'I am very sorry Sir,' said the old man, 'but I don't know any such man here in Benares'. The only ones in whom truth is to be found are Sikhs.

BOMBAY

Bombay was the final stop-off on this leg of the journey. In total, Hjalmar noted in his statistical tables (to which he so liked devoting himself), 2,488 objects were purchased during the Indian journey, and together with gifts the number of objects rose to 2,700. The statistics also include Hjalmar's washing in Bombay, and show what was required for a globetrotter and gentleman at the time: 18 collars, 20 cuffs, 7 day shirts, 1 nightshirt, 8 handkerchiefs, 3 pairs of cotton socks, 4 woollen ones and 2 sweaters.

Well ahead of time, by letter and telegram, Stolpe had made thorough plans regarding useful contacts during the course of the journey. At nearly every location of importance he was awaited by a contact, who would often offer food and accommodation and provide ethnographic objects. The imposing Doctor Stolpe was held to be an exciting guest and a pleasant companion who, when required, could sing a tune, a traveller from afar with many a tale to tell, experienced, energetic and purposeful. And thanks to the princes Oscar and Carl (who joined the party in Calcutta) the Vanadis expedition was, after all, graced by a royal lustre that made Stolpe especially attractive to both the Indian and the English elite. The fact that he travelled around the world with his own photographer and assistant doubtless also raised his stature.

The journey had been particularly successful for the two travelling companions. The practical result was extraordinary: nearly 10,000 ethnographic objects and 700 photographic glass plates were taken back to Sweden. A present-day observer might well be surprised that this venture could all be accomplished without major incidents or losses – the objects were fragile and the available means of transport primitive. Just as notable is the fact that nothing serious happened to the two travellers in what, at times, were extreme and exposed environments – no accidents, no bandits or violence, despite their equipment being spectacular and expensive, and the fact that those around them will of course have known that they carried cash and cheques in their pockets. Remarkably, they arrived home as healthy as when they had left, free of any diseases or tropical bacilli. The young Oscar Ekholm suffered minor ailments, and Hjalmar was at worst affected by tapeworm, which was successfully treated by the ship's doctor. That was all. And no major arguments between the two were ever noted.

They had parted company at the beginning of February, on leaving Delhi, after which Oscar rejoined the Vanadis in Bombay for the journey home. It was a good month later before Hjalmar left the same port. He contentedly recorded the exact time of departure: five o'clock on the afternoon of 20 March 1885. This finally marked the conclusion of the collecting element of the tour, and he was satisfied with what he had achieved. On arrival at Aden, where the threat of cholera infection from India was a real fear, strict quarantine regulations prevented Stolpe from going ashore – and the Swedish/Norwegian consul had to be conveyed to his vessel with the travel money that the Vanadis had left for him the month before. From Bombay, Hjalmar was forced to travel on credit, with the consulate's help. He ruefully noted: 'I ultimately found myself in a completely destitute position'.

Some weeks later, now travelling on the SS Mongolia, Hjalmar arrived in Brindisi in southern Italy, where he immediately despatched a telegram summoning Emmy: 'Directly Vienna expect you Dresden 14 April'. She replied that she had to take receipt of the keys for the leased apartment in Stockholm, and tried to tempt Hjalmar with an attractive new desk for her travel money. Hjalmar steamed on to Venice, and finally arrived in Vienna, where he erupted: 'Why are you doing this to me?'

The duel ended in Hjalmar, happy as a lark, meeting his wife off the train at Dresden. They had not seen each other for 16 months, but he observed that Emmy was completely unchanged, and described himself as follows: 'I've experienced the passage of time without it leaving any trace, other than the moonlight having perhaps become a little brighter'. His pate had certainly become lighter.

On 9 May 1885, the marine's cannons at Skeppsholmen saluted the frigate Vanadis when she dropped anchor off the Royal Palace. Oscar Ekholm was finally back home. Hjalmar Stolpe and his Emmy had already returned, beating the ship back when arriving by train from Copenhagen just a few days before.

The Vanadis expedition had come full circle and reached its final destination.

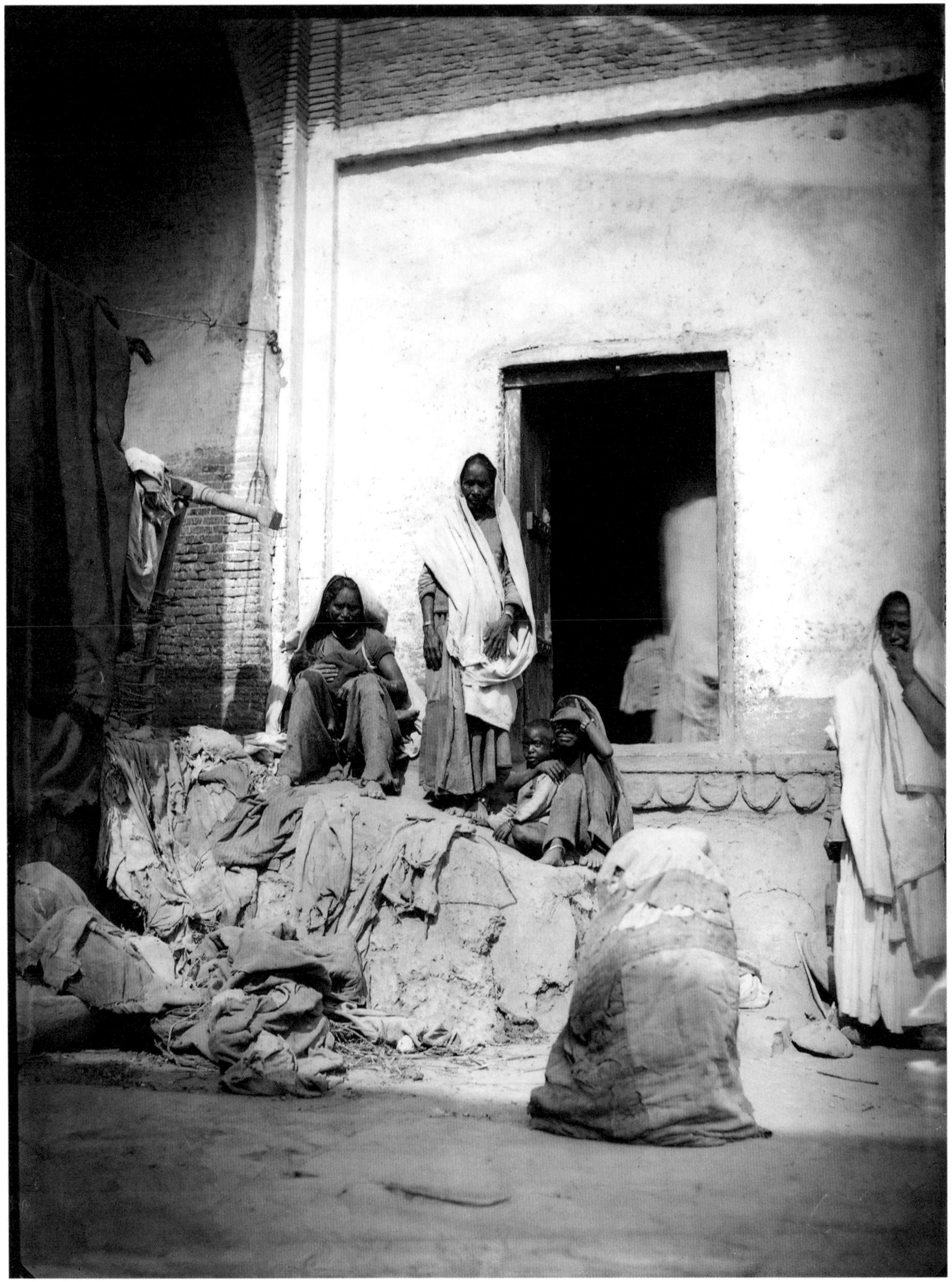

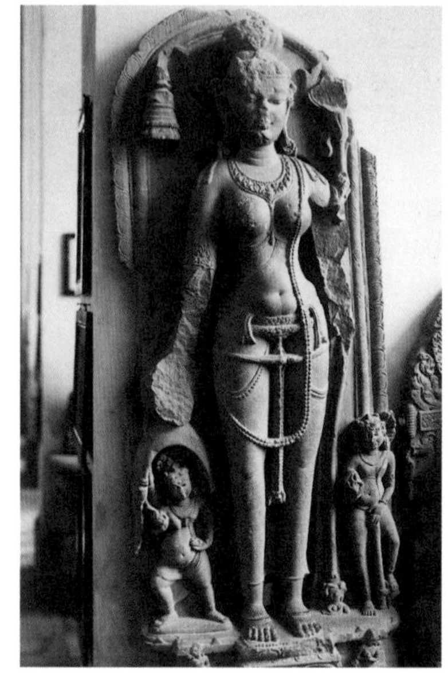

Left: 'Women and children, rags outside Hammam' is the caption in Hjalmar's list of photographs that he kept at Roesler's studio. They are probably waste pickers who collect cloths and rags, then boil them for reuse. It is a low-caste occupation that exists to this day. Oscar Ekholm took the picture in or near Agra.

Above: After Ekholm and Stolpe had left the Vanadis in Calcutta, they stayed in the city for a few weeks. In a letter home to his wife, Hjalmar wrote: 'We generally spent our time in museums, where a number of old stone sculptures were photographed. On the final day I also discovered an ethnographic department, but I got through it in 2 days.'

*Opposite page: In Benares Stolpe and Ekholm hired 'a big rowing boat, on whose cabin roof we sat observing the thousands of bathers along the shores of the Ganges.
We rowed slowly right through the city. Everywhere jetties teeming with the pious, who were to wash in the holy river. You could not imagine a more beautiful backdrop: a tall palace and a temple. ... All the broad stone steps leading down to the river are full of people of all ages, sexes, castes and statuses, dressed in all the colours of the rainbow – red, white and yellow. From the stone steps, on which a host of religious symbols (Mahadeo or Linga-Yoni) are to be found, narrow wooden jetties extend, at the end of which a devout worshipper usually sits, staring motionless at the water while mumbling prayers or attending to his ablutions. While the boat glides past I make notes on how everyone is behaving, and the photographer has his work cut out.'*

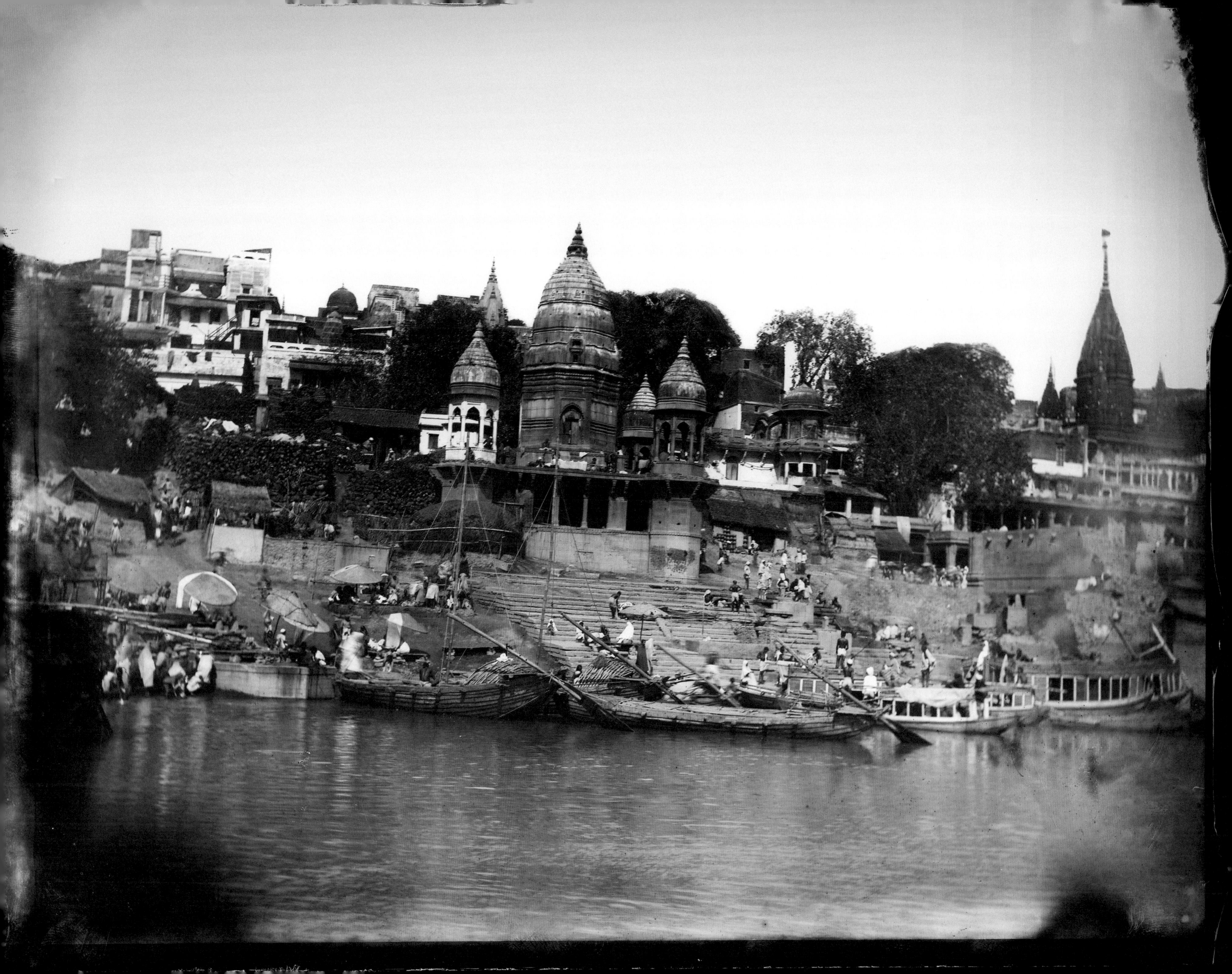

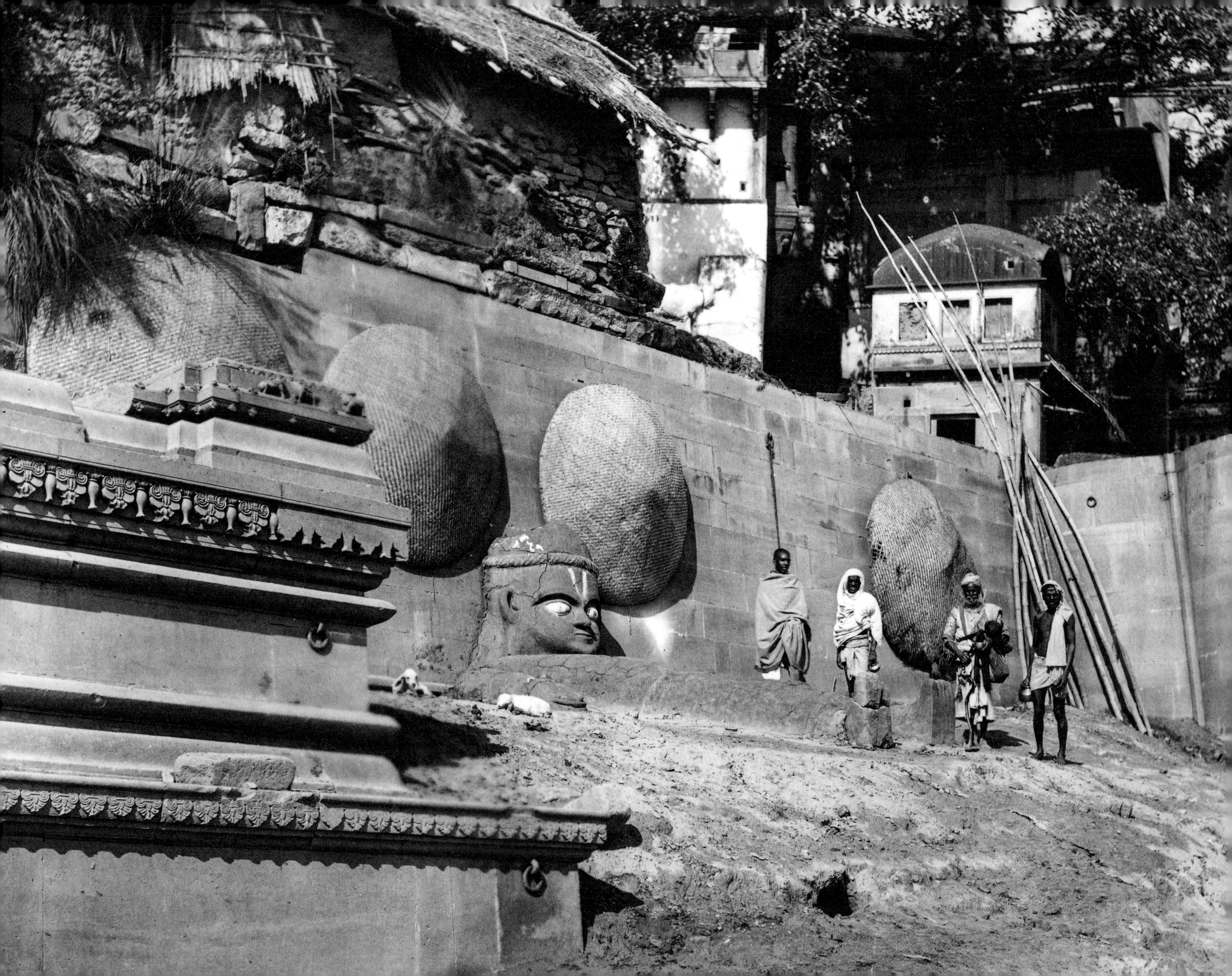

Opposite page: The pictures are from Benares and represent 'People at the cremation site'. The people are carrying out their ablutions in the holy waters of the Ganges, in order to wash away their sins.

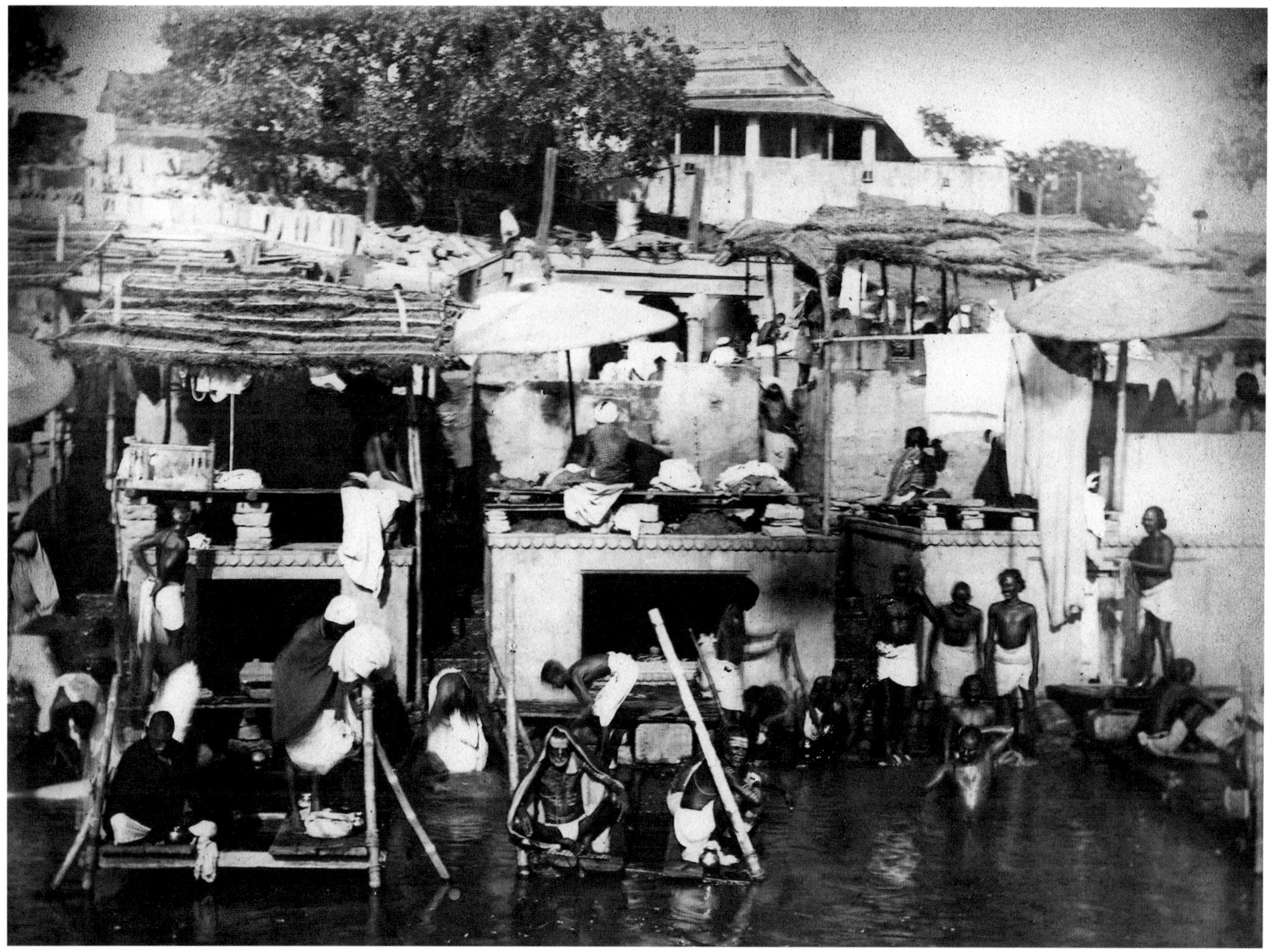

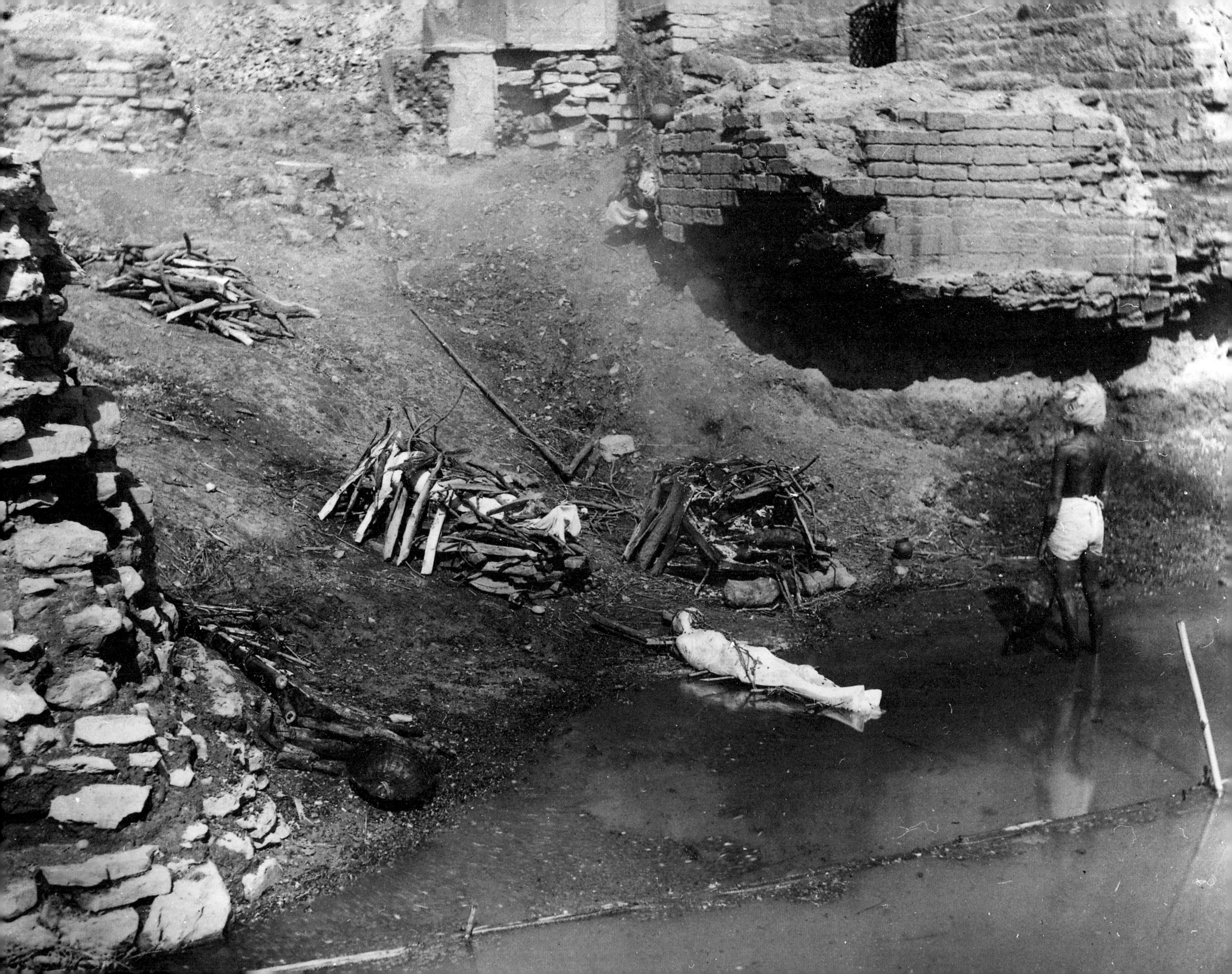

Opposite page: The city of Benares (now Varanasi) is around 3,000 years old, and one of the Hindus' holiest places, where huge hordes of pilgrims go to wash away their sins in the holy Ganges – often at the final stage of their life, so they can die on its banks. One day in January 1885 Oscar and Hjalmar were gliding along the river photographing cremation sites. At the water's edge 'lay the body of a woman on a bamboo bier, half in the river, and a man went and splashed water over her with a coal barrel. She was to have her ablutions before her immolation'. A few metres further up on the river bank a funeral pyre awaited her. When the Vanadis was at anchor in Calcutta at the mouth of the Ganges, it was noted (probably by Lieutenant Carl Edward Ulff) that every day there were 'numerous bodies going with the current, and on one occasion a corpse even got stuck on the Vanadis's anchor cable, whereupon some seamen had to go down and release the unsavoury guest'.

Below: According to Stolpe's notebook, this is a 'Priest under a sun shade'. The picture was probably taken outside the so-called Nepali Temple in Benares, dedicated to Shiva, one of Hinduism's most important gods, also known as Mahadeva. In a letter to his wife Hjalmar says: 'I became good friends with the Brahmin [priest] Burtug Nath, and photographed him sitting in front of a collection of stones out in the courtyard.'

Right: A snake charmer photographed by Oscar Ekholm in Benares.

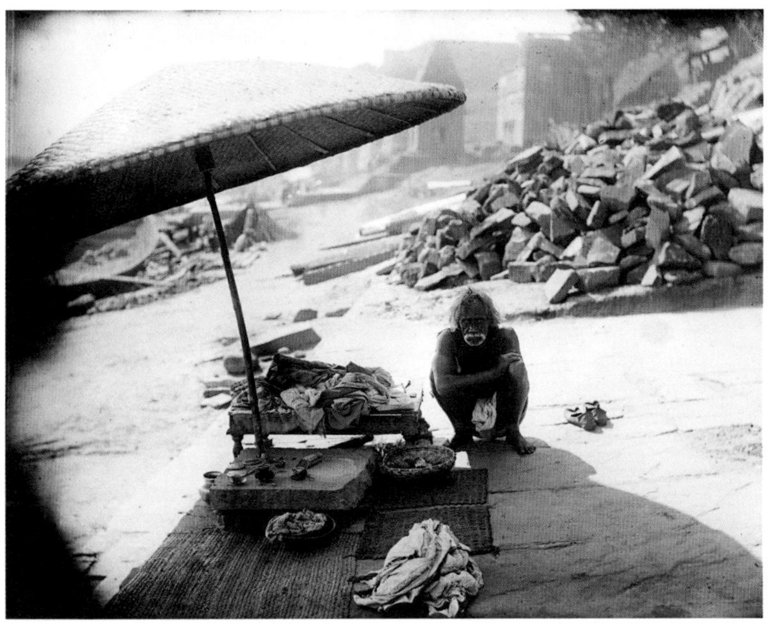

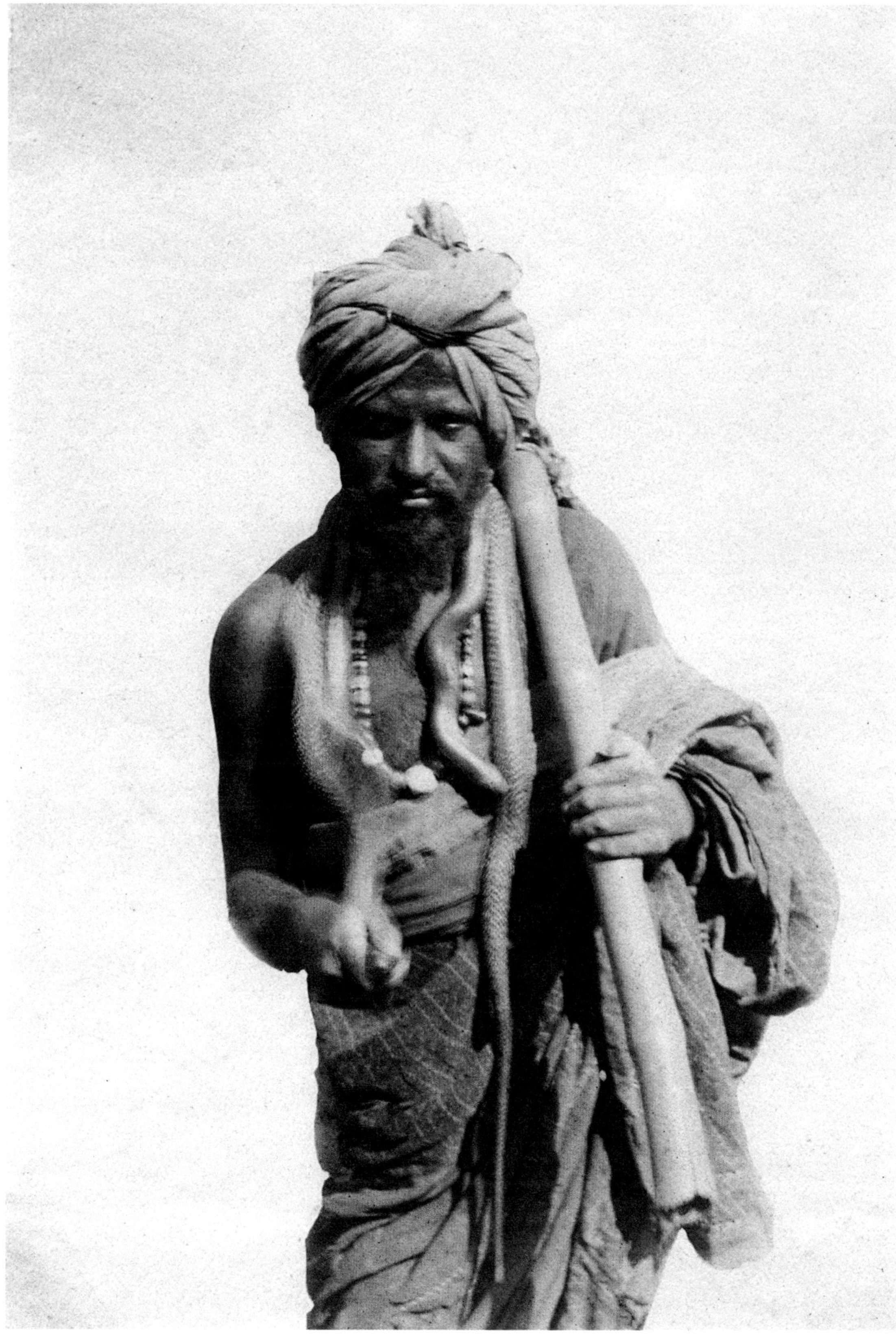

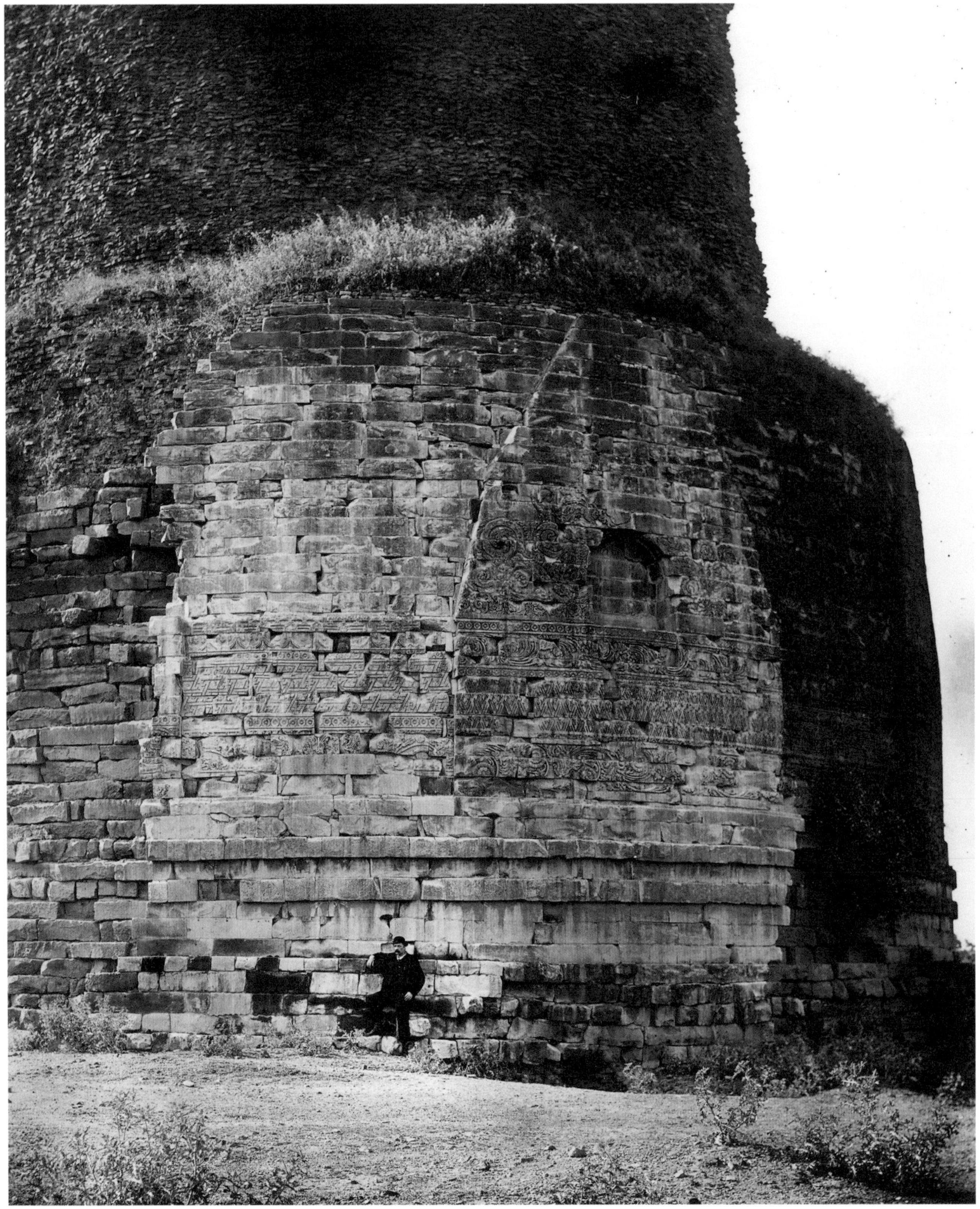

In Sarnath, an old Buddhist place of pilgrimage near Benares, a huge tower rises from the plain. It is a so-called stupa, which usually houses a Buddhist reliquary. Ekholm positioned Stolpe in front of the Dhamekh stupa, a cult building over a thousand years old that towers above the old temple area.

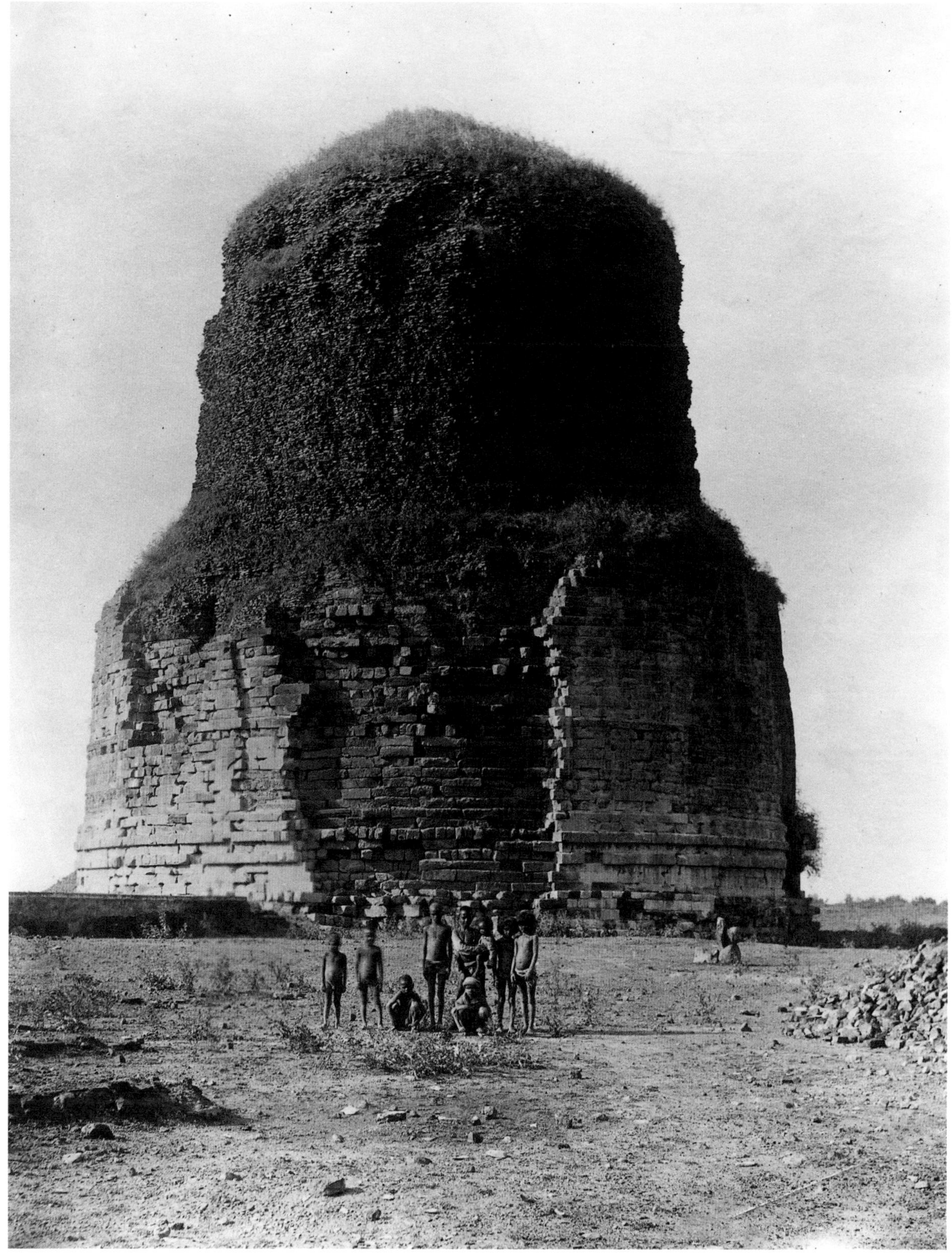

The Taj Mahal, near Agra, was built in the 17th century by the Great Mughal Shah Jahan, as a burial place for his favourite wife Mumtaz Mahal. The ambitious Ekholm climbed up onto the 'roof of the main portal' to take the picture, but Stolpe was only moderately impressed by this mausoleum: 'I've seen others that seem to me more interesting. The Taj looks too new. It was restored for the Prince of Wales's visit, and everything consequently looks completely new.' The Taj Mahal is now deemed to be one of the seven new wonders of the world. But the originally resplendent white of the marble building is starting to go yellowish green through air pollution. The mausoleum is often enshrouded in smog from Agra, which is considered to be the world's eighth-most polluted city.

Pages 190–191: When the motif was particularly attractive, Oscar Ekholm created a basis for panoramic images using two photographs that could be joined together. In Agra the photographic results of the visit to the mighty 16th-century castle were magnificent. Two small figures are to be seen on the roof: Hjalmar Stolpe with his notebook and bowler hat, and in front of him the interpreter and guide. In the background one has a glimpse of the Yamuna (Jumna) River, and in the haze on the horizon is one of the world's most famous buildings, the Taj Mahal.

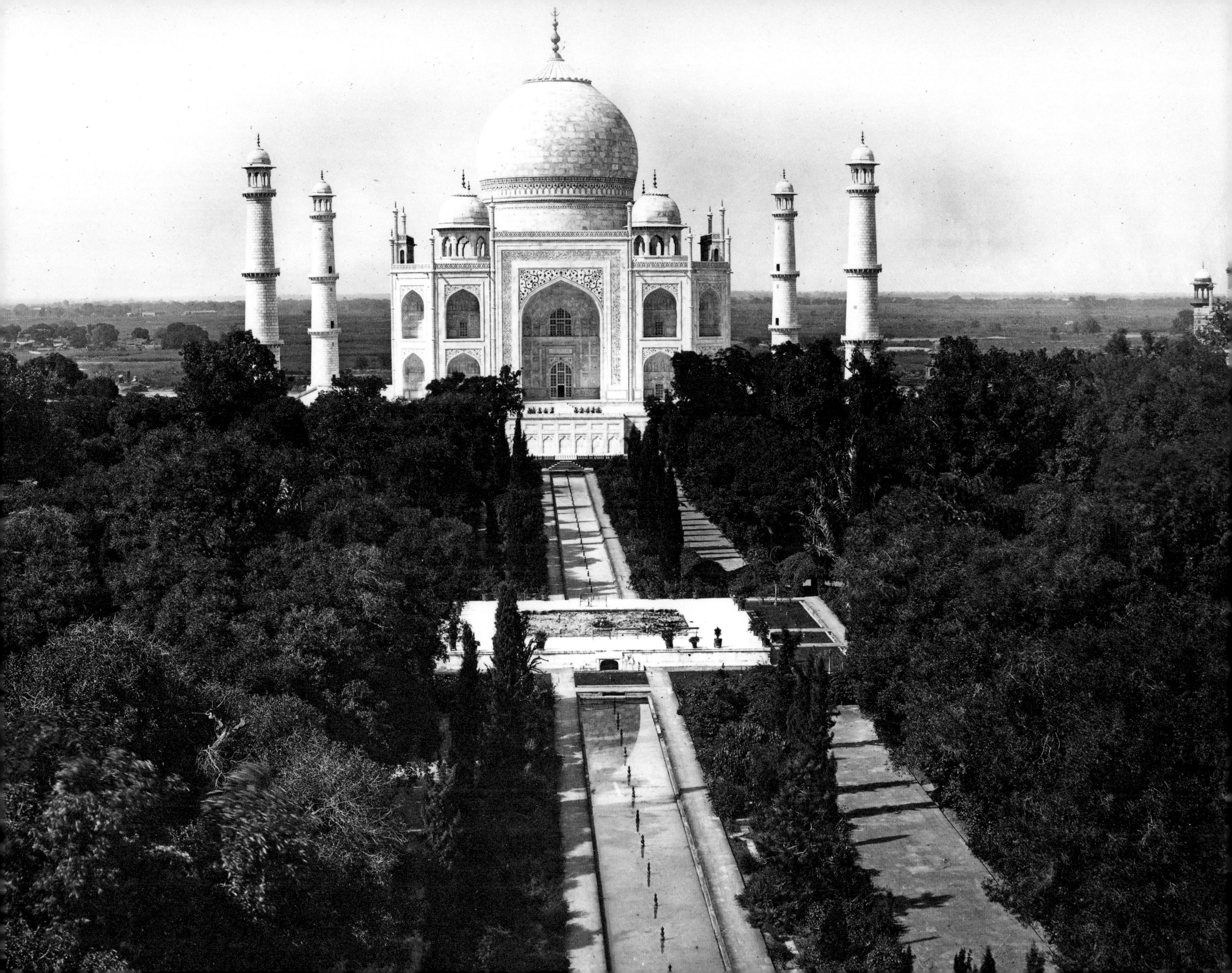

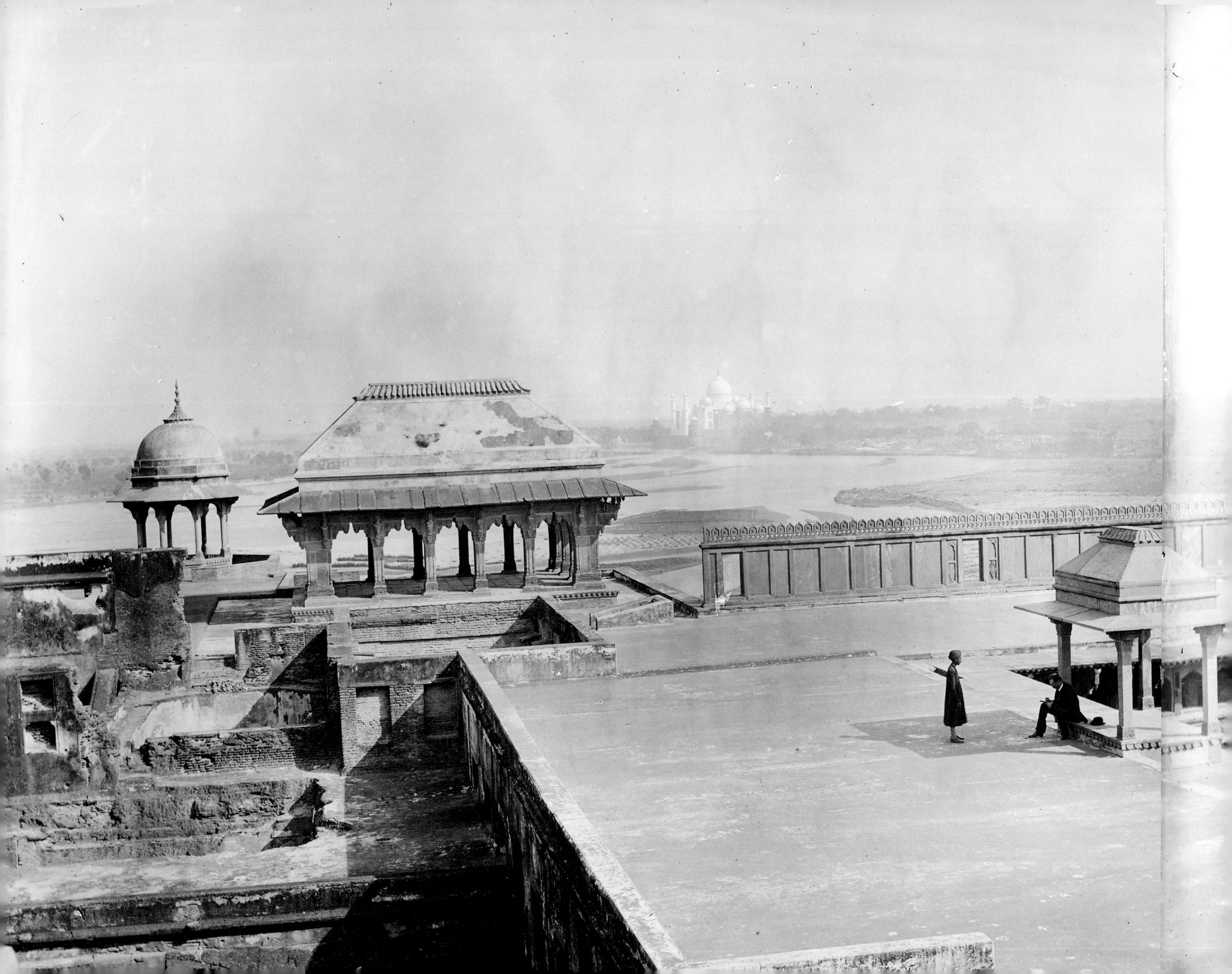

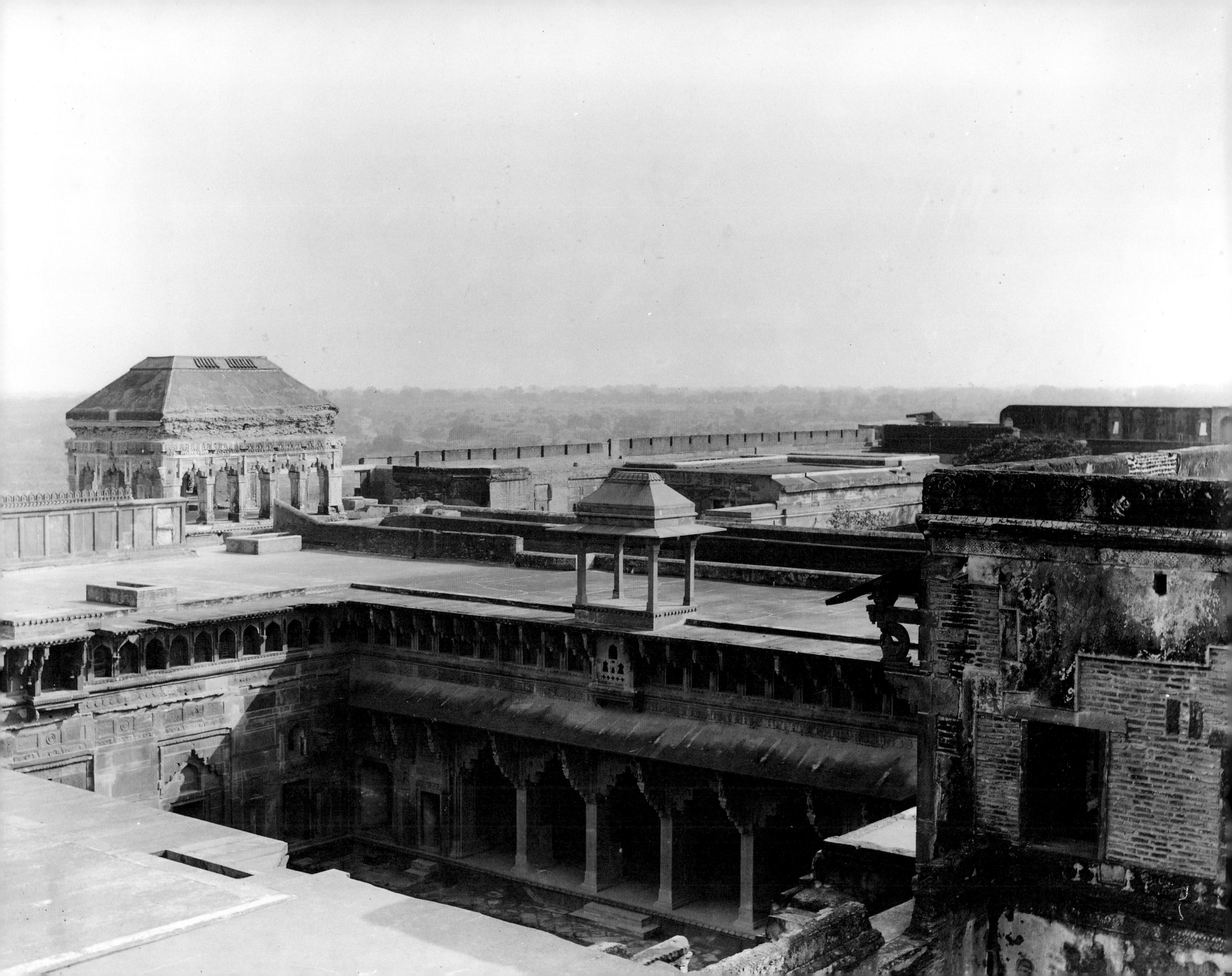

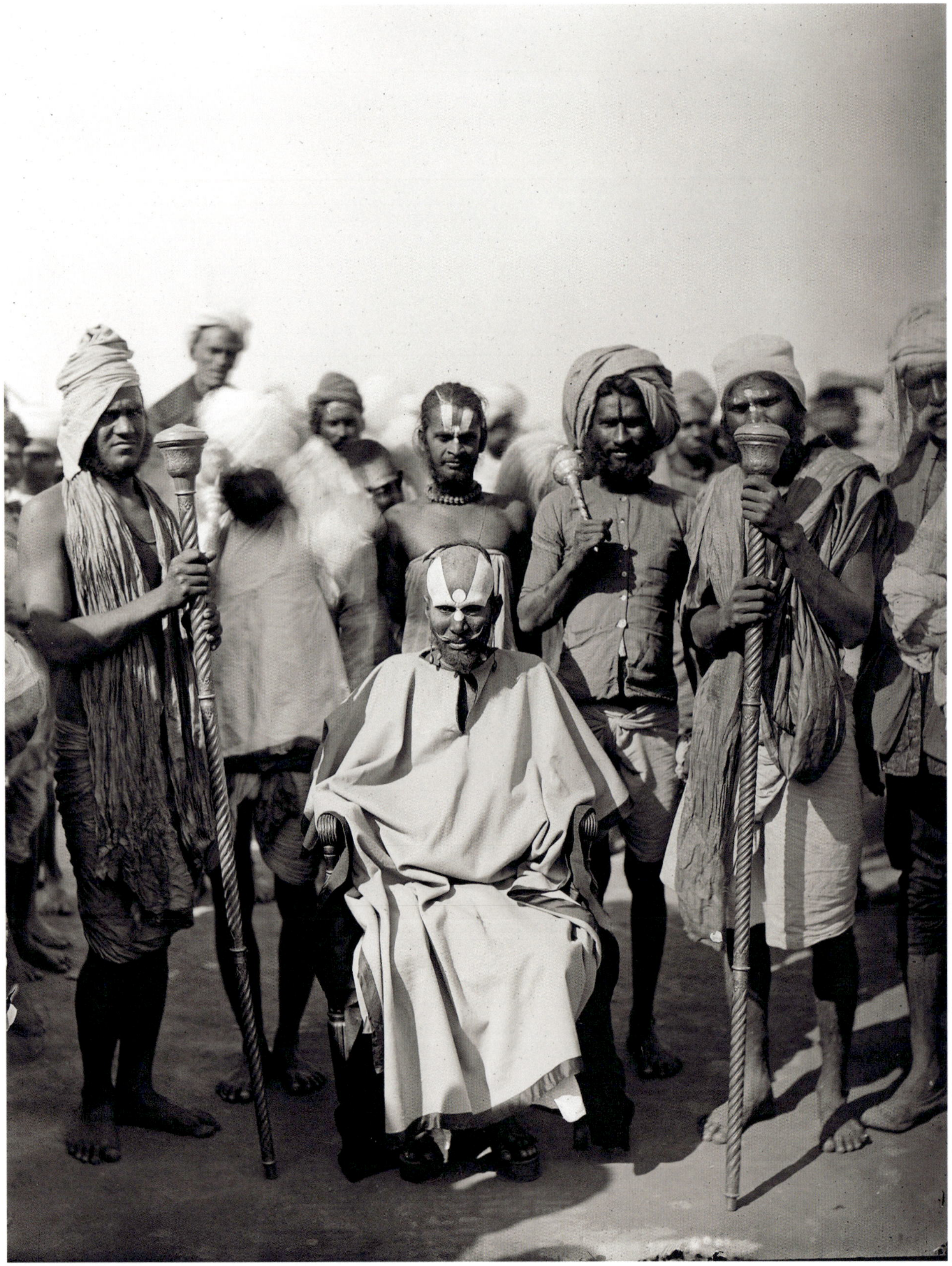

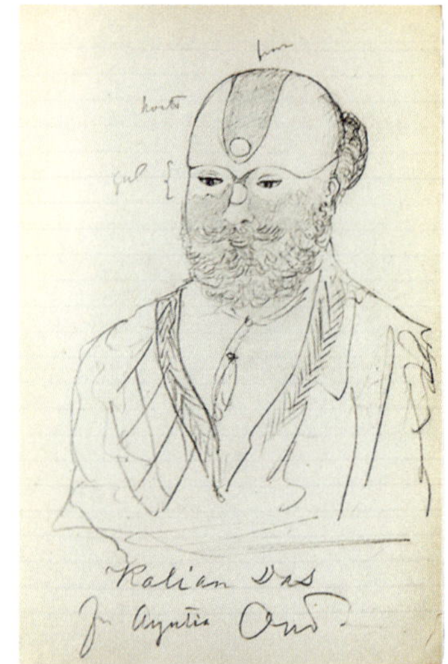

Kalian Das

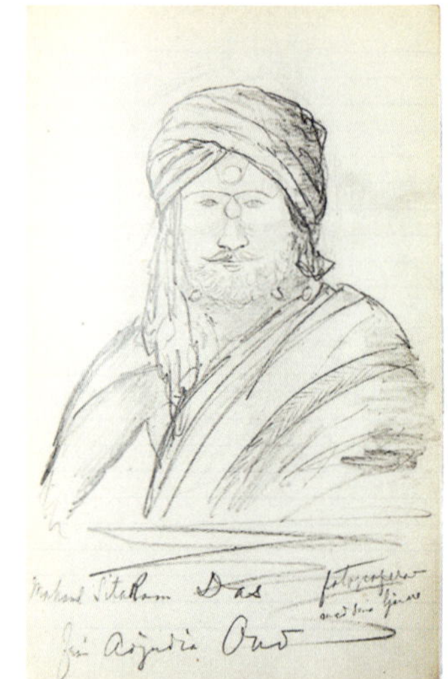

Mahant Sitaram Das

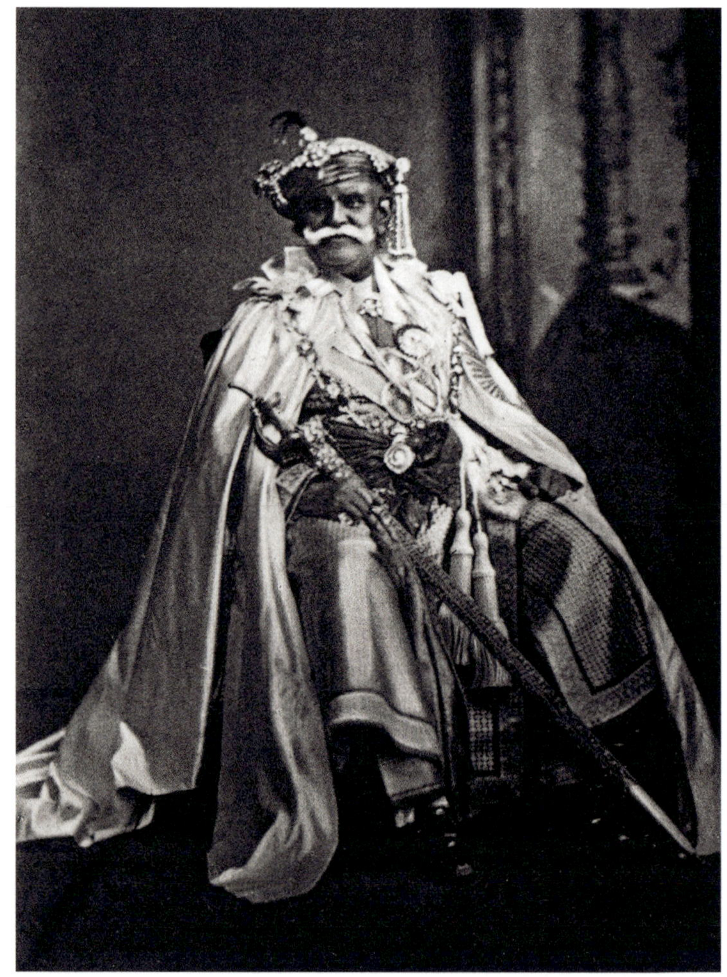

Opposite page, left and top right: In Allahabad, Hjalmar encountered two distinguished fakirs, i.e. not the sort who sit smeared with ash and stare at a fire. 'I did not have an interpreter, and had to use my own cunning. So I took out a pencil and drew pretty successful portraits of the two distinguished gentlemen, and showed them to them. Whilst I was busy drawing one of them it was amusing to observe the envy of the other. When I'd finished he looked worried, and his eyes followed the book and pen when I put them in my pocket. So I drew a portrait of him too.' Both of them 'had the mark of the Shiva sect magnificently painted right across their forehead.' After that, Oscar was to photograph the finest of them: 'But he insisted on sitting under a canopy. I at last got him out into the sunshine, and at the final moment pulled away the parasol a servant was holding over his holy head.' A 'snapshot' of how Sweden's first ethnographic documentary team collaborated during their voyage round the world. Below the portrait in his notebook Hjalmar wrote the name Kalian Das, and in the photograph one sees the holy man surrounded by his servants. The other – envious – fakir is depicted in a turban. They both came from Ayodhya in the principality of Oudh.

Opposite page, bottom right: Just before Stolpe and Ekholm arrived in Benares by train from Calcutta, Prince Oscar had been given a grand reception by the Raja of Benares. The 70-year-old prince was carried to the palace in a solid silver litter, and in all his finery, 'dressed in an ankle-length, gold-embroidered coat made of yellow silk, adorned with large jewelled buttons and other precious jewellery, and over his left shoulder he wore a large blue silk sash complete with a bediamonded grand star on his chest', according to the prince's adjutant Axel Ribbing. Stolpe doubtless basked in the wake of the royal visit. In the letters of recommendation he had with him he was presented as a professor, with the addition that he 'came out with the Swedish Princes'. Prince Oscar's brother Carl had joined in Calcutta.

Right: 'Seated man with bells round his ankle.' Thus the text in Hjalmar's list of photographs. Oscar Ekholm took the picture in Allahabad in January 1885.

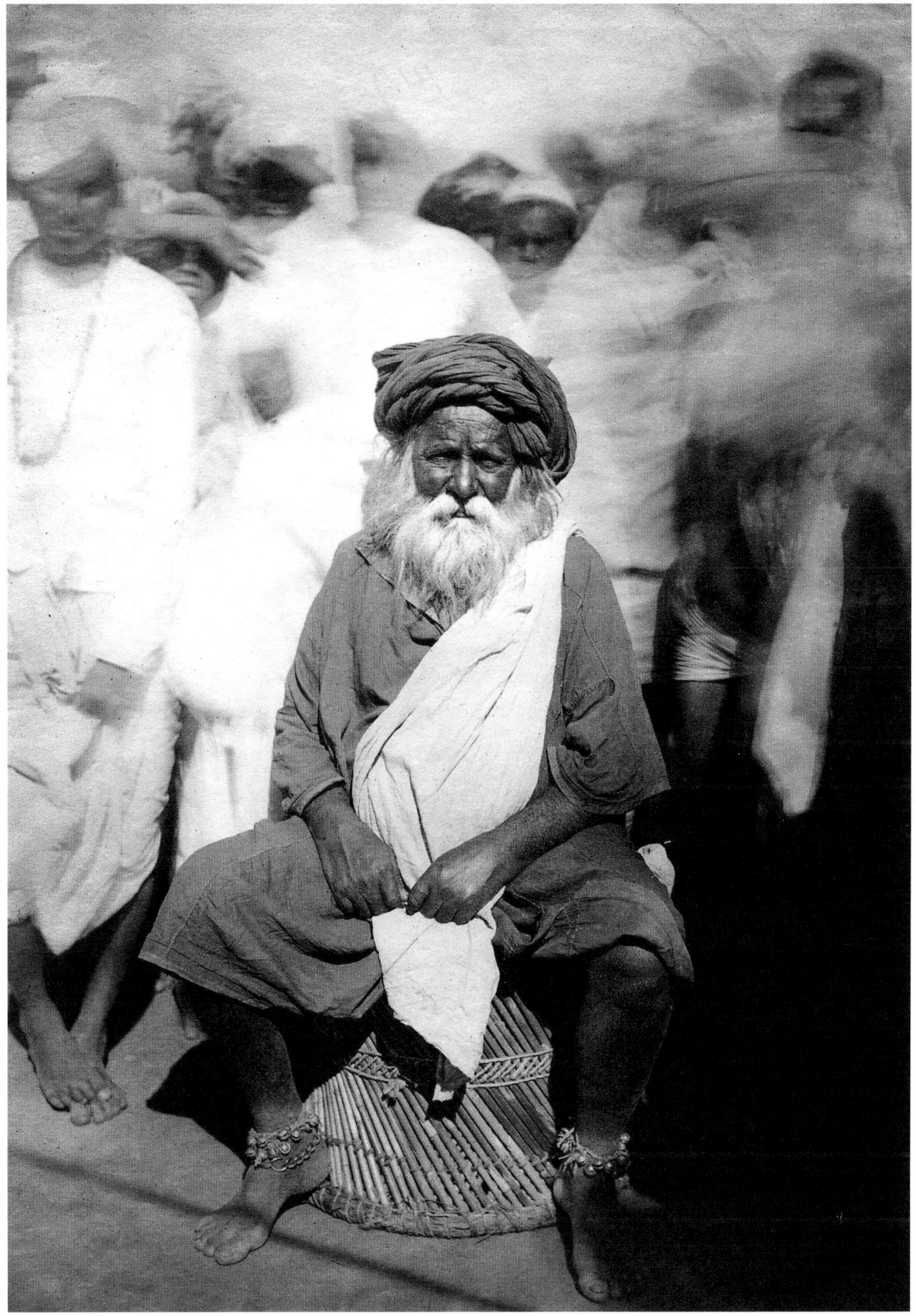

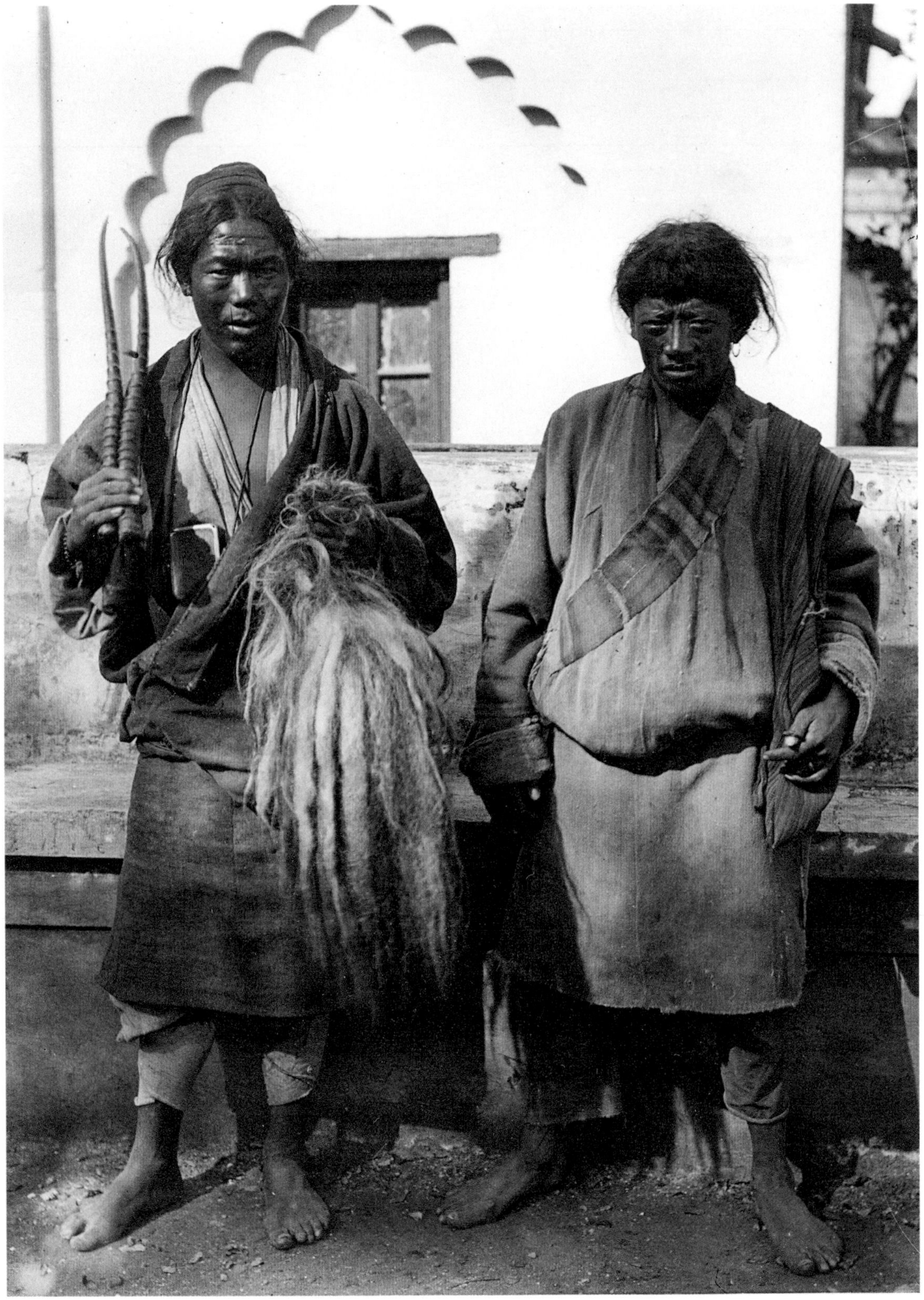

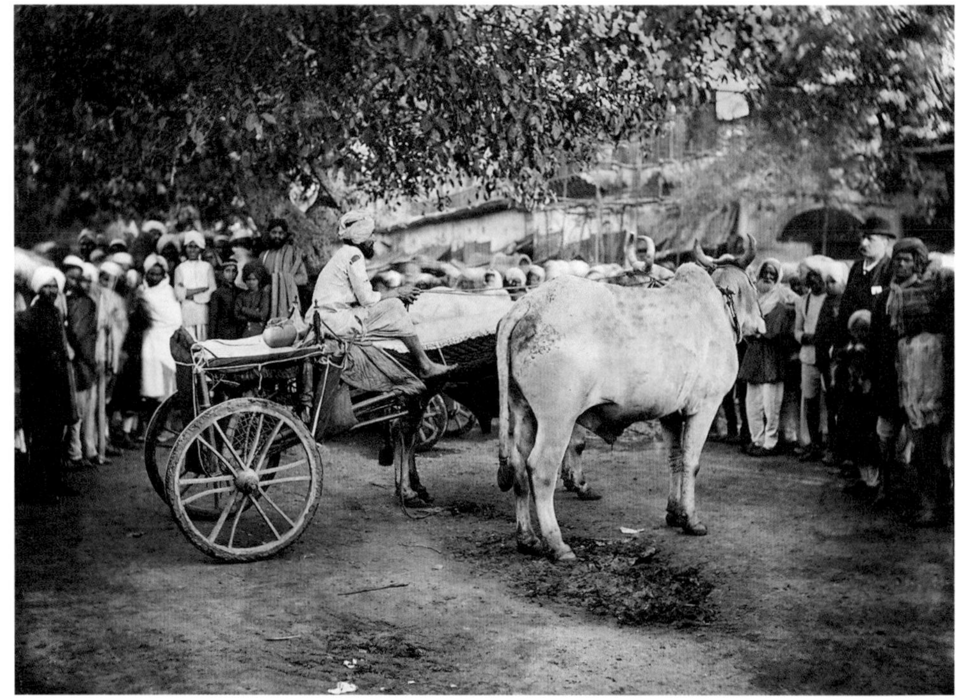

Left: 'Delhi. Two men from Olmora. One of them is holding 2 horns and a yak's tail.' The pointed horns come from an antelope and the tail is that of a yak – the long-haired bovine that lives in the Himalayas and is threatened with extinction. The strong tail was a commodity, and served as a religious symbol in India. It was also attached to British officers' helmets.

Above: Ekholm probably took this picture of a cart drawn by two oxen in Delhi, and Hjalmar Stolpe is to be seen amongst the observers, on the far right.

Opposite page: Under the heading Delhi, Stolpe writes in his list of pictures: 'Humayun's Mausoleum. Myself down in the courtyard.' For some years during the 16th century the Great Mughal (Mughal Emperor) Humayun was the ruler of northwestern India.

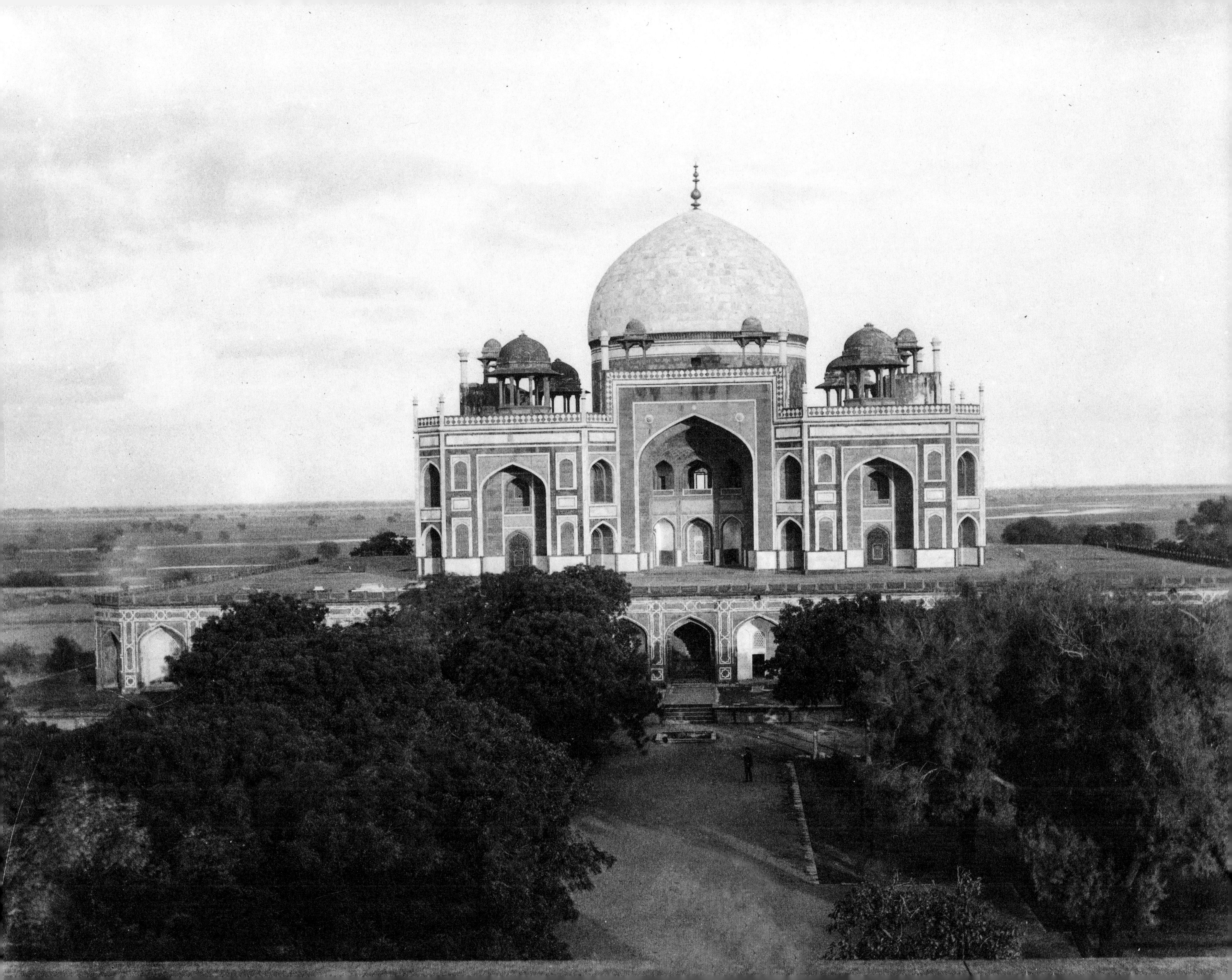

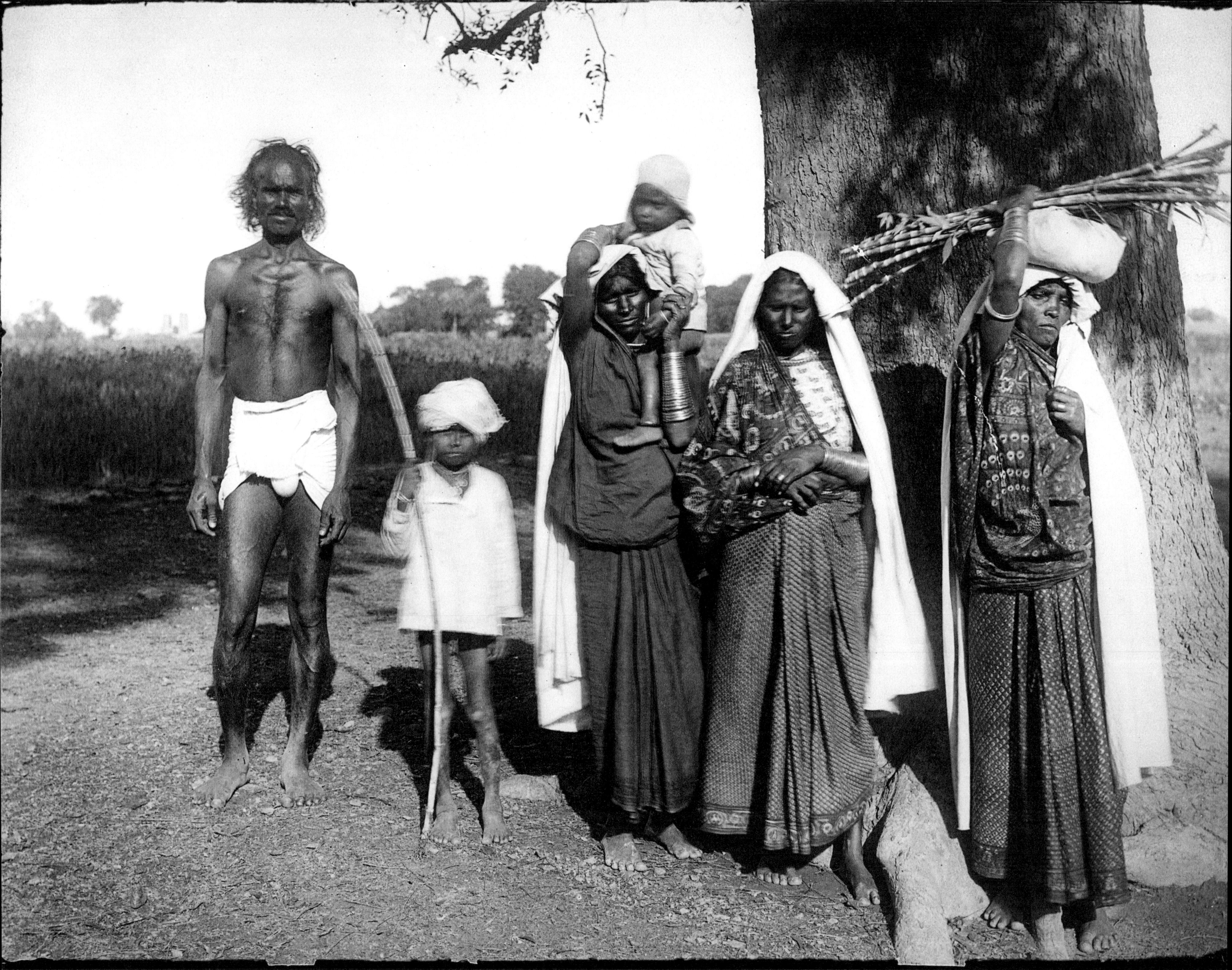

Opposite page: 'Countryfolk at Sarnath', the place of pilgrimage near Benares.

Below: After the visit to the Taj Mahal, Stolpe wrote to his wife: 'we came upon a family on a pilgrimage to Benares, riding on a very strange vehicle drawn by oxen. I had a shrewd guide who led them to believe I would photograph the Taj, which was to be seen behind them, and they stopped so as not to block our view. And that's how I got them. Just like that! The light's now gone off! Have to stop.' In his hotel room, Hjalmar's letter-writing was suddenly interrupted. Was it Buddha punishing him for his fib?

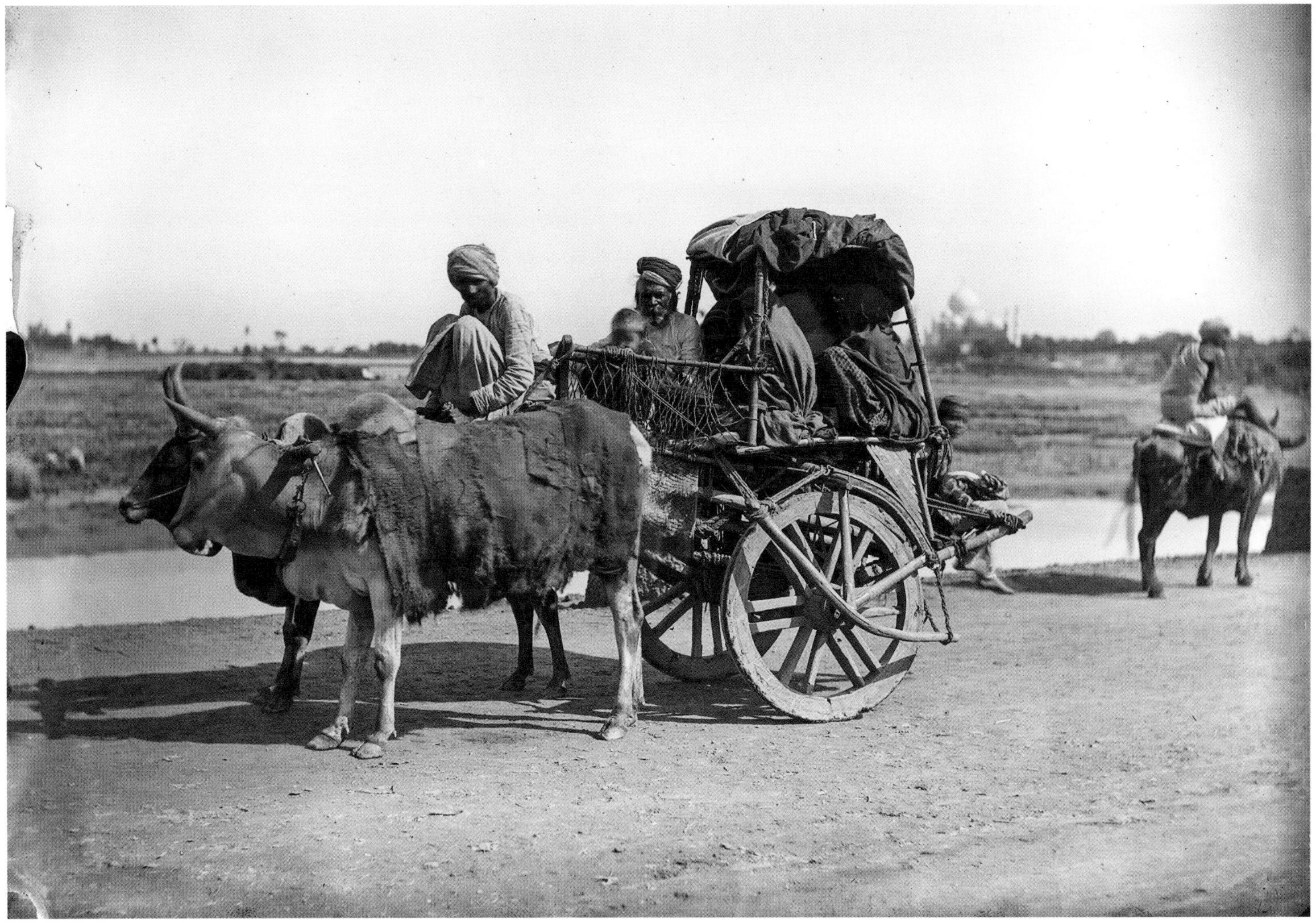

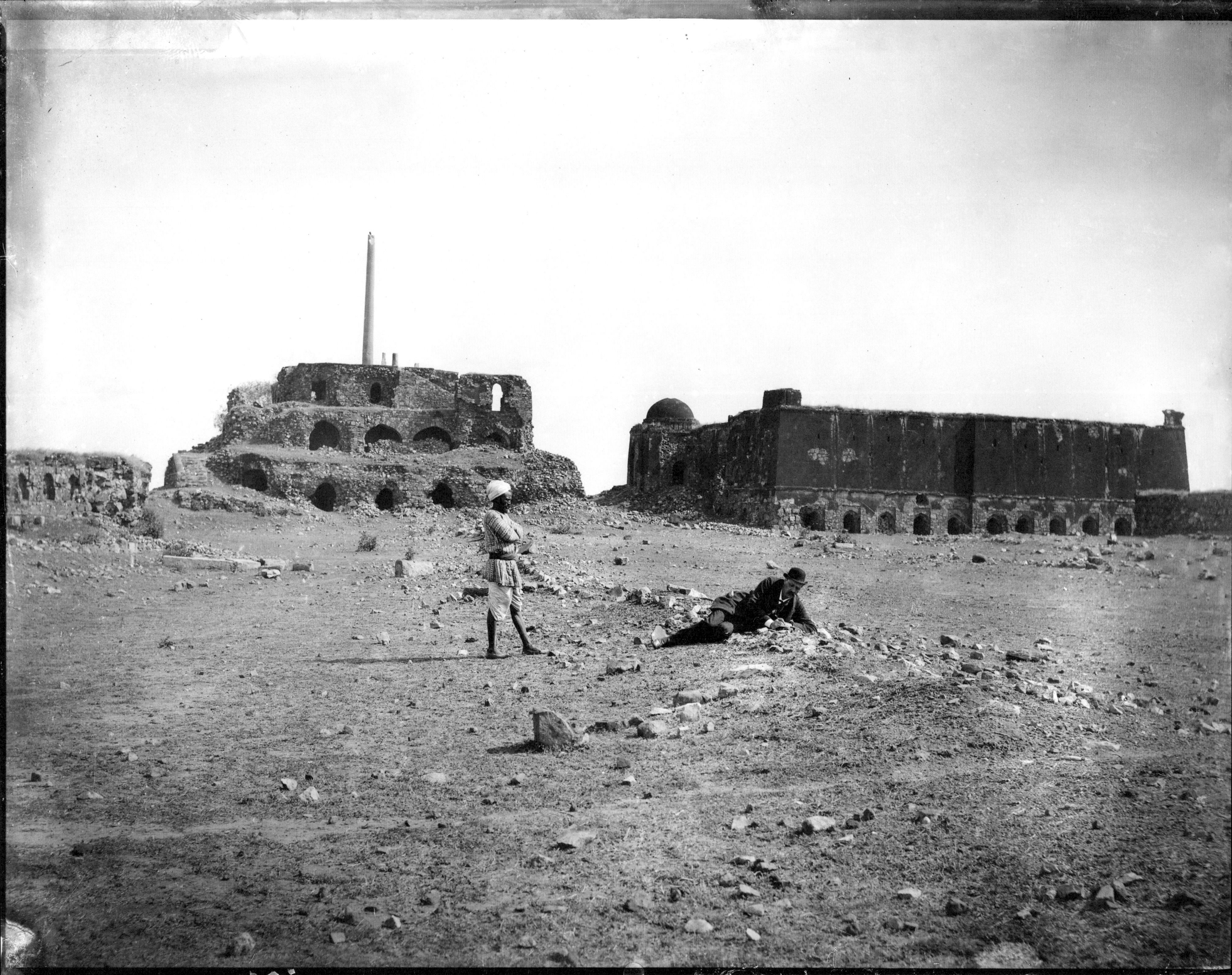

Opposite page: A somewhat odd picture of an ethnographer working in the field. Dressed in a dark suit and bowler hat, Hjalmar Stolpe is lying stretched out amongst stones somewhere in northern India, writing in his notebook. One fancies that the guide is casting a surprised glance. Maybe it was Ekholm wanting a different picture. And he got it, after all.

Right: This is Oscar Ekholm's last photograph from India. After 14 months together, Stolpe and Ekholm went their separate ways in Delhi on 1 February 1885. Ekholm's finances did not permit him to continue the journey, and he was sent to Bombay (now known as Mumbai) to meet up with the Vanadis and travel home. The picture was taken 'next to an old Mohammedan grave by the ruins of Firozabad fort – an old fort dating from the reign of Firoz Shah (1351–1388)'. Ekholm was commissioned with copying the picture from glass plate to paper whilst on the Vanadis and posting it to Emmy from Aden, with Stolpe's message: 'May it serve ... to show that I was thinking of you whilst on the banks of the Jumna.' On his chest Hjalmar has placed an anchor – a Christmas present Emmy sent with him, to be opened in Lisbon. She had been wondering why he was not wearing her gift. Over a month later the anchor picture reached Emmy in Norrköping: 'Thank you my Darling! You can just imagine how much it meant to me to see you in such a large format, though it was admittedly not the most successful of portraits. But I ... will always particularly cherish it!'

After Delhi, Hjalmar travelled to Lahore, declaring: 'I feel free and happy.'

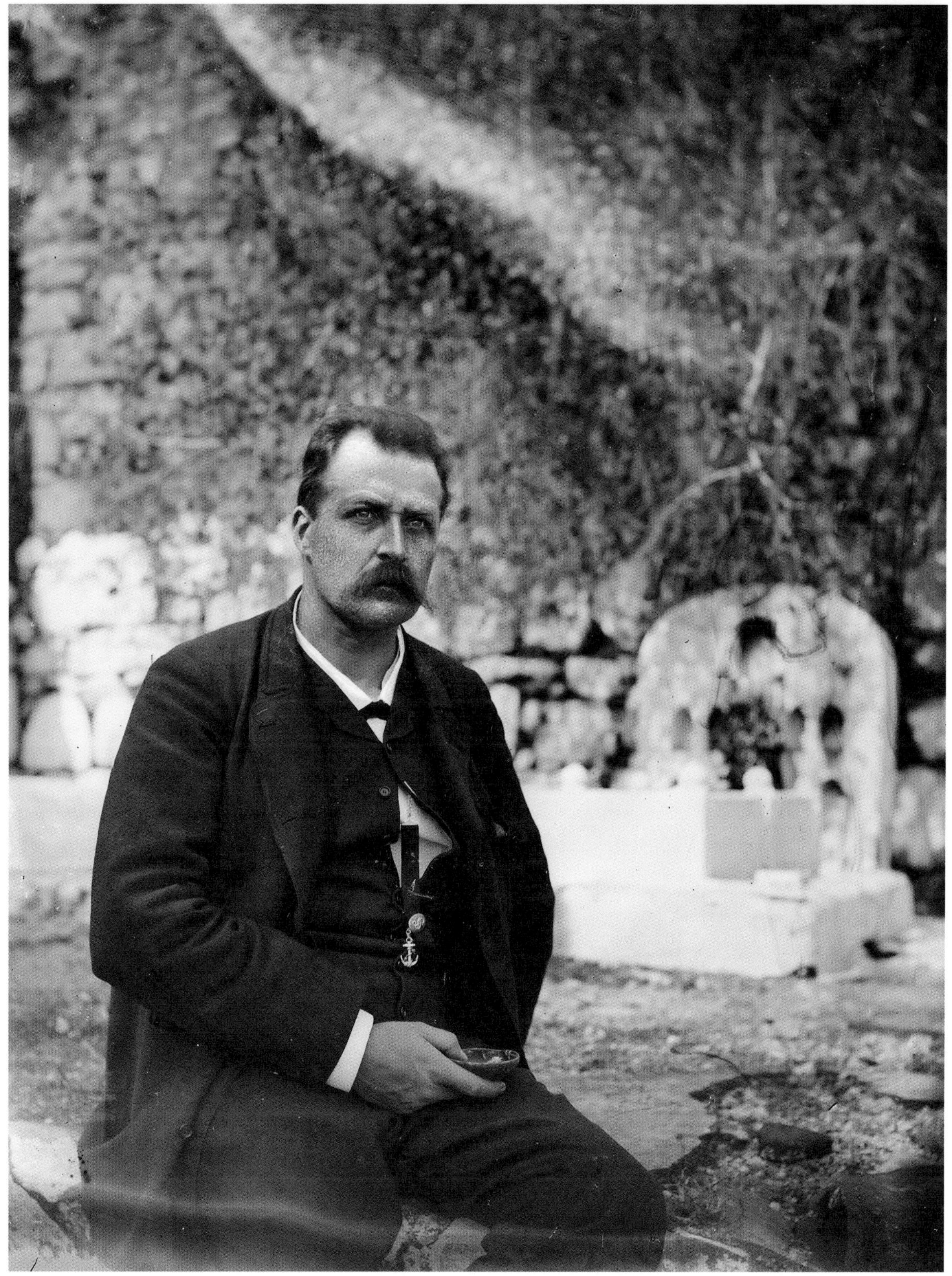

Left: 'Delhi. Interior of the palace in the castle. Myself by a pillar.' This is Stolpe's caption in one of his lists of pictures. As usual well dressed, and here nonchalantly leaning against a pillar, he is writing about the Delhi Palace in his notebook. Throughout the voyage around the world, in keeping with the requirements of the time, he played the part of an elegant European gentleman and scientist. In Delhi, after over a year's travelling he wanted to renew his wardrobe, but was at loggerheads with the tailor: 'I boxed the rascal's ears and was on the verge of walking away from it all.' Stolpe was infamous for his irascibility. But no other such indiscreet actions are otherwise known. He took his incorrectly sewn clothes, paid and went.

Opposite page: This building is situated in southern Delhi, and houses the Khirki Masjid mosque. The composition of the image can be seen as being typical of Ekholm. He had fun placing his boss at the top left of the steps, to contrast with the huge mosque. But you have to look carefully to pick out Hjalmar Stolpe, sitting there in a dark suit and bowler hat.

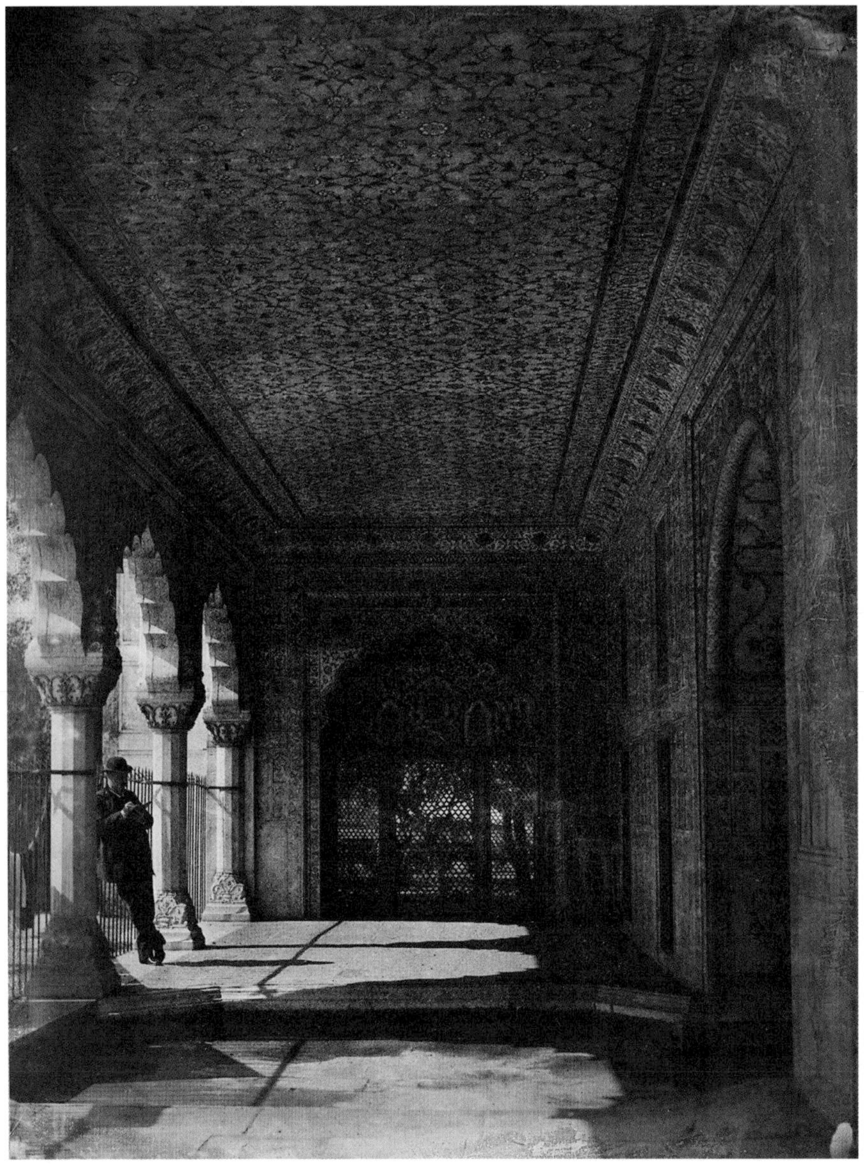

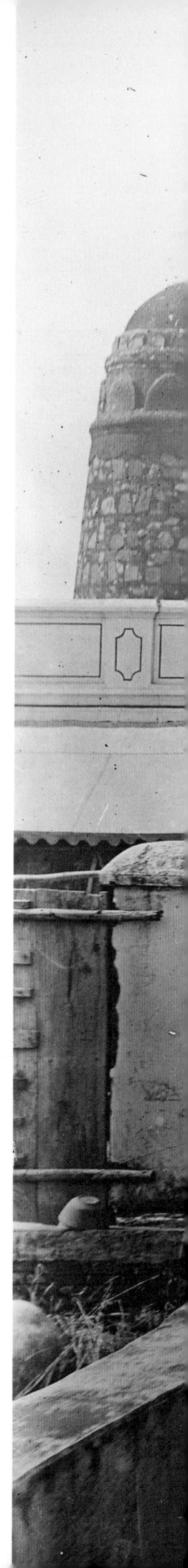

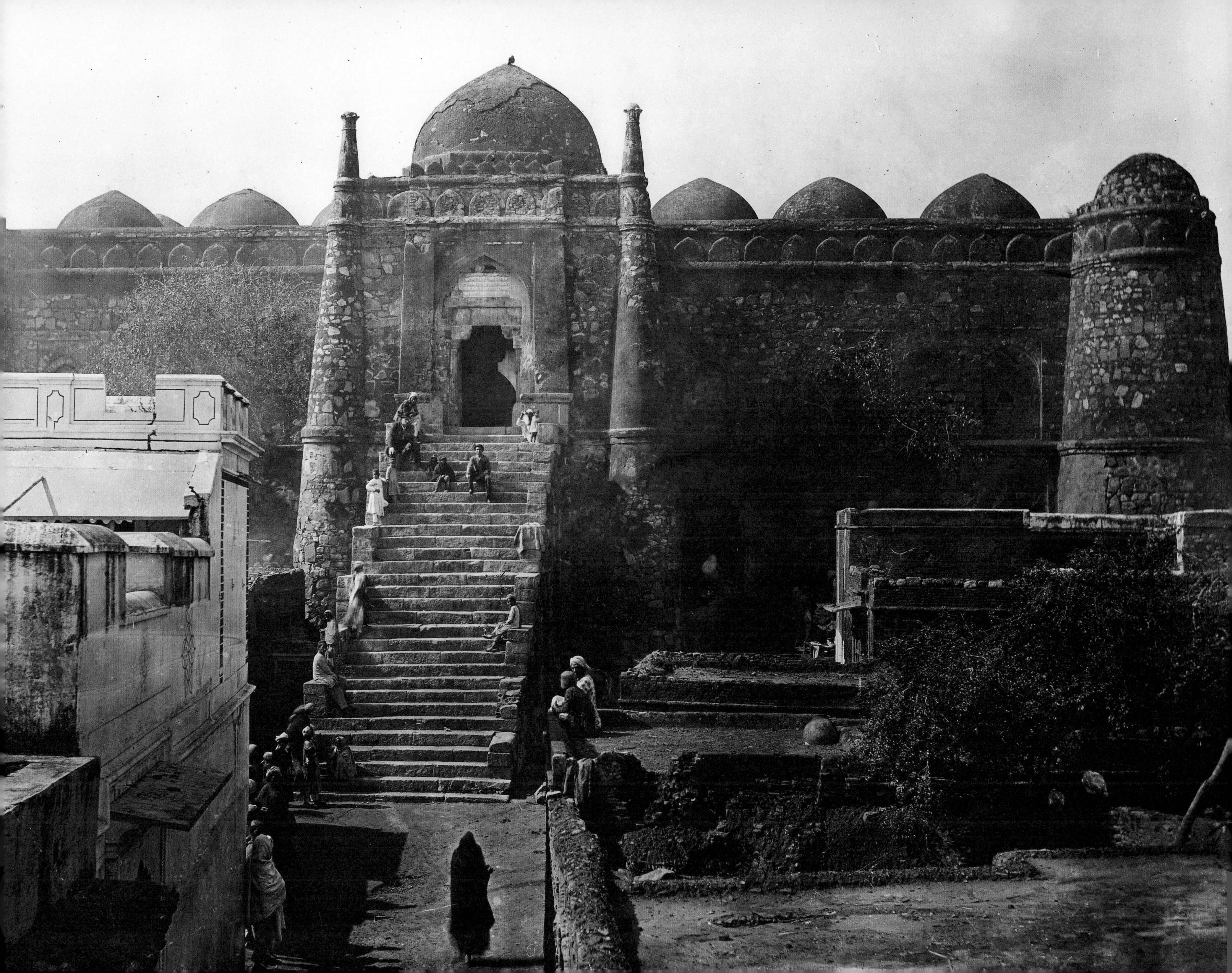

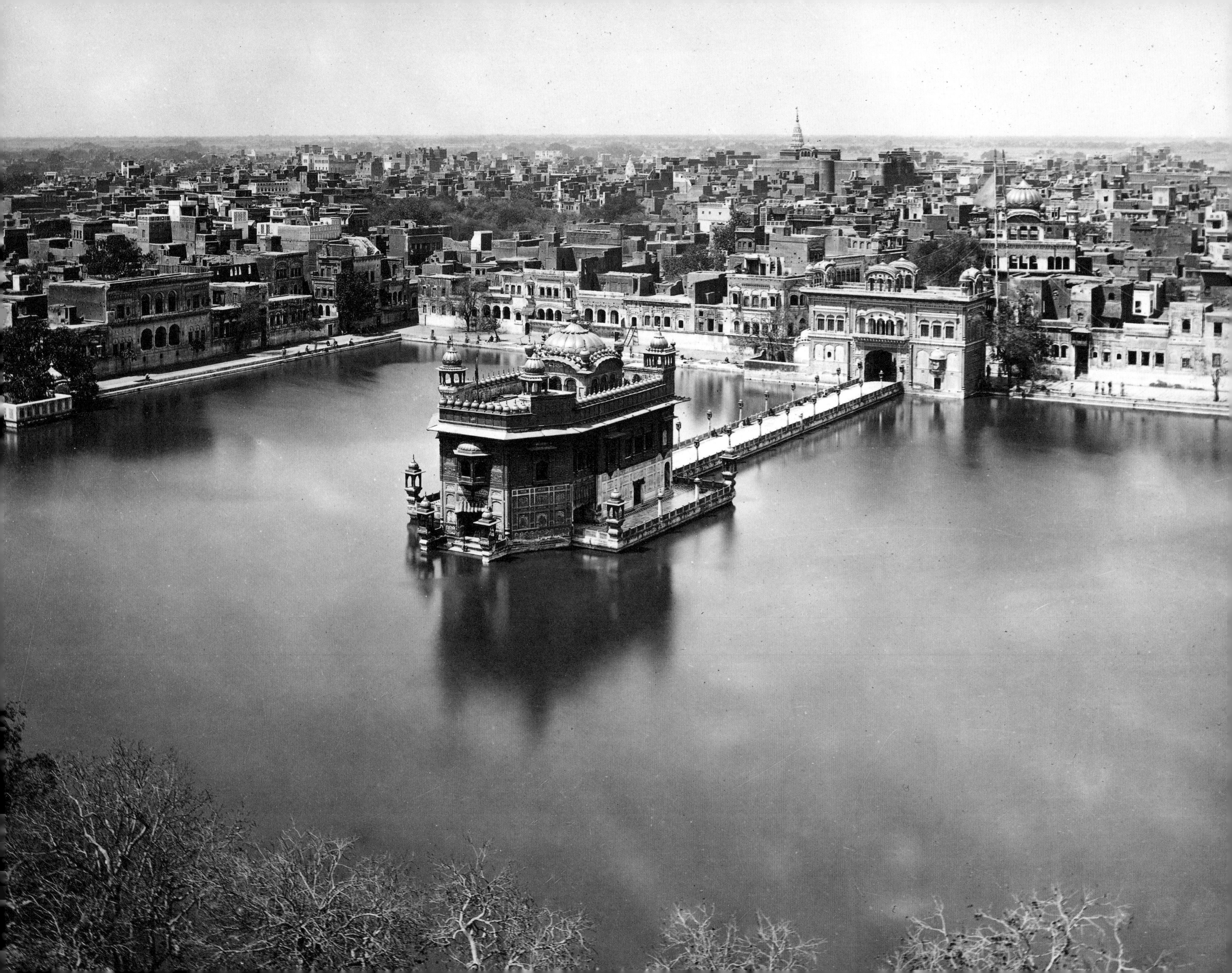

Opposite page: During the Vanadis voyage Oscar Ekholm's pictures were often supplemented with purchases from local photographers. One sees here James Craddock's magnificent photograph of the Golden Temple of Amritsar – the centre for the Sikh religion. The marble temple is situated in an artificial square lake (enlarged during the period up to 1581), where the pilgrims washed themselves. The Sikh sacred scripture is housed within the temple. Stolpe bought 30 or so photographs by Craddock when he visited his studio in Lahore on 28 February 1885.

Below: In the big inner courtyard of the Khirki Masjid mosque in Delhi two priests are busy washing each other's hair, in the presence of a high priest to the left.

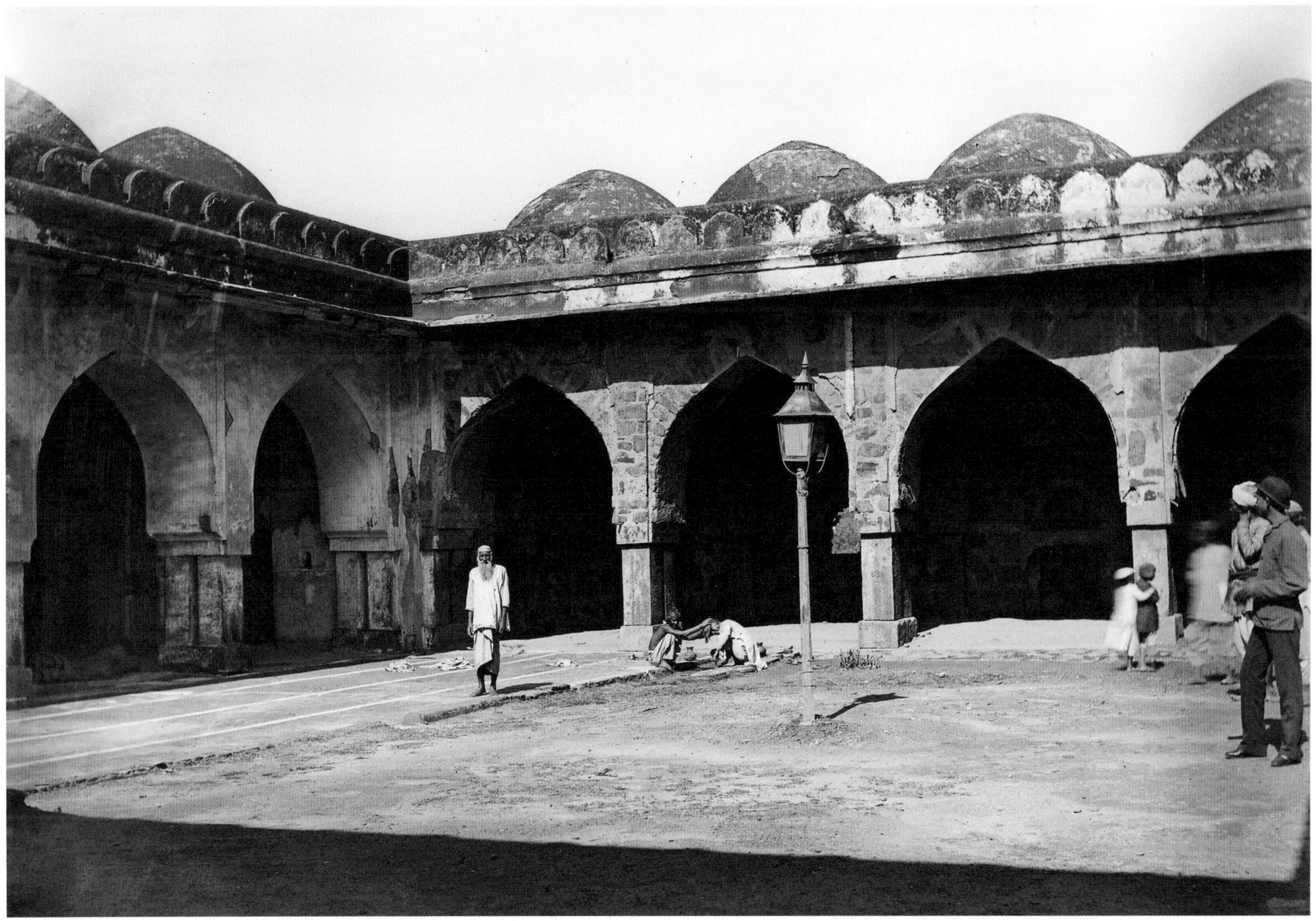

Right: On 27 March 1885 Oscar Ekholm (seen in the portrait) wrote to his brother Fredrik from the port of Valletta in Malta. His voyage round the globe had almost come to an end, and he said that he had 'today taken a plate of the Frigate and a group on deck comprising the Commander-in-Chief, Prince Oskar and some of the Officers and visiting Ladies and Gentlemen. I'm delighted to be hearing more noise than usual on deck today, as they're in the process of bringing the boats into action, because early tomorrow when I come up on deck we will no longer be in Malta – with every minute that passes I'll be closer to home. So farewell, brother – I'd hope that in a month there won't be many miles separating us.'

Below: 'The camel square in Aden.'

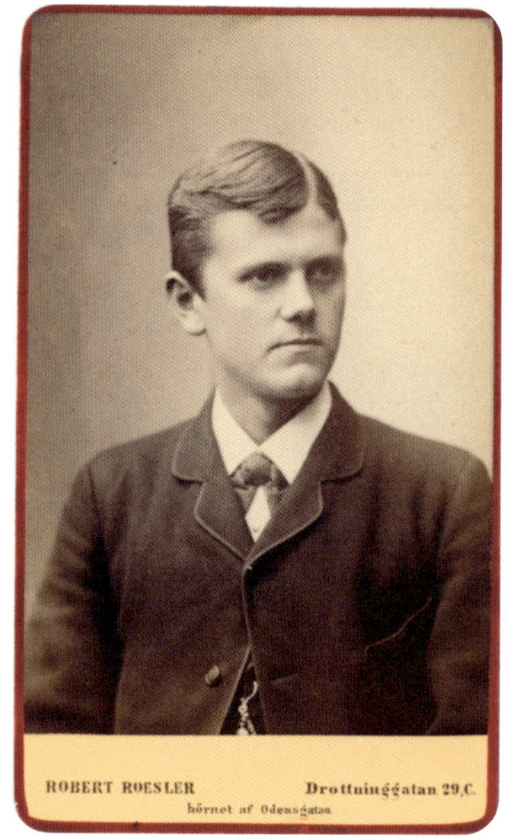

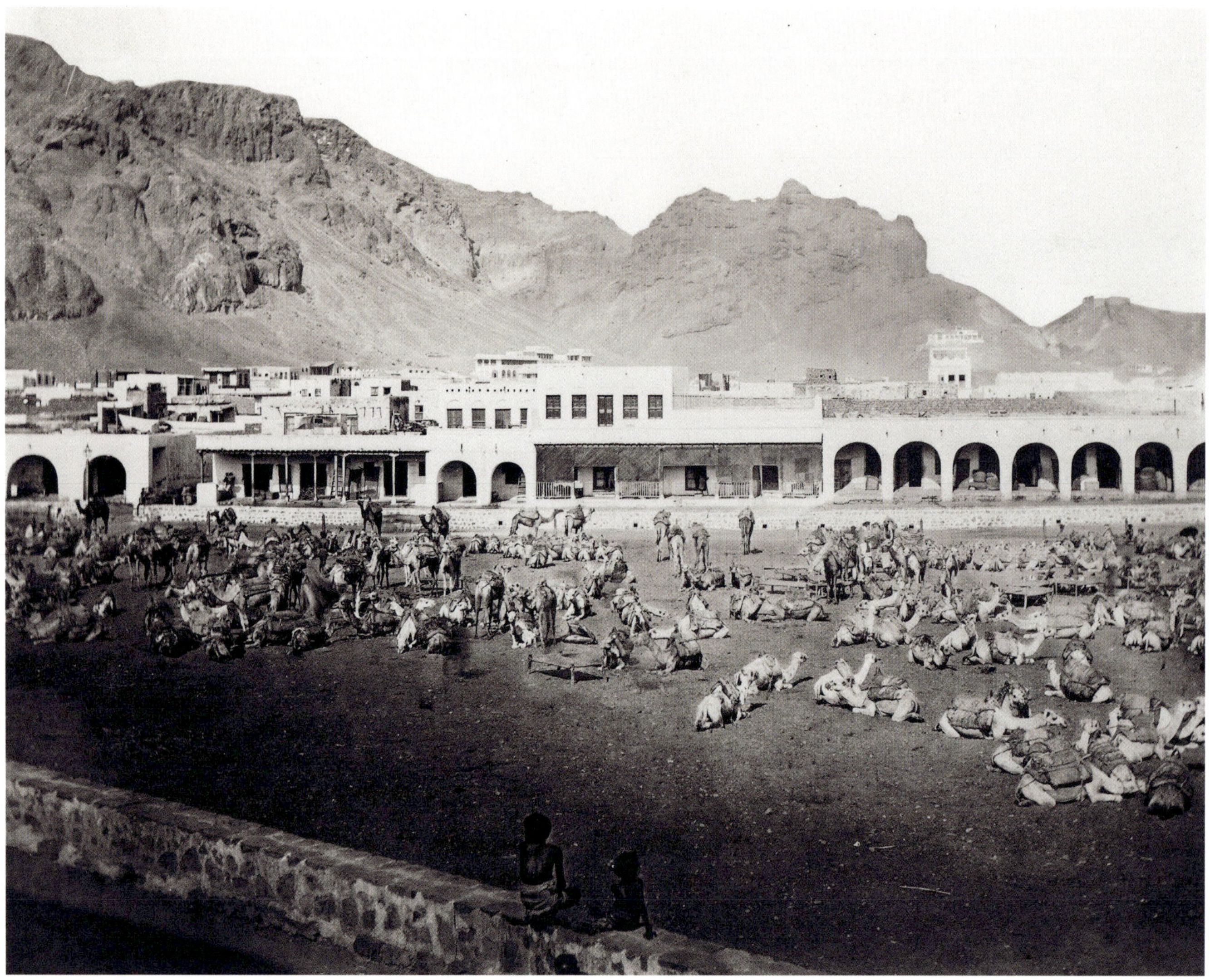

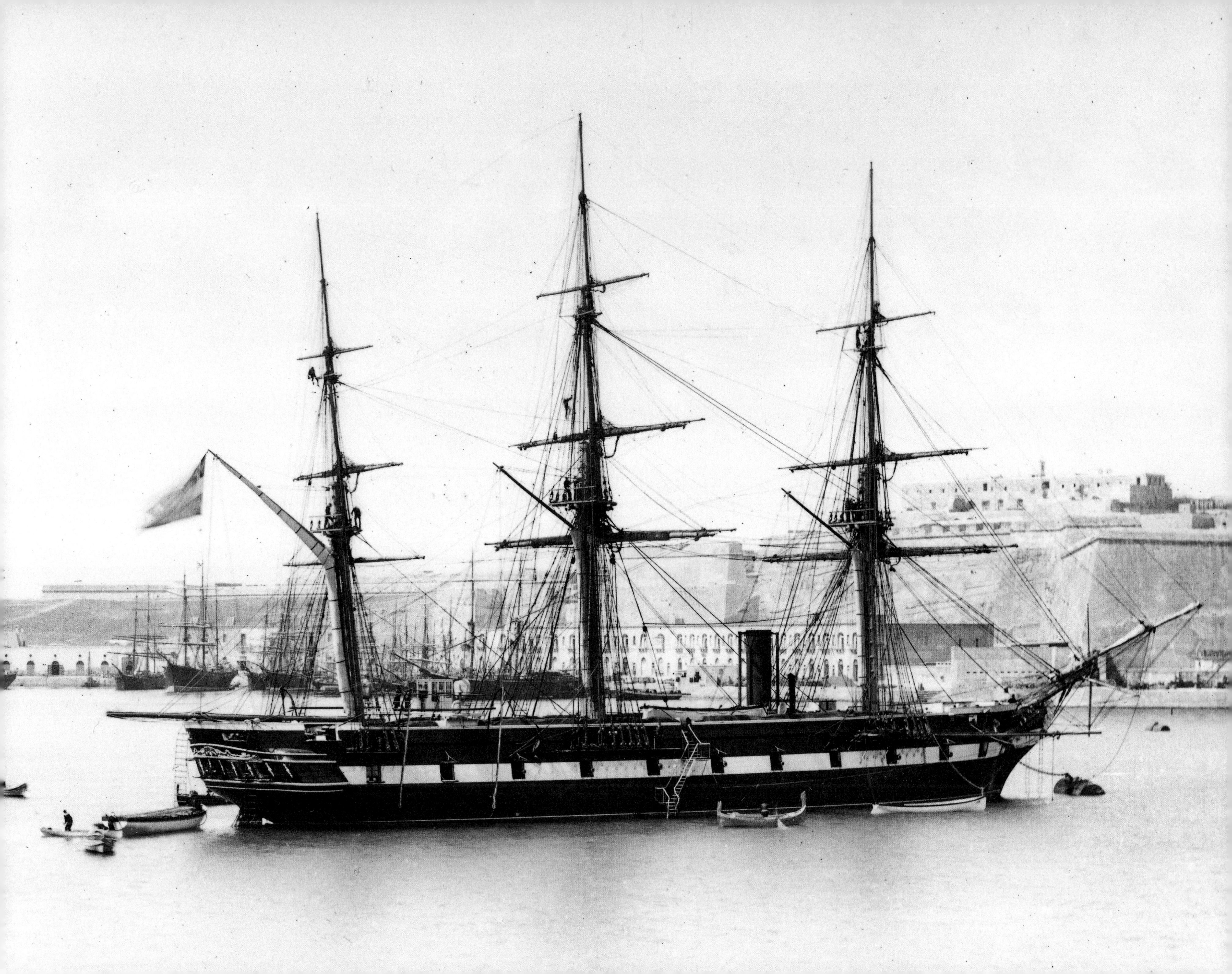

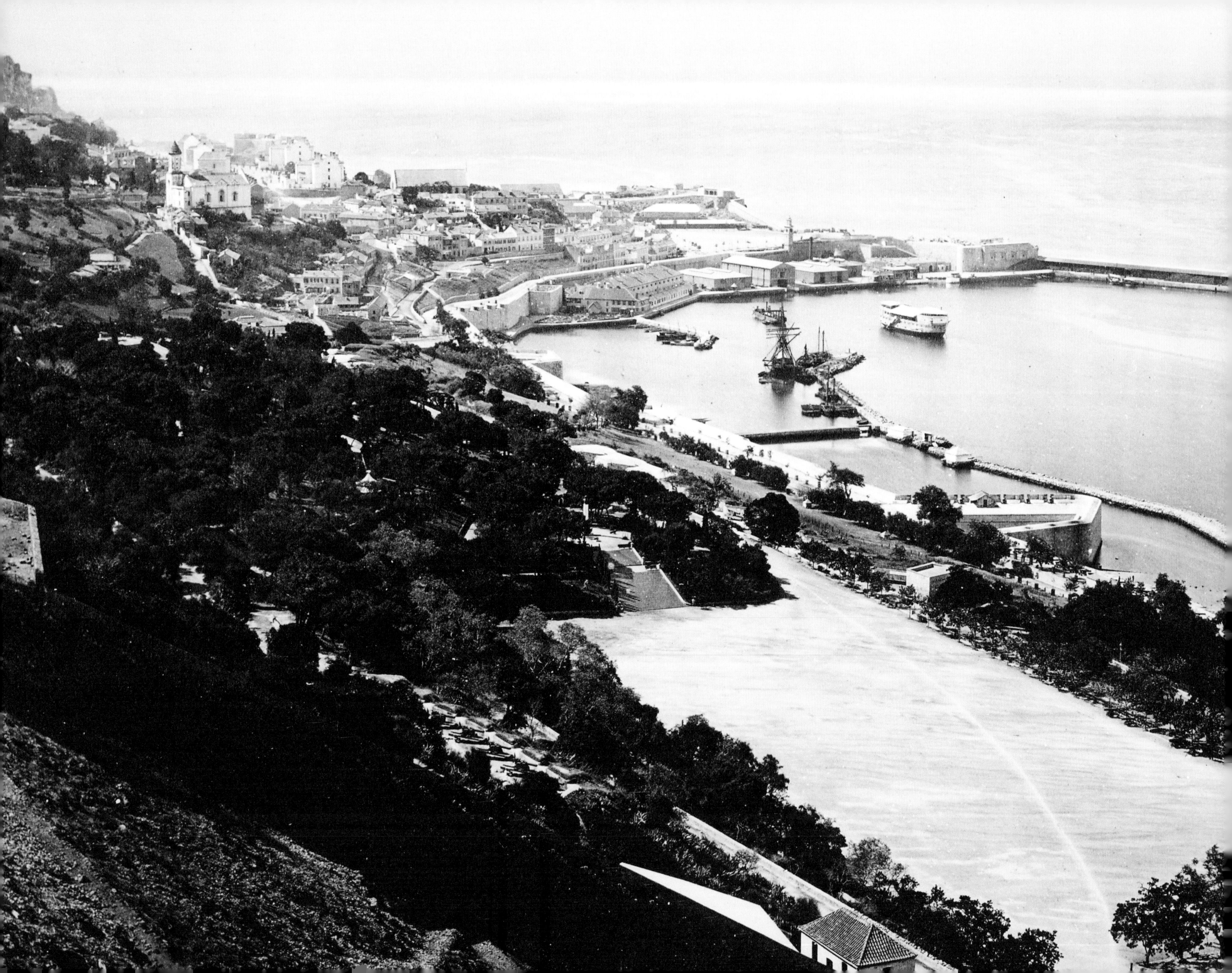

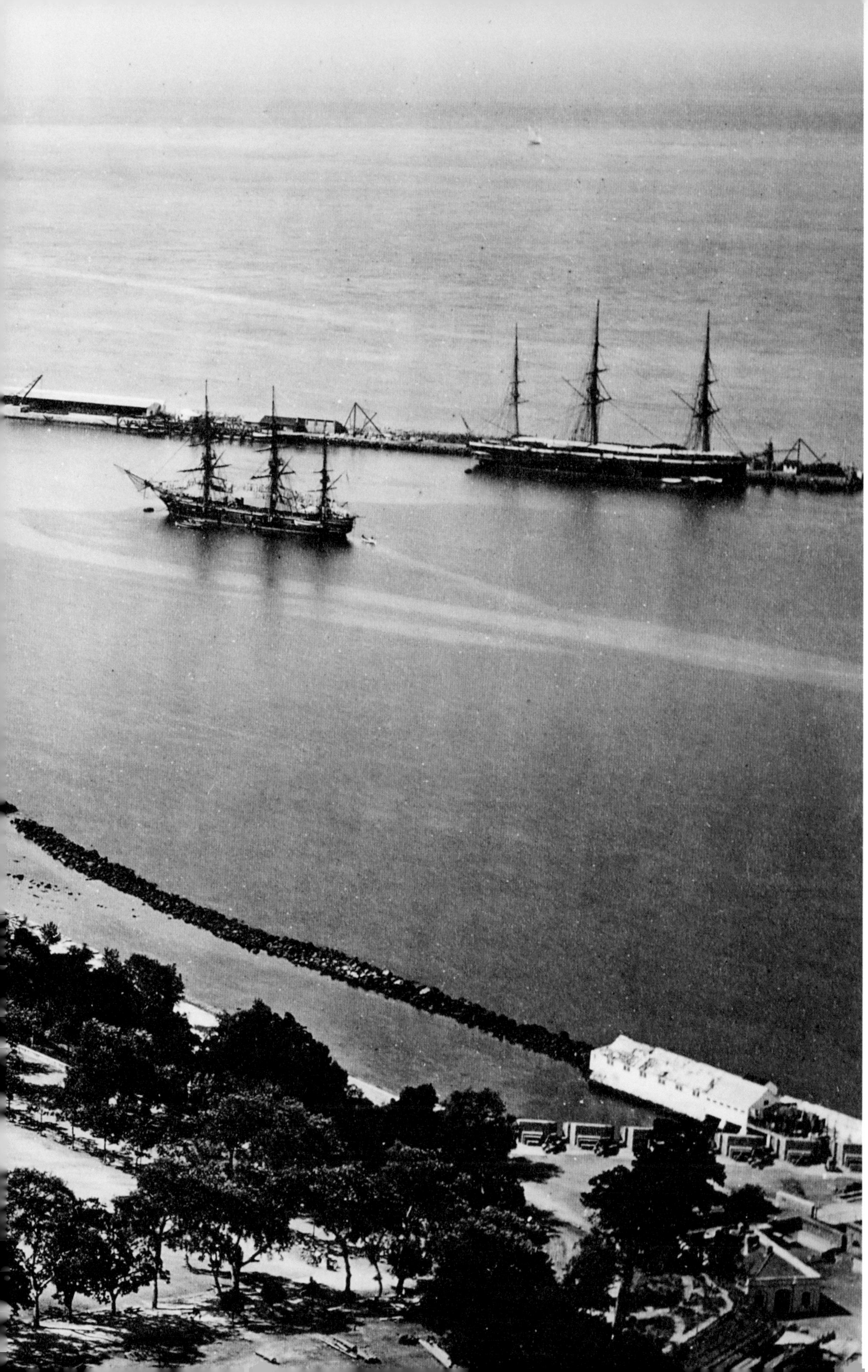

Left, and previous pages: For several days in April 1885 the Vanadis lay at anchor in Gibraltar, where the seaman E F Hultberg was buried. He had been up in the rigging taking in the sails, but fell onto the deck and died. During these days there was a political crisis involving England and Russia. The stop-off in Gibraltar was thus abruptly broken off, according to Boatswain Humbla: 'The captain [had] received a telegram from King Oscar II saying he should take the Vanadis home just as quickly as possible, under both sail and steam. The situation was that the Russian emperor wanted a coal-storage station on the island of Gotland for Russia's Baltic Fleet. A war between Russia and Sweden was not on the cards, but we were nevertheless to go home in the event of unrest.' Prince Oscar recorded in his diary that Sweden had 'equipped over 30 vessels to maintain neutrality', and that the Vanadis had been ordered to make her way to Stockholm as quickly as possible. Fårösund in the north of Gotland was fortified with a battery of cannons: '12 cm Fleet-type front-loading cannons'. But Sweden was never drawn into the crisis.

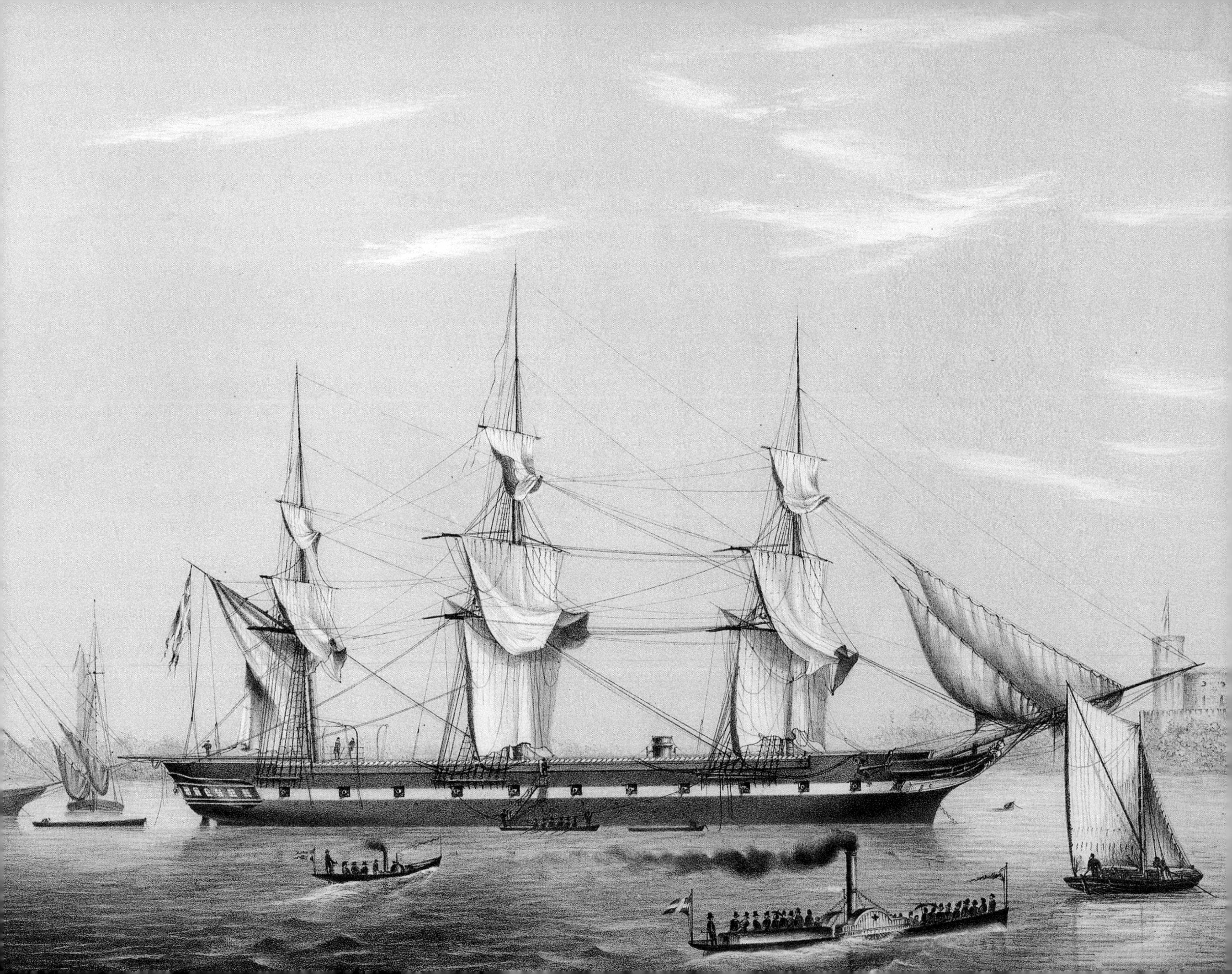

EXIT VANADIS

On 9 May 1885, the Vanadis anchored at Stockholms Ström. King Oscar II had gone to meet the frigate and his son Prince Oscar out in the archipelago. The newspaper *Stockholms Dagblad* reported: 'Having steamed its way between festively decorated vessels and along shores full of people, at about 4pm the Vanadis dropped anchor off Skeppsholmen, from where a salute was given and the shores of which boasted a welcoming crowd of people'.

There followed a royal dinner reception at the Royal Palace for the Vanadis's officers and their equals, including Hjalmar Stolpe. The rank and file were fed in the drill building at Skeppsholmen, and some men were put in the clink for past transgressions on board. Members of parliament were invited to view the ship, whose journey round the world had cost tax payers 556,252 kronor and 16 öre (approximately SEK 35 million in today's money).

Hjalmar Stolpe had a profound sense of things having come to a close when he returned to his job at the Museum of National Antiquities. He wrote: 'I'm sitting here describing ancient archaeological finds from the Iron Age, but my thoughts are in the Far East. And that will always remain the case'. But in his spare moments he unpacked the collections he had sent home, which had been stored in the cellar below the Palace of the Hereditary Prince on Gustav Adolfs Torg (the square where the Ministry for Foreign Affairs is located today). His aim was to put the impressive new ethnographic treasures on display for the benefit of the general public, as well as for doubting researchers for whom ethnography was still deemed a non-science. But his ultimate objective was to create an ethnographic museum under his own direction, and he saw the Vanadis collection as his most powerful line of argument.

The Vanadis exhibition was conceived out in the Pacific Ocean. Stolpe talked about the idea to his ship companion Prince Oscar, who willingly placed the Palace of the Hereditary Prince at his disposal. Thus, during the journey, most of the objects sent home were stored in the palace's cellars in wooden boxes, unopened pending Hjalmar's return. The mummies from Peru ended up in the cellar of the Museum of National Antiquities (the present-day National Museum), where the heating and naphthalene were to save them from disintegration.

Much later Hjalmar Stolpe detailed the scope of his collections:

> The collections inventory involves the following figures: for India over 3,000 items, for Siam around 500 and for Japan around 1,500. The collections from Singapore and Manila with environs run to just 130 items or so in total. The Peruvian collections run to over 1,200 items. For completeness' sake it should be stated that the material I collected from the Pacific Ocean area runs to about 1,000 items, and that from Tierra del Fuego and Chile amounts to around 120... The Vanadis expedition succeeded in bringing together over 7,500 items.

It should be added that, in letters and newspapers, Stolpe estimated the total number of *objects* at 10,000. The explanation for this discrepancy was that the inventory's individual *items* could involve several objects.

Work on the exhibition began in January 1886. The budget was tight, but upholsterers and carpenters were hired, while Emmy attached labels and Hjalmar worked free of charge and attended to private expenditure. Everything was unpacked, and he put a 'head and arms on Kintoki', the Japanese statue of a god guarding the entrance to the palace, whom he described as looking 'extremely grim'. Stolpe added: 'But it is a sad truth that the things I bought in Yokkaichi and sent back to Yokohama are lost, and remain so. I'm in particular aggrieved about the loss of a complete set of moulds and tools for a porcelain teapot, plus similar at various stages. I quite simply had the entire process of Japanese porcelain manufacture done and dusted – and now I have nothing!'

In the wake of the Vanadis's journey, serious financial problems emerged as a result of negligence on the part of the so-called Vanadis Committee, which was supposed to help Hjalmar Stolpe with the financing. Despite this he remained in a good mood, and slaved away night and day, as his wife described in a letter to her mother-in-law:

> He himself says, however, that only with misgivings does he think about leaving this work, which diverts him more than anything else. At midday last Thursday he was quite unexpectedly visited by the King and Prince Oscar. The King several times expressed his surprise at the number of items, but everything is still in such disarray that a closer inspection of the collection was not possible for Their Majesties. I was up with Hjalmar taking him his usual midday sandwiches at the time, and deem myself most fortunate to have escaped the Royal visit.

Emmy also recounted that during this busy and unsettled period her husband enthusiastically participated in social life and dinners: 'It had been a long time since I'd heard Hjalmar's voice sound like that – exactly as it was in the springtime of his

The Vanadis sits at anchor, drying its sails off Kastellholmen in Stockholm. Amongst the small surrounding vessels, the closest one is a jam-packed passenger boat powered by paddle wheels. The artist is Adolf Nordling (1840–88), and his lithograph was produced by J F Meyer o Co, Kongl. Hof-Stentryckeri, Stockholm.

youth! And he was just as handsome!' On Hjalmar's 45th birthday, on 23 April 1886, Emmy noted: 'Now Namman is singing so beautifully from The Creation – his voice is at its very best!'

On Friday 7 May the Vanadis exhibition opened with a private view for a distinguished group of guests. Those present included King Oscar II and Prince Oscar with retinue, the minister for foreign affairs Count Albert Ehrensvärd, heading up a group of cabinet members, admirals and other high-ranking officers, academics such as Adolf Erik Nordenskiöld and Hans Hildebrand, and directors-general and members of parliament. Hjalmar Stolpe had skilfully attended to the publicity, and the newspapers lavished column inches on the orderly exhibition, and on the exhibitor's diligence and tenacity. *Dagens Nyheter* stated that the exhibition was 'far more magnificent and opulent than one had imagined', adding:

> It is without a doubt one of the most interesting and successful cultural exhibitions Stockholm has had to offer, and one can just imagine how much work it involved, both in terms of collecting the thousands of objects and as regards organising them in the clear and tasteful manner displayed here, in both of which respects, as is well known, Doctor Hj. Stolpe is due almost sole credit. The exhibition is said to comprise over 6,000 items, though it should be noted that many of them involve several objects, thus the number of the latter is probably in excess of 10,000.

The public recognition came as music to Hjalmar's ears, who contentedly ambled through the palace's four halls where the objects were exhibited: in the first was Peru, with its mummies, 80 skulls and numerous clay vessels, in the second Japan and China, with its focus on the tea ceremony, then India's multitude of household utensils and costumes, and finally the South Sea Islands and their simple instruments.

Inside the entrance to Gustav Adolfs torg stood the god Kintoki, glowering angrily to drive away any evil spirits. On the wall beside the statue hung 'a picture representing the frigate Vanadis, which brought back all these exhibition objects [which was wrong], on the journey across the Pacific Ocean. The painting is surrounded by a set of Swedish flags'. The newspaper *Aftonbladet* reported: 'The arrangement of everything displays good taste and expertise. Numerous photographs, largely taken by the expedition's photographer, shed further light on the collection'.

The exhibition's status as an exclusive attraction was confirmed by the presence of foreign royalty, as Stolpe noted:

> Yesterday I had the Japanese prince at the exhibition for a short while. The only useful thing about this was that it appeared to draw a few people in. He was a man of few

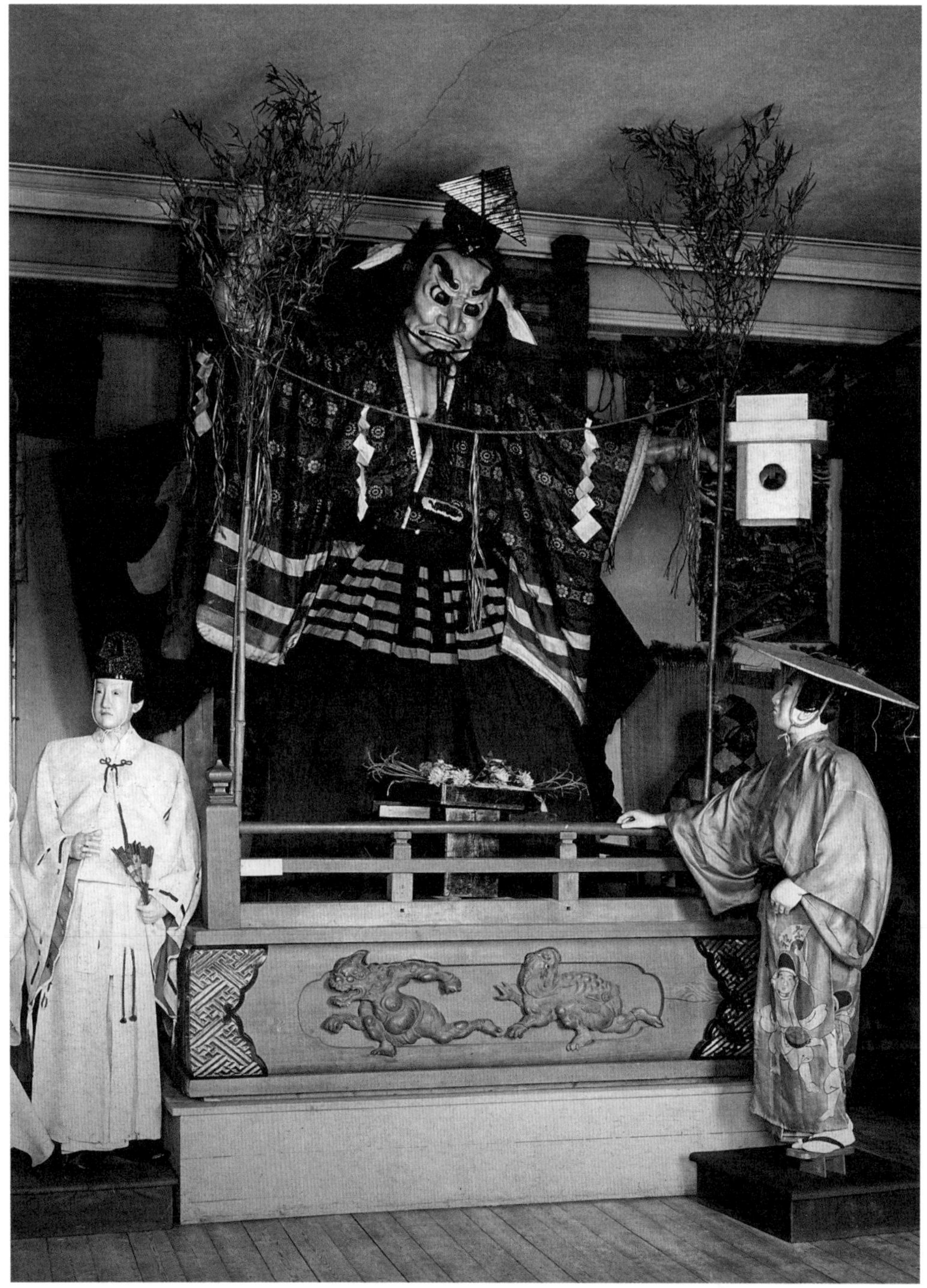

words, and the same applied to the whole party. Irritatingly enough none of them spoke any English – I'd just gone round the collection a little while before and thought through it in that language, but it transpired that the entire party only spoke French. To the prince's great surprise, however, I said a little to him in Japanese ... The tea ceremony was declared to be absolutely correct. Otherwise I was provided with no information at all. Such gentlemen know nothing. And they were in any case in a hurry.

Another visitor who bestowed lustre on the exhibition was the king of Portugal, who arrived together with Crown Prince Gustaf. Emmy related as follows to her daughter:

> Papa was woken up so he could be given a letter from the Royal Palace, in which he was informed that King Luiz of Portugal intended to visit the Vanadis exhibition as early as 10 o'clock. There was a great hurry here, believe me! Mama had to go with Papa to the exhibition to help him relieve the fellows of their nightwear (you remember the black cowls, don't you?), and by the time specified everything was under control. However, we were on the lookout for the Royal carriage to no avail until as late as 11 o'clock – only then did it become visible on Norrbro, with outriders and accompanied by three other carriages, in which his retinue were sitting.
>
> Papa had to rush down the stairs, and I was in my hidey-hole, behind the Chinese cabinet over by Papa's study – you know where I mean. There was a chink through which I was able to watch the eminent visitor – the King –and others. The Crown Prince was there too. I had great fun standing watching everyone without being seen myself! Papa spoke English when he was presenting the exhibition, and the King walked alongside him, listening attentively and looking interested.

The Vanadis exhibition ran from May to December 1886, transferring to Gothenburg the following year. Entrance tickets cost 25 öre, and in the catalogue (50 öre) Hjalmar Stolpe described the objects and their background. He pointed out that 'only a small proportion of the over 700 photographs' taken during the journey could be displayed, and that they had been taken by a 'photographer accompanying the expedition'. No name was given. By contrast, the Gothenburg patron who had provided 10,000 kronor in sponsorship for the photographer did get a mention: 'It is a welcome obligation for me to gratefully call attention to the great service Mr [Pontus] Fürstenberg has done to science through this generous appropriation'. We do not know how Oscar Ekholm felt about this slight. He had up until then been commended by Stolpe, but in 1886 – the year of the exhibition – their relationship appears to have deteriorated. Ekholm's newly opened photographic studio was on the verge of going into liquidation. And before Ekholm was allowed to leave Stockholm for Sundsvall, Stolpe demanded that he return the photographic equipment and the remaining glass-plate negatives.

Neither do we know how Oscar Ekholm's Vanadis pictures were judged at the time. With one important exception. In Copenhagen in the early 1890s, Kristian Bahnson was preparing the publication of his work *Etnografien fremstillet i dens Hovedtraek* (An Outline Presentation of Ethnography). He was the administrator and, in practice, the director of the Danish Museum of Ethnography – one of the oldest in the world, created in 1841. He was in need of illustrations, and contacted his friend Hjalmar, who sent him a number of Vanadis photographs. Bahnson quickly replied that he was overwhelmed by the gems he had beheld. They far exceeded his expectations – a collection of pictures that 'in terms of their clarity and picturesque effect are among the very best I have ever seen'.

It should be mentioned that Stolpe and Bahnson were close friends. But his young Danish colleague was one of the few to have the courage to question Stolpe's theses regarding the interpretation of 'primitive' peoples' ornamentation – the scientific contribution by which he himself and his European colleagues set the greatest store. Bahnson was only able to use a small number of the Vanadis pictures, Stolpe having taken too long to send them. But they contrast strongly with Bahnson's other illustrations – lively, on-the-spot depictions amidst lifeless woodcuts. Neither did his Danish friend state that the name of the photographer was Oscar Ekholm. The acknowledgement in the book goes to Hjalmar Stolpe. After the Vanadis exhibition, Bahnson's book and Stolpe's lectures in various contexts in the 1890s, it was to be a good half-century before Ekholm's photographs were rediscovered.

While the Vanadis exhibition was in progress, the frigate itself continued to sail the world's oceans. But in 1893, after 30 years of glorious service, the downgrading of the Vanadis commenced. She was first converted into a receiving ship, then in 1911 to a barracks ship, with Skeppsholmen as her moorings. The Vanadis sank there in October 1941, and was eventually salvaged and transferred to the Fjäderholmarna islands near Stockholm. Two years later she was bought for 11,950 kronor by the firm P. Näsman Co, who reckoned on getting ten tons of copper and 18 tons of scrap iron out of the hull. She was broken up at Smedsudden shipyard in Stockholm. But the well-built Vanadis put up resistance to the very last, and had to be exploded using dynamite – a final salute for the old lady of the seas.

After more than eight decades of life, time was finally up for the good ship – Exit Vanadis.

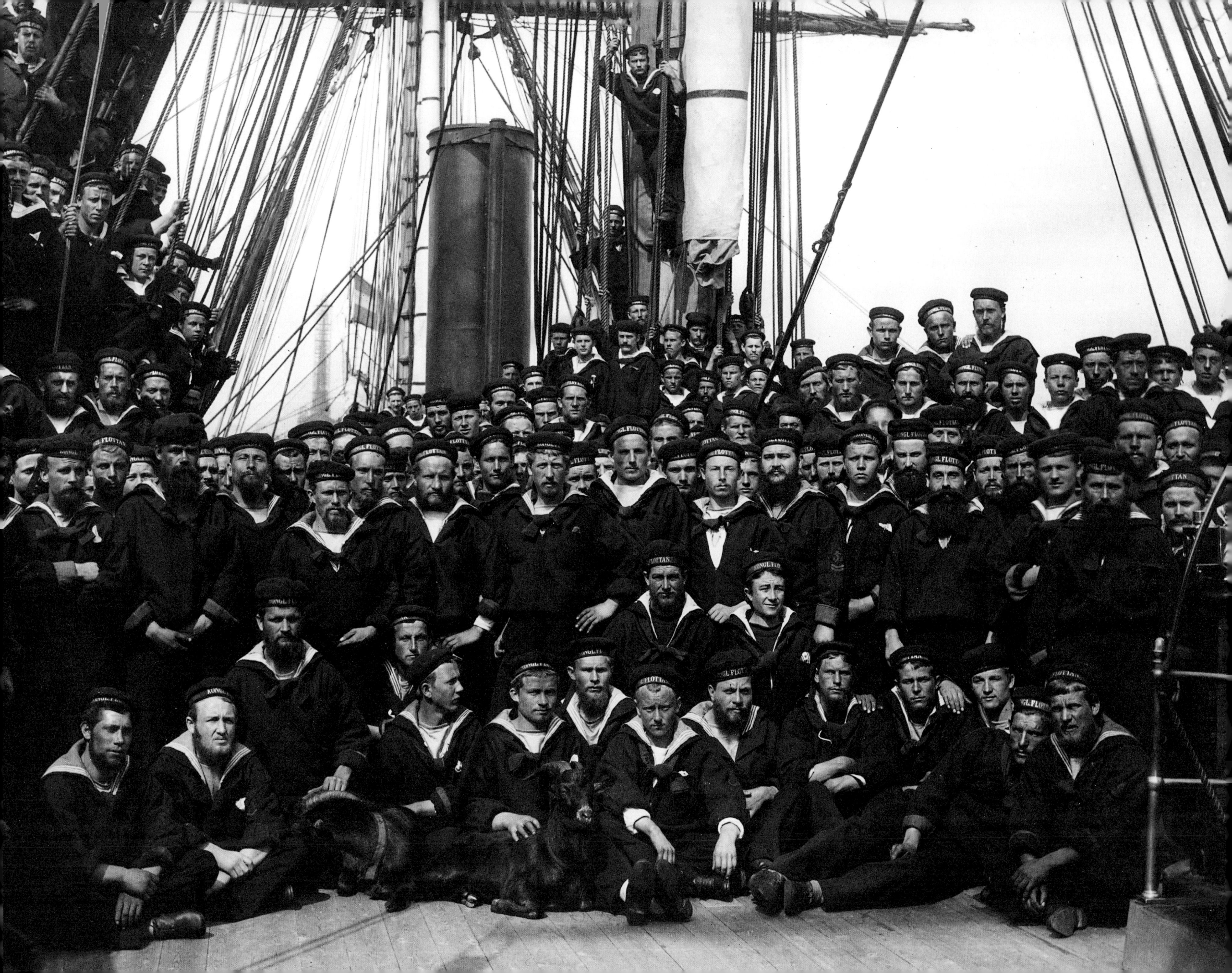
KONGL. FLOTTAN

Previous page: The mighty Kintoki was a hero from Japanese mythology who had been elevated to the status of a god. Stolpe described his 'wild appearance and ostentatious clothing, partly silk, throwing peas at the demons to drive them away'. He added: ' The whole thing is an excellent museum piece, though at 5 1/2 metres colossally high. It will be accommodated in the Prince's palace, however, and will draw people into the exhibition.' The Vanadis exhibition opened in 1886 in what is now the Department of Foreign Affairs building at Gustav Adolfs Torg, and Kintoki stood in the vestibule. He was a processional statue who was drawn by bulls on the occasion of the annual Kanda Miojin festival, which is celebrated in the Tokyo district of Kanda on 15 September.

On 9 May 1885 the Vanadis dropped anchor off Skeppsholmen in Stockholm. Some of the nearly 340-strong crew can be seen here in front of the funnel. On the deck near to the camera one can with some difficulty make out the two goats, Markus and Lukas. They had been caught the year before on Nukuhiva in the Pacific Ocean during a hunt on the island, when the seaman Otto Larsson fell down a cliff and lost his life.

Right: The Vanadis as a receiving ship at Skeppsholmen in Stockholm.

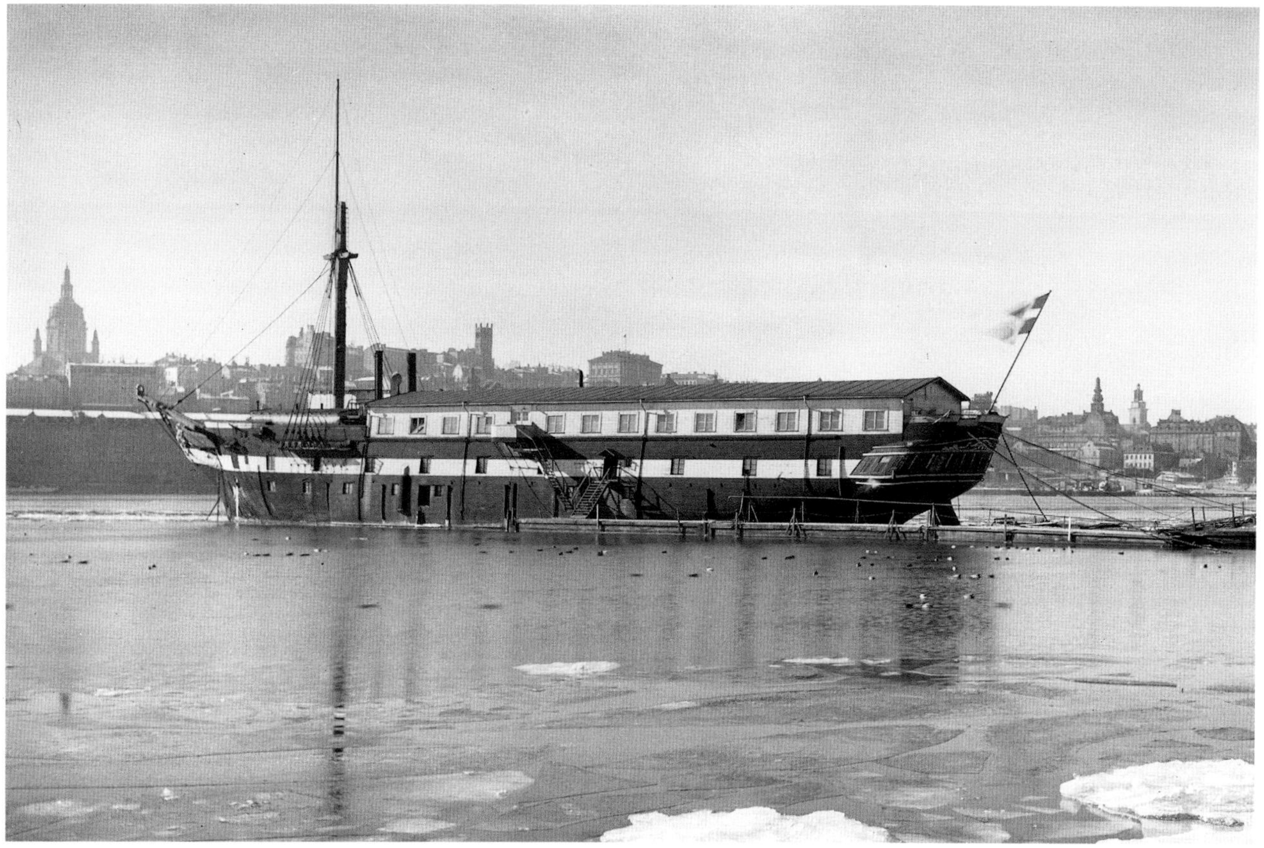

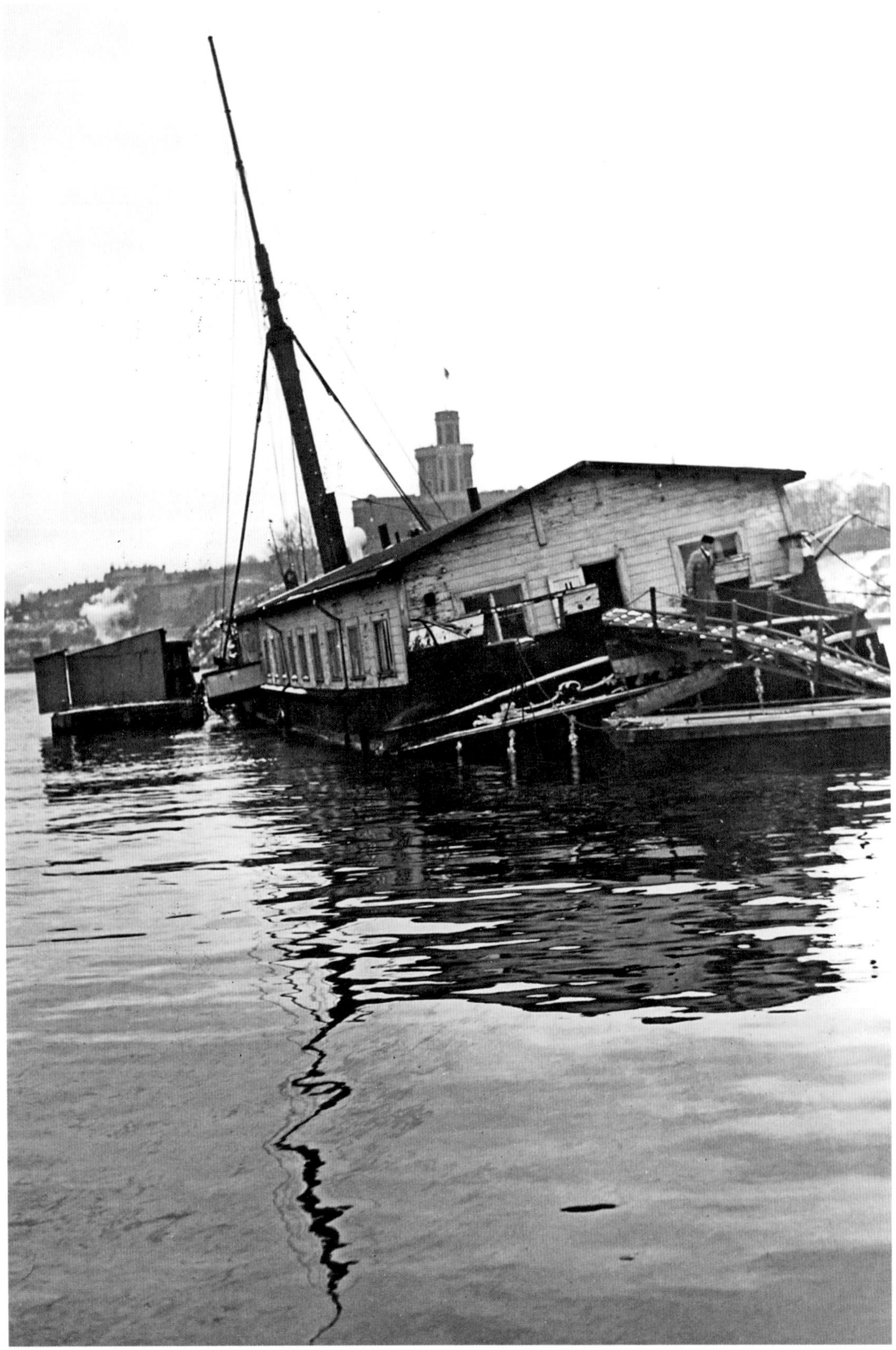

On 29 October 1941 the press reported that the Vanadis had sunk at her Skeppsholmen moorings. She was gradually salvaged and transferred to the Fjäderholmarna islands near Stockholm. In 1943 she was sold for SEK 11,950 to the company P. Näsman & Company, who reckoned on getting ten tons of copper and eighteen tons of scrap iron out of the hull.

The picture in which the Vanadis is being towed was probably taken that year, off Kastellholmen, with Beckholmen to the left and Fåfängan visible in the background.

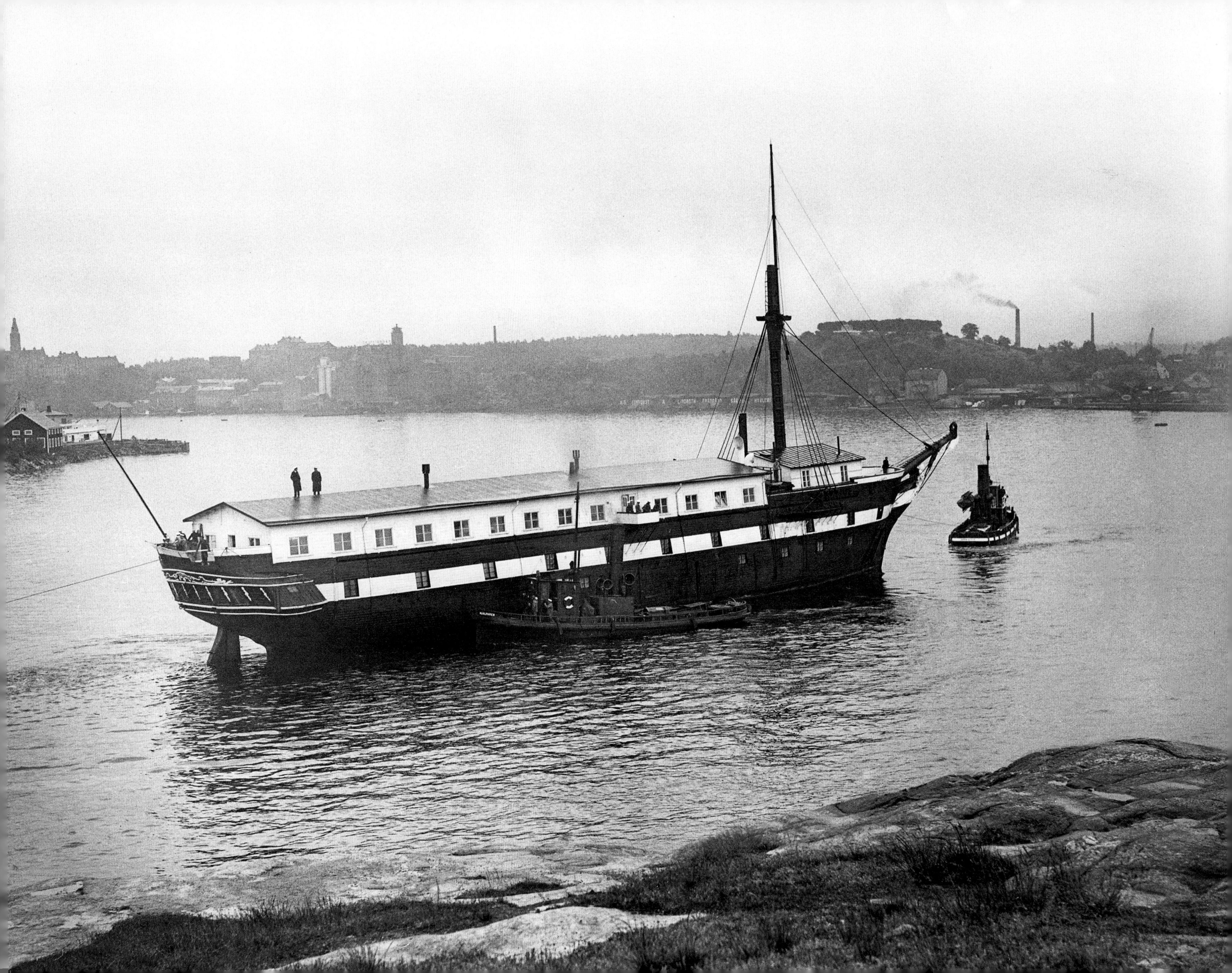

SOURCES AND LITERATURE

The family archive left by Emmy and Hjalmar Stolpe forms the basis of my account of the Vanadis expedition. It has been kept in Copenhagen since 1908, when Emmy died. The archive is dominated by letters exchanged between the couple and with their relatives, friends and colleagues at home and abroad. There are also notebooks, drawings, financial accounts, certificates, newspaper cuttings and concepts for Hjalmar's printed writings. In total there are several thousand letters and other documents.

But the archive also includes 400 paper copies of the glass negatives the photographer Oscar Ekholm exposed during the Vanadis voyage. He copied most of them in the little photographic laboratory on board the Vanadis and during the long stopovers in Japan and India. They constituted a portable photographic archive for Hjalmar Stolpe, who was able to record the motif and location on the back of the pictures. Stolpe was an enthusiastic follower of the nascent art of photography, and during the course of the journey he bought a large number of photos. He himself calculated that his collection numbered a thousand or so photographs, including the so-called *cartes de visite*, i.e. small portraits of relations and friends (and enemies).

Stolpe's photo collection was the point of departure for my plans for the photo book *The Vanadis Expedition*, which now also includes a selection from Ekholm's 700 glass negatives that are archived at the Museum of Ethnography in Stockholm. In addition to these two collections, several archives contain photographs linked to the Vanadis voyage. The Royal Library, for example, holds the photo collection of the ship's priest Immanuel Kiellman-Göranson (466 copies), and the Maritime Museum houses the collections that belonged to the frigate's officers Arvid Lindman and Oscar Bernadotte. Uppsala University Library also owns a collection of Vanadis photographs, as does Västervik Museum.

Emmy and Hjalmar Stolpe's family archive was placed at my disposal by their late grandson Björn Fabricius, and forms the basis of my biography *Kungen av Birka. Hjalmar Stolpe – arkeolog och etnograf* (*The King of Birka. Hjalmar Stolpe – archaeologist and ethnographer*) (2015). Fabricius wanted the archive to end up at the Royal Library in Stockholm, and now that *The Vanadis Expedition* has rolled off the press it is my desire that this wish be granted.

Kungen av Birka contains a complete Stolpe bibliography. In the present book the source references are restricted to the years of the frigate's journey and the subsequent Vanadis exhibition in Stockholm. The texts by Hjalmar Stolpe and other players are reproduced in a form as close as possible to the original, which means the authors are accountable for the factual content. The sources are listed below by chapter.

N.B. The comparisons between the value of the Swedish krona (SEK) in the 1880s and those of today have been made with the help of Professor Rodney Edvinsson's site Historicalstatistics.org – Portal for Historical Statistics. The amounts stated in the text are rounded off and give an idea of the monetary value during the Vanadis journey.

THE VANADIS EXPEDITION

Archive

Ekholm, Oscar. Letters to his brother Fredrik 2/12 and 26/12 1883. Private ownership.

Stolpe, Hjalmar. Letters to Carl Johan Stolpe 7/5 1883; Emmy Stolpe 4/12, 10/12, 23/12 1883, 22/1, 28/2 1884; Albert Nydahl 1/4 1890.

Literature and press

Alexandersson, Gustav. 'J V Svensons Motorfabrik' [J V Svensson's Engine Factory]. In *Nackaboken 1978. Årsbok för Nacka kommun* [Nacka Book 1978. Yearbook for Nacka Municipality]. Nacka: 1978.

Arenhill, Emma. 'Biografiska data om Oscar Ekholm och hans släkt' [Biographical data on Oscar Ekholm and his family]. Unprinted.

Bengtson, Bo och Cedergren, Hugo (ed.). *Prins Oscar Bernadotte. En minnesbok av 32 författare* [Prince Oscar Bernadotte. A commemorative volume by 32 authors]. Uppsala: Lindblads, 1953.

Erikson, Bo G. *Kungen av Birka. Hjalmar Stolpe – arkeolog och etnograf* [The King of Birka. Hjalmar Stolpe – archaeologist and ethnographer]. Stockholm: Atlantis, 2015.

'Fotografier af fornlemningar. Om artisten och fotografen C. F. Lindberg' [Photographs of ancient remains. About the artist and photographer C. F. Lindberg]. In *Kongl. Vitterhets-, historie- och antiqvitetsakademiens månadsblad* [Monthly Magazine of the Royal Swedish Academy of Letters, History and Antiquities]. Eleventh year, 1882. Stockholm: 1883.

Hägg, Christer. *Fregatten Eugenies världsomsegling* [The frigate Eugenie's circumnavigation]. Stockholm: Magnus Ullman, 1999.

Hägg, Erik. *Örlogsliv. Minnen från ungdomsresor med svenska flottans segelfartyg och de första officersåren i Karlskrona 1885–1896* [Naval life. Memories of journeys as a young man with the Swedish fleet's sailing vessels and the first years as an officer in Karlskrona 1885–1896]. Stockholm: Sv. litteratur [Swedish literature], 1943.

'Instruktion för Filosofie Doktorn Hj. Stolpe i hans egenskap af deltagare såsom etnograf i en längre sjöexpedition med Kongl. Fregatten Vanadis. Den 14 November 1883. På Kongl. Vetenskaps Akademiens vägnar, C. Skogman, D. Lindhagen'. [Instructions for Hj. Stolpe, Doctor of Philosophy, in his role as participant as an ethnographer in an extended maritime expedition with the Royal Frigate Vanadis. 14th November 1883. On behalf of the Royal Swedish Aademy of Sciences, C. Skogman, D. Lindhagen]. Unprinted.

Jansson, Alfred. *Prins Oscar Bernadotte* [Prince Oscar Bernadotte]. Stockholm: Norstedts, 1948.

Jensen, Ola Wolfhechel. *På spaning efter det förflutna. En historia om arkeologiskt fältarbete i Sverige 1600–1900* [In search of the past. A history of archaeological field work in Sweden 1600–1900]. Forms part of Antiquarian Series 54. Stockholm: Royal Swedish Academy of Letters, 2018.

Lagerholm, Folke. *Deltagarna i fregatten Vanadis jordomsegling 1883–1885. Förteckning med biografiska uppgifter intill 1928 års utgång* [The participants in the frigate Vanadis's circumnavigation of 1883–1885. List of biographical information up to the end of 1928]. Karlskrona: K. L. Svensson, 1929.

Lehmann, Erik. *Med fregatten Vanadis på världsomsegling 1883–1885. F.d. båtsman Humblas dagboksanteckningar och minnen samlade och utgivna av Erik Lehmann* [With the frigate Vanadis on the circumnavigation of 1883–1885. Former boatswain Humbla's diary entries and memories collected and published by Erik Lehmann]. Västervik: Västerviks-Posten, 1927.

Manneby, Hans. *Båtsman Humbla. Anteckningar från Vanadis världsomsegling 1883–85* [Boatswain Humbla. Notes from the Vanadis's circumnavigation of 1883–85]. Stockholm: Rabén & Sjögren in collaboration with the Swedish National Maritime Museum, 1978.

Melander, Richard. *Jorden rundt under svensk örlogsflagg. Ögonblicksbilder från fregatten Vanadis verldsomsegling 1883–1885 samlade af S. Natt och Dag och i brefform återgifna af Richard Melander* [Round the globe under a Swedish naval flag. Glimpses from the frigate Vanadis's circumnavigation of 1883–1885 collected by S. Natt och Dag and in the form of letters reproduced by Richard Melander]. Stockholm: Bonniers, 1887.

Mårtensson, C. G. *Vid firandet af trettioårsdagen (den 4 maj 1915) af vår hemkomst till Karlskrona från fregatten Vanadis jordomsegling åren 1883–1885* [On the celebration of the thirtieth anniversary (4 May 1915) of our return to Karlskrona from the frigate Vanadis's circumnavigation of 1883–1885]. Karlskrona: Amiralitetsboktryckeriet [Admiralty Printing Press], 1915.

Norberg, Erik (ed.). *Karlskronavarvets historia.* [The History of Karlskrona Shipyard]. *Part I 1680–1866* and *Part II 1866–1992.* Karlskrona: Karlskronavarvet [Karlskrona Shipyard] (Kockums), 1993.

Stolpe, Hjalmar. *Studier i amerikansk ornamentik. Ett bidrag till ornamentens biologi* [Studies of American ornamentation. A contribution to the biology of ornaments]. Stockholm: 1896.

The journal *Flottans män* [Men of the Fleet], 2013, No. 1.

Widén, Albin. 'Fregatten Vanadis världsomsegling. Med utdrag ur Prins Oscar Bernadottes dagböcker' [The frigate Vanadis's circumnavigation. With extracts from Prince Oscar Bernadotte's diaries]. From *Aktuellt och historiskt. Meddelanden från försvarsstabens krigshistoriska avdelning* [Matters current and historical. Communications from the defence staff's Dept of War History]. Stockholm: Försvarsmakten [Armed Forces], 1962.

CIRCUMNAVIGATIONS

Literature

Hägg, Christer. *Fregatten Eugenies världsomsegling*. Stockholm: Magnus Ullman, 1999.

Östberg, Carl Axel. *De första världsomseglingarna under svensk flagg – kapten Nils Werngren berättar. Författat och sammanställt av Carl Axel Östberg* [The first circumnavigations under a Swedish flag, as told by Captain Nils Werngren. Written and compiled by Carl Axel Östberg]. Karlskrona: Abrahamson, 1989.

SOUTH AMERICA

Archive

Ekholm, Oscar. Letter to his brother Fredrik, 25/2 1884. Private ownership.

Stolpe, Hjalmar. Letters to Emmy Stolpe 21/1, 3/2, 28/2, 18/3, 20/3, 31/3, 8/4, 14/4 1884.

Stolpe, Hjalmar. Notebook, Vol. 1, 1884: Karlskrona–Valparaíso. Museum of Ethnography. Stockholm. Unprinted.

Literature and press

Jansson, Alfred. *Prins Oscar Bernadotte.* Stockholm: Norstedts, 1948.

Stolpe, Hjalmar. 'Bref från Vanadis. Från Stockholms Dagblads korrespondent' [Letters from the Vanadis. From Stockholms Dagblad's correspondent]. From *Stockholms Dagblad* 5/8, 6/8, 18/8 1884.

Stolpe, Hjalmar. *Vägvisare genom Vanadis-utställningen i Arffurstens palats 1886* [Guide to the Vanadis exhibition at the Palace of the Hereditary Prince 1886]. Stockholm: 1886.

Stolpe, Hjalmar. 'Om vårt Etnografiska Museum, särskildt om dess afdelning II' [About our Museum of Ethnography, in particular its Dept II]. From *Ymer – tidskrift utgifven af Svenska sällskapet för antropologi och geografi* [Ymer – journal published by the Swedish Society for Anthropology and Geography], fifteenth year 1895. Stockholm: 1896.

Westergren, Georg. *Dikter, gamla minnen. Bref från Sydamerika, reseskildringar af en uppbördsläkare* [Poems, old recollections. Letters from South America, travel writings by a military physician]. Stockholm: Hæggström, 1877.

THE PACIFIC OCEAN

Archive

Ekholm, Oscar. Letters to his brother Fredrik 12/6, 21/8 1884. Private ownership.

Lindman, Arvid. Dagbok förd under Fregatten Vanadis verldsomsegling 1883–85 [Diary kept during the Frigate Vanadis's circumnavigation 1883–85]. Part 1. Maritime Museum. Stockholm. Unprinted.

Loggbok för H. M. Fregatt Vanadis, 1883–85 [Logbook for H.M. Frigate Vanadis, 1883–85]. Krigsarkivet [Military Archives of Sweden]. Stockholm. Unprinted.

Naht [tjänsteman vid Hernsheim & Co]. Dagbok, juli 1884. Av prins Oscar överlämnad till Sjöhistoriska museet 1939 [Naht (official at Hernsheim & Company). Diary, July 1884. Left to the Maritime Museum by Prince Oscar in 1939]. Unprinted.

Stolpe, Hjalmar. Letters to Emmy Stolpe 18/6, 22/6, 6/7, 21–22/7, 30/7, 12/8 1884; to Georg Lindhagen, August 1884.

Stolpe, Hjalmar. Notebook, Vol. 4, 1884: Honolulu 20 June to 10 July. Jaluit 26 July to 2 Aug. 1884. Museum of Ethnography. Stockholm. Unprinted.

Literature and press

Bengtson, Bo och Cedergren, Hugo (ed.). *Prins Oscar Bernadotte. En minnesbok av 32 författare* [Prince Oscar Bernadotte. A commemorative volume by 32 authors]. Uppsala: Lindblads, 1953.

Callmer, Christian. *Abraham Fornander. Thirteen letters to Erik Ljungstedt edited with introduction and notes by Christian Callmer.* Part of Scripta minora Regiae Societatis humaniorum litterarum Lundensis 1972/1973:2. Lund: Royal Scientific Society of Letters in Lund, 1973.

Finsch, Otto. 'Die Marshall-Inseln' [The Marshall Islands]. From *Die Gartenlaube – Illustriertes Familienblatt* 1886 [The Arbour – Illustrated Family Paper 1886], No. 2. Leipzig: 1886.

Greenwell, Jean. 'Doctor Georges Phillip Trousseau. Royal Physician'. From *Hawaiian Journal of History*. Honolulu: Hawaiian Historical Society, 1991.

Kaeppler, Adrienne Lois and Van Tilburg, Jo Anne. *The Iconic Tattooed Man of Easter Island.* Melbourne: Mana Press, 2018.

Manneby, Hans. *Båtsman Humbla. Anteckningar från Vanadis världsomsegling 1883–85.* Stockholm: Rabén & Sjögren in

collaboration with the Swedish National Maritime Museum, 1978.
Mörner, Birger. *Söderhavet. Del I. De stora färderna. Vid rikets port. Polynesien* [The South Pacific. Part I. The major voyages. At the gateway to the kingdom. Polynesia]. Part of the series *De geografiska upptäckternas historia* [The history of geographical discoveries]. Uppsala: Lindblads, 1923.
Stolpe, Hjalmar. 'Från Vanadis. Bref till Stockholms Dagblad' [From the Vanadis. Letters to Stockholms Dagblad]. From *Stockholms Dagblad* 13/10, 14/10, 15/10 1884.
Stolpe, Hjalmar. 'Om undersökningar af polynesiska grafhålor, hvilka af honom utförts under Vanadisexpeditionen. Föredrag, 17/2 1888' [On examinations of Polynesian grave holes carried out by him during the Vanadis expedition. Lecture, 17/2 1888]. From *Ymer – tidskrift utgifven af Svenska sällskapet för antropologi och geografi*, eighth year 1888. Stockholm: 1889.
Stolpe, Hjalmar. 'Om vårt Etnografiska Museum, särskildt om dess afdelning II'. From *Ymer – tidskrift utgifven af Svenska sällskapet för antropologi och geografi*, fifteenth year 1895. Stockholm: 1896.
Stolpe, Hjalmar. *Über die Tätowirung der Oster-Insula* [On the Easter Islanders' tattoos]. Part of *Abhandlungen und Bericht des Königlichen Zoologischen und Anthropologischen Museums zu Dresden* [Treatises and Report by the Royal Zoological and Anthropological Museum in Dresden]. Festschrift No. 6. Berlin: R. Friedländer, 1899.
Stolpe, Hjalmar. *Vägvisare genom Vanadis-utställningen i Arffurstens palats 1886*. Stockholm: 1886.
Våra minnen [Our recollections]. Commemorative volume about the Vanadis's journey, with preface Christmas 1885 by the princes Oscar, Carl and Eugen. With contributions by Fredrik Adelborg, Nils Sundström, Carlo di Terramont, Axel Ribbing and Carlo Landberg. Stuttgart: 1886.

JAPAN

Archive

Ekholm, Oscar. Letter to his brother Fredrik, 17/9 1884. Private ownership.
Stolpe, Hjalmar. Letters to Emmy Stolpe 28–30/8, 1/10, 26/10, 2/12 1884.

Literature and press

Bengtson, Bo och Cedergren, Hugo (ed.). *Prins Oscar Bernadotte. En minnesbok av 32 författare*. Uppsala: Lindblads, 1953.
Kiellman-Göranson, Immanuel. 'Bilder från Fregatten Vanadis färd från åren 1883–85. Reseberättelser i *Blekinge läns tidning*'. [Photographs from the Frigate Vanadis's voyage from the years 1883–85. Travel writings from the newspaper *Blekinge läns tidning*] Collected in 'Minnes-Bok' [Commemorative Volume], National Library of Sweden, Stockholm.
Stolpe, Hjalmar. 'Från Vanadis-utställningen' [From the Vanadis exhibition]. From the weekly paper *Ny Illustrerad Tidning*, 23/10 1886.
Stolpe, Hjalmar. 'Från Vanadis-utställningen. II. Sagan om Sakata Kintoki, den stora japanska processionsstatyen' [From the Vanadis exhibition. II. The story of Sakata Kintoki, the great Japanese processional statue]. From the weekly paper *Ny Illustrerad Tidning*, 20/11 1886.
Stolpe, Hjalmar. 'Hvad är och hvad bör ett etnografiskt museum vara?' [What is a museum of ethnography and what should it be?] From the weekly paper *Hvar 8 dag*, 19/4 1903.
Stolpe, Hjalmar. 'Om vårt Etnografiska Museum, särskildt om dess afdelning II'. From *Ymer – tidskrift utgifven af Svenska sällskapet för antropologi och geografi*, fifteenth year 1895. Stockholm: 1896.
Stolpe, Hjalmar. *Vägvisare genom Vanadis-utställningen i Arffurstens palats 1886*. Stockholm: 1886.

THE PHILIPPINES AND THAILAND

Archive

Ekholm, Oscar. Letter to his brother Fredrik 29/11 1884. Private ownership.
Stolpe, Hjalmar. Letters to Emmy Stolpe 26/10, 6/11 1884; Carl Johan Stolpe 22/12 1884.

Literature and press

Stolpe, Hjalmar, 1895. Lecture held at the Swedish Society for Anthropology and Geography, 14/12 1894. From *Ymer – tidskrift utgifven af Svenska sällskapet för antropologi och geografi*, fifteenth year 1895. Stockholm: 1896.
Stolpe, Hjalmar. Summary of lecture given at Stockholms borgarskola. From the daily paper *Svenska Dagbladet*, 23–24/3 1900.
Våra minnen. Stuttgart: 1886.

INDIA

Archive

Ekholm, Oscar. Letter to his brother Fredrik 27/3 1885. Private ownership.
Stolpe, Hjalmar. Letters to Emmy Stolpe 19–23/12, 30/12 1884. 21/1, 27/1, 1/2, 8–15/2, 23/2, 2/3, 11/3 1885.
Stolpe, Hjalmar. Notebook, Vol. 6, 1885: Hong Kong, Manila, Bangkok, Singapore. Notebook, Vol. 9: India III. Museum of Ethnography. Stockholm. Unprinted.

Literature and press

Stolpe, Hjalmar. 'Om vårt Etnografiska Museum, särskildt om dess afdelning II'. From *Ymer – tidskrift utgifven af Svenska sällskapet för antropologi och geografi*, fifteenth year 1895. Stockholm: 1896.
Våra minnen. Stuttgart: 1886.

EXIT VANADIS

Archive

Stolpe, Emmy. Letters to Charlotta Stolpe 3/4, 8/5 1886; Estrid Stolpe 24/8 1886.
Stolpe, Hjalmar. Letters to Emmy Stolpe 1/7, 12/8 1885.

Literature and press

The daily paper *Dagens Nyheter*, 8/5 1886.
The daily paper *Stockholms Dagblad*, 10/5 1885.
Stolpe, Hjalmar. 'Hvad är och hvad bör ett etnografiskt museum vara?' From the weekly paper *Hvar 8 dag*, 19/4 1903.
Stolpe, Hjalmar. *Vägvisare genom Vanadis-utställningen i Arffurstens palats 1886*. Stockholm: 1886.
The daily paper *Svenska Dagbladet*, 7/9 1935, 6/3 1939, 29/10 and 11/11 1941, 11/7, 27/7 and 13/8 1943, 20/11 1994.

TABLE OF IMAGES

Used abbreviations:

EM – Etnografiska museet – the Museum of Ethnography, Stockholm
HS – Hjalmar Stolpe's private collection
KB – Kungliga biblioteket – The Royal Library, Stockholm
SMTM – Statens maritima och transporthistoriska museer – the Swedish National Maritime and Transport Museums
SSM – Stockholms stadsmuseum – the Stockholm City Museum, Stockholm

THE VANADIS EXPEDITION

CIRCUMNAVIGATIONS

SOUTH AMERICA

THE PACIFIC OCEAN

JAPAN

134 Photo: Oscar Ekholm. HS.
135 Photo: Oscar Ekholm. HS.
136 Photo: Oscar Ekholm (both photos). HS.
137 Photo: Oscar Ekholm. HS.
138 Photo: Oscar Ekholm. EM.
139–148 Unknown photographer. Purchased photos. HS.
149 Photo: Oscar Ekholm. EM.
151 Photo: Oscar Ekholm. HS.
152–153 Photo: Oscar Ekholm. HS.

THE PHILIPPINES AND THAILAND

154 Photo: Oscar Ekholm. Kiellman-Göranson's collection. KB.
158 Photo: Oscar Ekholm (both photos). HS.
159 Photo: Oscar Ekholm. HS.
160 Photo: Oscar Ekholm. HS.
161 Photo: Oscar Ekholm. HS.
163 Photo: Oscar Ekholm. EM.
164 Photo: Oscar Ekholm (both photos). HS.
166 Photo: Oscar Ekholm. HS.
167 Litter. Photo: Oscar Ekholm. HS. / King Chulalongkorn. Unknown photographer. EM.
168 Photo: Oscar Ekholm. HS.
169 Unknown photographer. Purchased photo. HS.

INDIA

170 Photo: Oscar Ekholm. HS.
174 Photo: Oscar Ekholm. EM.
178 Photo: Oscar Ekholm. HS.
180 Waste pickers. Photo: Oscar Ekholm. EM. / Stone sculptures. Photo: Oscar Ekholm. HS.
181 Photo: Oscar Ekholm. EM.
182 Photo: Oscar Ekholm. HS.
183 Photo: Oscar Ekholm. HS.
184 Photo: Oscar Ekholm. HS.
185 Photo: Oscar Ekholm (both photos). EM.
186 Photo: Oscar Ekholm. HS.
187 Photo: Oscar Ekholm. EM.
188–189 Photo: Oscar Ekholm. HS.
190–191 Photo: Oscar Ekholm. HS.
192 Fakir. Photo: Oscar Ekholm. EM. / Drawings. Photo: Bengt Wanselius. / The Raja of Benares. Unknown photographer. Purchased photo. HS.
193 Photo: Oscar Ekholm. EM.
194 Two men. Photo: Oscar Ekholm. HS. / The cart. Photo: Oscar Ekholm. EM.
195 Photo: Oscar Ekholm. HS.
196 Photo: Oscar Ekholm. HS.
197 Photo: Oscar Ekholm. EM.
198 Photo: Oscar Ekholm. EM.
199 Photo: Oscar Ekholm. EM.
200 Photo: Oscar Ekholm. EM.
201 Photo: Oscar Ekholm. EM.
202 Photo: James Craddock. EM.
203 Photo: Oscar Ekholm. HS.
204 Photo: Oscar Ekholm. Kiellman-Göranson's collection. KB. / Portrait. Photo: Robert Roesler. Private collection.
205 Photo: Oscar Ekholm. HS.
206–207 Photo: Oscar Ekholm. Kiellman-Göranson's collection. KB.

EXIT VANADIS

208 Lithograph by Adolf Nordling. Kungl. Hovstaterna.
210 Unknown photographer. EM.
212 Photo: Oscar Ekholm. HS.
213 Unknown photographer. SMTM.
214 Unknown photographer. TT Bild.
215 Photo from *Aftonbladet*. SSM.

INDEX

Please note that since Hjalmar Stolpe and Oscar Ekholm are mentioned so often in the book, their names have been left out of the index.

ACKNOWLEDGEMENTS

Thanks to her persistent trawling of the archives, Emma Arenhill found information on Oscar Ekholm and his relations. The clues led to unknown letters and portraits held by Oscar's distant relations, the siblings Anna and Axel Nelson. I am extremely grateful for the permission to use the material.

For years Magnus Johansson from the Museum of Ethnography in Stockholm has been my friend and has lent me a helping hand; his efforts have been untiring, and he has displayed a passionate interest in the museum's photographic treasures. I am also indebted to the museum's photographer Johan Jeppsson.

Håkan Wahlquist, the Museum of Ethnography's curator responsible for Asia, has also provided generous support. His knowledge and encouragement have been invaluable, as has his reading of the manuscript.

Ulla Wagner has a past as director of the Museum of Ethnography, and is the author of exciting texts on Japan.

Olaf Winter is an ethnographer and doctor of archaeology at the Australian National University, where he has researched the islands of the Pacific Ocean. I am enormously grateful to him for reading the manuscript.

I have had the privilege of working with Nils Johan Tjärnlund as a publishing editor, and once again wish to thank him for his unstinting editorial work, enhanced by his humour!

Over the years the Axel and Margaret Ax:son Johnson Foundation for Public Benefit has provided generous financial support for my work on the Vanadis Expedition project.

And last but foremost, my heartfelt thanks to my wife Madeleine von Rohr, who for decades has put up with an interminable number of files.

THE VANADIS EXPEDITION

An Ethnographic Voyage Around the World 1883–1885

Original title: *Expedition Vanadis. En etnografisk världsomsegling 1883–1885*

Published by Bokförlaget Stolpe, Stockholm, 2021

Edited by: Nils Johan Tjärnlund
Translated from Swedish by: William Sleath
Copy editor: Andrew Mackenzie
Picture editor: Bengt Wanselius
Design: Patric Leo
Layout: Patric Leo, Petra Ahston Inkapööl
Prepress: Italgraf Media, Stockholm, Sweden
Print: Narayana Press, Denmark, via Italgraf Media, 2021
First edition, first printing

ISBN 978-91-8906-920-6

BOKFÖRLAGET STOLPE

AXEL AND MARGARET AX:SON JOHNSON FOUNDATION
FOR PUBLIC BENEFIT